T0393837

Dynamical Projectors Method in Hydro and Electrodynamics

Dynamical Projectors Method in Hydro and Electrodynamics

Sergey Leble and Anna Perelomova

CRC Press is an imprint of the
Taylor & Francis Group, an **informa** business

Printed on acid-free paper

International Standard Book Number-13: 978-1-138-03560-7 (Hardback)

Library of Congress Cataloging-in-Publication Data

Names: Leble, S. B. (Sergeæi Borisovich), author. | Perelmova, Anna, author.
Title: The dynamical projectors method: hydro and electrodynamics / Sergey Leble and Anna Perelmova.
Description: First edition. | Boca Raton, FL: CRC Press, Taylor & Francis Group, 2018. | "A CRC title, part of the Taylor & Francis imprint, a member of the Taylor & Francis Group, the academic division of T&F Informa plc." | Includes bibliographical references and index.
Identifiers: LCCN 2017050734 | ISBN 9781138035607 (hardback: acid-free paper) | ISBN 9781351107990 (e-book)
Subjects: LCSH: Magnetohydrodynamics–Mathematics. | Hydrodynamics–Mathematics. | Electrodynamics–Mathematics. | Projection.
Classification: LCC QC151.7 .L44 2018 | DDC 532/.5–dc23
LC record available at https://lccn.loc.gov/2017050734

Visit the Taylor & Francis Web site at
http://www.taylorandfrancis.com

and the CRC Press Web site at
http://www.crcpress.com

Dedication

To our school teachers
Viktor Nekrasov (School 19, Arkhangel'sk)
Rudolf Bega (School 2, Moscow)

Epigraphs

Песнь **двадцать вторая**
Я тени принимаю **за тела**
Поистине, нередко внешний знак
Приводит ложным видом в заблужденье,
Тогда как суть погружена во мрак.
Veramente più volte appaion cose
che danno a dubitar falsa matera
per le vere ragion che son nascose.
Purgatorio Canto XXII - La Divina Commedia
Appearances will often, it is true,
give rise to false assumptions, when the truth
to be revealed is hidden from our eyes
Aleksandr Pushkin
Dante Alighieri
"Who'd Fully Stopped You..."
1823
Who'd fully stopped you, billows, proud?
Who'd dressed with chains your movement, mightiest?
Into a pool, so mute and lightless,
Who'd changed such a rebellious flood?
Whose magic wand had thrown aside
My hope, sorrow and gladness
And, with a doziness of laziness,
My ever-boiling soul had lulled?
So, blow, winds, rise, waters-dreamers,
Distract the pestilent stronghold!
Where are you, thunderstorm of freedom?
Wing the slaved waters' manifold!
Aleksandr Pushkin
Кто, волны, вас остановил,
Кто оковал ваш бег могучий,
Кто в пруд безмолвный и дремучий
Поток мятежный обратил?
Чей жезл волшебный поразил
Во мне надежду, скорбь и радость
И душу бурную
Дремотой лени усыпил?

Взыграйте, ветры, взройте воды,
Разрушьте гибельный оплот!
Где ты, гроза — символ свободы?
Промчись поверх невольных вод.
Александр Пушкин

Contents

List of Figures

Preface

Generally, the way to a somewhat explicit formula that describes the evolution of disturbances in a medium implies a number of crucial simplifications. Such names of this procedure are "derivation" or "heuristic arguments." The results often are very impressive and begin to live their own history. A reader can find celebrated examples in the frames of the linearized statements of problems in every textbook on mathematical physics. Among them, we may list the D'Alambert (wave, string) equation, Laplace-Poisson, and heat (diffusion) equations. Derivation of these equations is also a necessary element of the textbooks. Performing the procedure, one may neglect some terms but very rarely meets attempts of their mathematical approval. Recently, a boom in mathematical physics was observed in respect with similar standard "minimal" nonlinear equations such as the so-called integrable ones. The list of them includes a lot of entries and still grows. We would mention the most known of them: there are Korteweg-de Vries (KdV) and nonlinear Schrodinger equations (NLSE), which describe the dynamics of one-dimensional wave packets in the long wavelength domain (KdV case) and in a small vicinity of a (carrier) frequency (NLSE case). In a sense, these equations are universal in small amplitude and dispersion realm. Considering the form of these equations, one understands that their one-component and one-dimensional cases require a lot of explanation to embed the resulting solutions in a general statement of problem, which is based on the original multicomponent and multidimensional description. Special accent is made in respect to a perturbation by nonlinearity in the context of the expansion of evolution operator in a series with respect to powers of the small magnitude parameter.

Both KdV and NLS equations contain only the first derivative in time, which imposes the only direction of a wave's propagation. This principally differs from the previously mentioned D'Alembert equation. It is clear that considering an initial stage of evolution of 1D small magnitude perturbation, we observe general propagation in both directions. This directly points out the complex content of the potential development of general initial perturbations. Some kind of superposition should be taken into account in mathematical description. The phenomenon is well studied in the theory of 1D string evolution, but it needs some efforts to translate the apparatus to the multicomponent field. In this book, we suggest a systematic realization of such a program, which makes visible all ingredients and stages of the implementation of the "minimal" equations into the general statements of problems in hydrodynamics and electrodynamics. We consider mainly two kinds of the problems: initial (Cauchy) and boundary regime problems.

A method of separating disturbances in a medium into specific components is developed. A hydrodynamic system is locally split into coupled nonlinear equations of interacting modes. Linearization provides independent modes with an individual evolution. The corresponding projector operators allow us to formulate initial-value problems for each mode and to introduce physical basis in order to follow interacting disturbances as they develop. The one-dimensional problem is properly examined as an example. The entropy and progressive components (modes) are introduced, and dynamic equations that account for nonlinear interactions are derived. In particular, weak nonlinearity and dispersion introduce Burgers/KdV-like systems for directed waves interacting with a mean flow.

In the general (three-dimensional) case, five eigenvectors of a linear thermoviscous flow over a homogeneous background are derived in the quasi-planar geometry of the flow. The corresponding projectors are evaluated and applied in order to get the nonlinear evolution equations for the interacting entropy, vortical, and acoustic modes. A dynamic equation on streaming caused by an arbitrary type of acoustic excitation is derived as well. The correspondence to the known results on streaming caused by quasi-periodic acoustic sources, is traced. The acoustic force of streaming is calculated also for an impulse sound, including a mono-polar one. General hydrothermodynamic perturbations in a two-dimensional boundary layer is split into eigen modes: a Tollmienn-Schlichting (TS) wave and two acoustic ones. The definition of modes is performed by means of local relations, which are extracted from the linearized general system over the equilibrium flow in a boundary layer. Each such connection defines every individual subspace and the corresponding projector. After subsequent projecting to the eigen subspace of an Orr-Sommerfeld equation for a TS wave and the corresponding procedure for acoustics, the equations rearrange to the one-dimensional system that describes evolution along the mean stream. A new mechanism of nonlinear resonant excitation of the TS wave by sound is proposed and modeled by means of four-wave interaction. Further, application of the projecting method is applied in problems relating to propagation and interaction of acoustic and gravity waves in a stratified medium.

The same method is systematically applied to dynamics of electromagnetic disturbances of a medium with a given dispersion and magnetic properties, including metamaterials. The separation of total disturbance is provided on the basis of linear dispersion relations that are introduced in either the frequency domain (for a boundary regime) or in the wavevector domain (in a Cauchy problem). The problems are in a certain sense "diagonalized" in a linear flow and partially diagonalized in a nonlinear flow. Linear and nonlinear problems are considered in terms of propagation of polarized directed waves and their interaction in a dispersive medium. The influence of restricting boundaries such as waveguides is considered from a practical point of view. These are the dielectric (optic) fibers and metal tubes that conduct radio frequency waves. As for wave propagation in open space, projecting helps to establish the general role of external sources and to take into account a symmetry of perturbations.

Authors

Sergey Leble was born in Sretensk, Russia, in 1945. He graduated from the secondary school in Arkhangelsk in 1962 and from Leningrad State University in 1968 with a specialty in theoretical physics. He conducted post-doctoral studies in the subdivision of nuclear theory and elementary particles at Leningrad State University, where he completed the PhD thesis entitled "Theory of elementary particles in spaces with curvature" under the supervision of Dr. Igor Terentyev in 1974. The full doctoral thesis, entitled "Waveguide propagation of nonlinear waves in layered media," was awarded by Leningrad State University in 1987. He has served as a professor since 1988, head of the Theoretical Physics department at Kaliningrad State University since 1989, and a professor at the Gdansk University of Technology in Poland from 1995–2015. Currently, he is a professor at Baltic Federal University in Kaliningrad, Russia. He has taught general and specialized courses in all areas of theoretical physics. An author of chapters in collective monographs and of approximately 200 scientific works in the field of theoretical and mathematical physics, concepts of scientific activity have lent his major monographs their titles: S.B. Leble. *Waveguide propagation of nonlinear waves in stratified media.* Leningrad University, 1988; S.B. Leble. *Nonlinear Waves in Waveguides* (Springer-Verlag, 1991); E. Doktorov, S. Leble. *Dressing method in mathematical physics* (Springer, 2007).

Anna Perelomova was born in Moscow in 1965. After graduating from high school in 1982, she began studying in the Department of Acoustics at the Physical Faculty of Moscow State University, where she specialized in radiophysics and received a degree with honors in 1989 under the supervision of Dr. Oleg. Rudenko. Her doctoral thesis, under the supervision of Drs. Ivan Nemchinov and Valerij Shuvalov "Acoustic-gravitational waves induced by a powerful source," was awarded in 1994 by the Institute of Physics of the Earth in Moscow. Anna Perelomova defended a habilitation thesis "Interaction of acoustic and non-acoustic modes: theoretical aspects of nonlinear acoustics" in 2006 for the Department of Physics at the university named after A. Mickiewicz in Poznan,

Poland. Since 2011, she has served as a professor in the Department of Technical Physics and Applied Mathematics at the Gdansk University of Technology in Poland and is the author of approximately 90 publications in the field of nonlinear hydrodynamics and acoustics.

1 Introduction

This book discusses the results published in papers devoted to the technique of derivation of model evolution equations in terms of fundamental systems such as hydrodynamics and electrodynamics. It also naturally includes initial or boundary conditions that fix a type of a medium perturbation. Such universal approach is absent in monograph or textbook literature. The contents of the book provide readers with a practical tool that allows them to divide a space of solutions into subspaces with specific behavior marked by such terms as direction of wave propagation, frequency range, and, that is important to stress, characteristics of a medium. The characteristics are expressed in terms of standard parameters of a matter, such as permeability in case of electrodynamics or dissipation terms (equilibrium state and attenuation) in hydrodynamics.

One such way is the introduction of a combinations of field variables, named as modes, which correspond to their specific behavior. The simplest example that appears already in one-dimensional (1D), or planar, dynamics may be described as unidirectional waves; that is, waves with a fixed direction of propagation. In the case of celebrated wave equation (the string equation in 1D) we have the d'Alembert formula that gives the general solution of the initial (Cauchy) problem, with the natural inclusion of the directed waves. A choice of initial condition that corresponds to a wave, propagating, say, rightwards, is readily realized *very easily* with the help of a technique discussed in this book, namely the dynamic projecting. This technique splits the initial or boundary conditions *space* to the modal subspaces of the directed waves. Readings of dispersion and/or dissipation (terms with higher derivatives) are automatically taken into account within the algorithmic formalism, described already in the example of introduction and with general detalization in Chapter 2. Account of nonlinearity leads to interaction of the modes that are also automatically specified. The simple case of nonlinearly directed wave leads to the celebrated Korteweg-de Vries and nonlinear Schrödinger equations.

The general technique is universal, the steps are quite similar in both huge fields of possible applications, in hydro- and electrodynamics. A transition to two- and three-dimensional (3D) problems complicates calculations, but lies in the same road, and may be either realized mechanically or one can effectively include the symbolic computation programs because all the steps are written analytically. A good chance to demonstrate a power of projecting the method relates to a case of atmosphere-ocean waves physics, whose exponential equilibrium stratification admits, however, a transition to equations with constant coefficients and separate directed acoustic and internal waves.

An account of combined dissipation and nonlinearity in 3D hydrodynamics leads to effective description of such important phenomena as heating and streaming. It is a direct excitation of so-called entropy and rotation modes (non-wave or near-zero frequency ranges). The application of the dynamic projecting technique allows to use both periodic and nonperiodic acoustic sources for control of such processes.

The impulse excitation is also included! Cumulatively, the combined dispersion and nonlinearity allows to effectively describe electromagnetic pulses up to very short ones (video) on the basis of generalization of Shafer-Wayne equations that includes counter-propagating waves with a polarization account.

Among others, there are problems that appear secondary, but, in our opinion, they are of great importance. We mean diagnostics problems. The question that naturally arises in physics, geophysics and other applied sciences, sounds: How to recognize a wave (or non-wave perturbation) in an overall field? The application of the projecting technique provides the answer for high-quality requirements, because the set of the projectors is complete, and each of them is strictly attached to a specific perturbation under consideration.

One more task that is very "hot" in applied modeling of the abundant problems of hydro- and electrodynamics is building numerical codes. The division of the components of the field with different scales and directions of propagation would be very effective. Especially, it relates to the nonlinear problems in which interaction of modes "spoils" the linear picture, and one needs effective tool for "cleaning" of perturbations when a program simulates time or space evolution. The dynamic projecting technique may be effectively used in such applications.

A numerical evaluation of basic physical parameters is an effective tool to present the results of a modeling in the form of tables or plots. Such a representation of results is a necessaary ingredient of theoretical physics [1] and can be effectively applied in various scales from nuclear to universe ones. In many cases, there are difficulties of such direct modeling if scales of different orders contribute to the phenomenon under simulation. In such situations, it is effective to separate the description into parts with different characteristic scales. Mathematically, we split a space of solution into corresponding subspaces in order to follow the following underlying calculations [2,3,5]. The division of the solution space can be realized by means of projecting operators, which correspond to the eigenvectors of the evolution operator of the problem under consideration [2]. Such a procedure leads generally to pseudo-differential operators as elements of the matrix evolution operator. Further, essential simplifications may be the corresponding we restrict ourselves by some frequency or wavelength range, expanding the kernel of projecting operators in the vicinity of the given frequency carrier. This restriction allows us to cut the series of corresponding pseudo-differential operators at some level, restricting the matrix elements by polynomials of some operator, including derivatives. Otherwise, in more complicated conditions, e.g. for the so-called short, or ultrashort, pulses, we use some wide-range approximation of the kernel. In wave theory, such a procedure is known as an approximation of dispersion relation, valid in some frequency domain. In nonlinear models, we obtain a definition of modes that may interact during evolution. This allows us to follow the wave perturbations and identify the modes content as well as the concomitant non-wave perturbations. From a mathematical point of view, we apply either a Freshet expansion of the evolution operator or some model approximation of the operator.

So, from a modeling perspective, we suggest a combined analytic—numerical scheme of integration of a multicomponent system of partial differential equations with two kinds of initial-boundary conditions. The problem may concern an initial (Cauchy) boundary condition of boundary regime problems (for an introduction see [6]).

Briefly, the idea of this approach may be described by the simplest example, as follows. Consider an evolution problem as the system of two equations with constant coefficients:

$$\frac{\partial u(x,t)}{\partial t} - a\frac{\partial u(x,t)}{\partial x} - b\frac{\partial v(x,t)}{\partial x} = 0, \tag{1.1}$$

$$\frac{\partial v(x,t)}{\partial t} - c\frac{\partial u(x,t)}{\partial x} - d\frac{\partial v(x,t)}{\partial x} = 0, \tag{1.2}$$

A compact form of (1.1) and (1.2) is

$$\psi_t = L\psi, \tag{1.3}$$

with matrices

$$\psi = \begin{pmatrix} u \\ v \end{pmatrix} \text{ and } L = \begin{pmatrix} a\partial_x & b\partial_x \\ c\partial_x & d\partial_x \end{pmatrix}, \tag{1.4}$$

introduces the evolution operator L and a state ψ of a system. The Fourier transformation in x

$$u(x) = \frac{1}{\sqrt{2\pi}}\int_{-\infty}^{\infty} \tilde{u}(k)e^{ikx}dk,\ v(x) = \frac{1}{\sqrt{2\pi}}\int_{-\infty}^{\infty} \tilde{v}(k)e^{ikx}dk, \tag{1.5}$$

may be written as

$$\psi = F\tilde{\psi}. \tag{1.6}$$

In compact notation of derivatives by index, it yields the system of ordinary differential equations

$$\tilde{\psi}_t = F^{-1}LF\tilde{\psi} = \tilde{L}\tilde{\psi}, \tag{1.7}$$

where the k-representation of the evolution operator takes the form

$$\tilde{L} = ik\begin{pmatrix} a & b \\ c & d \end{pmatrix}. \tag{1.8}$$

The matrix eigenvalue problem

$$\tilde{L}\phi = \lambda\phi \tag{1.9}$$

introduces two subspaces, which we represent by the matrix of solutions Φ

$$\tilde{L}\Phi = \Phi\Lambda \tag{1.10}$$

with diagonal matrix $\Lambda = \text{diag}\{\lambda_1, \lambda_2\}$. We normalize the eigenvectors such that their first components are units. It is easy to check that if $\lambda_1 \neq \lambda_2$, the inverse matrix exists and

$$\Phi^{-1}\tilde{L} = \Lambda\Phi^{-1}. \tag{1.11}$$

Let us establish a matrix $\tilde{P}_i$ so that $\tilde{P}_i\,\Phi = \Phi_i$ are eigenvectors of the evolution matrix in Eq. (1.7). Moreover, the standard properties of orthogonal projecting operators

$$\tilde{P}_i \cdot \tilde{P}_j = 0, \quad \tilde{P}_i^2 = \tilde{P}_i, \quad \sum_i \tilde{P}_i = I, \tag{1.12}$$

where I is the unit operator, are implied. By means of (1.11), one can prove that the $\tilde{P}_i$ are expressed as the direct matrix product

$$\tilde{P}_i = \Phi_i \otimes (\Phi^{-1})_i, \tag{1.13}$$

where Φ_i is the ith column and $(\Phi^{-1})_i$ is the ith row of the matrix of solutions Φ (1.10) [8] and the identity

$$\tilde{L}\tilde{P}_i = \tilde{P}_i\tilde{L} \tag{1.14}$$

holds. Explicit form of the eigenmatrix Φ in the mentioned normalization is given by

$$\Phi = \begin{pmatrix} 1 & 1 \\ \phi_1 & \phi_2 \end{pmatrix}. \tag{1.15}$$

The bottom components ϕ_i are established from (1.10)

$$\phi_i = \frac{\lambda_i - a}{b}, \tag{1.16}$$

$$\lambda_{1,2} = \frac{ik}{2}[a + d \pm \sqrt{\Delta}] \tag{1.17}$$

if $\Delta = (a-d)^2 + 4bc > 0$, the eigenvalues $\frac{\lambda_i}{i}$ are real and the equations are hyperbolic, which corresponds to the wave propagation. It is the elliptic case, if $\Delta < 0$, whereas the parabolic one, if $\Delta = 0$, requires individual consideration (see Section 2.1.3). The idempotents in k-representation take the form

$$\tilde{P}_1 = \frac{1}{\phi_1 - \phi_2}\begin{pmatrix} 1 \\ \phi_1 \end{pmatrix} \otimes \begin{pmatrix} -\phi_2 & 1 \end{pmatrix} \tag{1.18}$$

and

$$\tilde{P}_2 = \frac{1}{\phi_1 - \phi_2}\begin{pmatrix} 1 \\ \phi_2 \end{pmatrix} \otimes \begin{pmatrix} \phi_1 & -1 \end{pmatrix}. \tag{1.19}$$

Similarly, evaluating the direct product and plugging (1.16) yields

$$\tilde{P}_1 = \frac{1}{\sqrt{\Delta}}\begin{pmatrix} \frac{a-d+\sqrt{\Delta}}{2} & b \\ c & \frac{d-a+\sqrt{\Delta}}{2} \end{pmatrix}, \tilde{P}_2 = \frac{1}{\sqrt{\Delta}}\begin{pmatrix} \frac{d-a+\sqrt{\Delta}}{2} & -b \\ -c & \frac{a-d+\sqrt{\Delta}}{2} \end{pmatrix}. \tag{1.20}$$

In x-representation, which follows from the inverse Fourier transformation, the matrix L and the projectors $\tilde{P}_i$ are as follows:

$$F\tilde{L}F^{-1} = L, \tag{1.21}$$

$$F\tilde{P}_i F^{-1} \equiv P_i \tag{1.22}$$

and also commute:

$$F\tilde{L}F^{-1}F\tilde{P}_i F^{-1} = F\tilde{P}_i F^{-1}F\tilde{L}F^{-1}. \tag{1.23}$$

In this "happy case" the projecting operators in k-representation are constant matrices that do not depend on k. Hence, their x-representations coincide with these established ones (1.20). Projecting the evolution Eq. (1.7) gives two independent equations, which in fact are the first lines of

$$(P_i\psi)_t = LP_i\psi, \tag{1.24}$$

obtained by simple application from the left; the commutation (1.23) is taken into account. We make use of the direct calculations:

$$P_1\psi = \frac{1}{\sqrt{\Delta}}\begin{pmatrix} \frac{a-d+\sqrt{\Delta}}{2} & b \\ c & \frac{d-a+\sqrt{\Delta}}{2} \end{pmatrix}\begin{pmatrix} u \\ v \end{pmatrix} =$$

$$\frac{1}{\sqrt{\Delta}}\begin{pmatrix} bv + u\left(\frac{1}{2}a - \frac{1}{2}d + \frac{1}{2}\sqrt{\Delta}\right) \\ cu + v\left(\frac{1}{2}d - \frac{1}{2}a + \frac{1}{2}\sqrt{\Delta}\right) \end{pmatrix} = \begin{pmatrix} \Pi \\ \frac{2c\Pi}{a-d+\sqrt{\Delta}} \end{pmatrix}, \tag{1.25}$$

$$P_2\psi = \frac{1}{\sqrt{\Delta}}\begin{pmatrix} \frac{d-a+\sqrt{\Delta}}{2} & -b \\ -c & \frac{a-d+\sqrt{\Delta}}{2} \end{pmatrix}\begin{pmatrix} u \\ v \end{pmatrix} \tag{1.26}$$

$$= \frac{1}{\sqrt{\Delta}}\begin{pmatrix} u\left(\frac{1}{2}d - \frac{1}{2}a + \frac{1}{2}\sqrt{\Delta}\right) - bv \\ v\left(\frac{1}{2}a - \frac{1}{2}d + \frac{1}{2}\sqrt{\Delta}\right) - cu \end{pmatrix} = \begin{pmatrix} \Lambda \\ -\frac{a-d+\sqrt{\Delta}}{2b}\Lambda \end{pmatrix}.$$

The first rows of these relations, by means of the new variables

$$\begin{aligned} \Pi &= \tfrac{1}{\sqrt{\Delta}}\left(bv + u\left(\tfrac{1}{2}a - \tfrac{1}{2}d + \tfrac{1}{2}\sqrt{\Delta}\right)\right), \\ \Lambda &= \tfrac{1}{\sqrt{\Delta}}\left(u\left(\tfrac{1}{2}d - \tfrac{1}{2}a + \tfrac{1}{2}\sqrt{\Delta}\right) - bv\right), \end{aligned} \tag{1.27}$$

yield in the inverse relations

$$\begin{aligned} u &= \Pi + \Lambda, \\ v &= -\tfrac{1}{2b}\left(a - d + \sqrt{\Delta}\right)\Lambda + \tfrac{1}{2b}\left(d - a + \sqrt{\Delta}\right)\Pi, \end{aligned} \tag{1.28}$$

The original system splits into the system of independent equations by means of (1.24).

$$\begin{aligned} \Pi_t &= \tfrac{1}{2}\left(a + d + \sqrt{\Delta}\right)\Pi_x, \\ \Lambda_t &= \tfrac{1}{2}\left(a + d - \sqrt{\Delta}\right)\Lambda_x \end{aligned} \tag{1.29}$$

Such a system of independent equations naturally describes one-dimensional evolution. Let us take the typical case for physical applications $a = d = 0$ (see also Section 7.1.2). Now $\Lambda_{1,2} = \pm ik\sqrt{bc}$, $v_{1,2} = \pm ik\sqrt{\frac{c}{b}}$, and therefore, (5.62) is significantly simplified as $\left[\left(a+d+\sqrt{a^2-2ad+d^2+4bc}\right)\right]_{a=0,d=0} = \sqrt{4bc}$, correspondingly. The referring variables now read

$$\begin{aligned} \Pi &= \tfrac{1}{2}u + \tfrac{1}{2}\sqrt{\tfrac{b}{c}}v, \\ \Lambda &= \tfrac{1}{2}u - \tfrac{1}{2}\sqrt{\tfrac{b}{c}}v, \end{aligned} \tag{1.30}$$

and the dynamic equations take the forms

$$\Pi_t = \sqrt{bc}\Pi_x, \tag{1.31}$$

$$\Lambda_t = -\sqrt{bc}\Lambda_x. \tag{1.32}$$

They are equivalent to the classic string equation (its transformation is mentioned in [4]) of acoustic or electromagnetic waves (see, e.g., [9]) and many others. In this case, we have oppositely propagating waves; namely the mode Π is rightward and Λ is leftward progressive waves. The Cauchy problem has an elegant formulation in the context of initial conditions $u(x,0) = f(x), v(x,0) = g(x)$, so that

$$\Pi(x,0) = (P_1\psi(x,0))_1 = \frac{1}{2}f + \frac{1}{2}\sqrt{\frac{b}{c}}g,\ \Lambda(x,0) = (P_2\psi(x,0))_1 = \frac{1}{2}f - \frac{1}{2}\sqrt{\frac{b}{c}}g. \tag{1.33}$$

We have taken the first components of the relations (1.25) and (1.26) at initial time $t = 0$. The D'Alembert formula follows directly from solutions of the equations with conditions (1.33). In the elliptic case ($\Delta < 0$), the square root is imaginary; then, in the simple case when $a = d = 0$, $\Delta = 4bc$, the equations form

$$\Pi_t = i\sqrt{-bc}\Pi_x, \tag{1.34}$$

$$\Lambda_t = -i\sqrt{-bc}\Lambda_x, \tag{1.35}$$

which reflects that the reference variables are complex conjugate $\Lambda = \Pi^*$. It is important to notice that the variables

$$\begin{aligned} u &= \Pi + \Lambda, \\ v &= -\sqrt{\tfrac{c}{b}}(\Lambda - \Pi), \end{aligned} \tag{1.36}$$

are real. They are convenient in the boundary problem of Laplace/Poisson equation at half-plane solution.

The specific parabolic case ($\Delta = 0$) corresponds to the unique eigenvalue

$$\lambda = \frac{ik}{2}(a+d), \tag{1.37}$$

the link between the components is found to be

$$cL\phi^1 = ik\begin{pmatrix} a & b \\ c & d \end{pmatrix}\begin{pmatrix} s \\ \phi_1 \end{pmatrix} = \begin{pmatrix} iask + ibk\phi_1 \\ ick + idk\phi_1 \end{pmatrix} \tag{1.38}$$
$$= \lambda\begin{pmatrix} s \\ \phi_1 \end{pmatrix} = \begin{pmatrix} s\lambda \\ \lambda\phi_1 \end{pmatrix},$$

which reads also as $\frac{\lambda s}{ik} - as = b\phi_1$. Plugging in the eigenvalue yields $\phi_1 = \frac{s}{b}\left(\frac{\lambda}{ik} - a\right) = -\frac{1}{2}s\frac{a-d}{b}$ or, for $s = 1$, $\phi_1 = -\frac{1}{2}\frac{a-d}{b}$.

We establish the second projecting matrix by use of the orthogonal to ϕ_1 vector ϕ_2, that is,

$$\begin{pmatrix} 1 & \phi_2 \end{pmatrix}\begin{pmatrix} 1 \\ \phi_1 \end{pmatrix} = 0. \tag{1.39}$$

This gives

$$\phi_2 = -\frac{1}{\phi_1}.$$

Using the basic formulas for the projectors (2.142) and (1.19), one arrives finally at

$$P_1 = \frac{1}{\phi_1 + \frac{1}{\phi_1}}\begin{pmatrix} \frac{1}{\phi_1} & 1 \\ 1 & \phi_1 \end{pmatrix} = \frac{a-d}{a^2 - 2ad + 4b^2 + d^2}\begin{pmatrix} 4\frac{b^2}{a-d} & -2b \\ -2b & a-d \end{pmatrix},$$

$$P_2 = \frac{1}{\phi_1 + \frac{1}{\phi_1}}\begin{pmatrix} \phi_1 & -1 \\ -1 & \frac{1}{\phi_1} \end{pmatrix} = \frac{a-d}{a^2 - 2ad + 4b^2 + d^2}\begin{pmatrix} a-d & 2b \\ 2b & 4\frac{b^2}{a-d} \end{pmatrix}.$$

The simplification $d = 0$, $a^2 = -4bc$ leads to the matrix operator P_1:

$$P_1 = \frac{1}{a^2 + 4b^2}\begin{pmatrix} 4b^2 & -2ab \\ -2ab & a^2 \end{pmatrix},$$

$$P_1\psi = \frac{1}{a^2 + 4b^2}\begin{pmatrix} 4b^2 & -2ab \\ -2ab & a^2 \end{pmatrix}\begin{pmatrix} u \\ v \end{pmatrix} = \frac{1}{a^2 + 4b^2}\begin{pmatrix} 4b^2u - 2abv \\ a^2v - 2abu \end{pmatrix}.$$

Making use of the first row, we determine the reference variable

$$\pi = -2b\frac{av - 2bu}{a^2 + 4b^2}.$$

Applying the projector to the original equation, we obtain

$$LP_1\psi = \partial\begin{pmatrix} a & b \\ c & 0 \end{pmatrix}\frac{1}{a^2 + 4b^2}\begin{pmatrix} 4b^2u - 2abv \\ a^2v - 2abu \end{pmatrix}.$$

The first row

$$b\partial\frac{a^2v-2abu}{a^2+4b^2}+a\partial\frac{4b^2u-2abv}{a^2+4b^2}=-ab\partial\frac{av-2bu}{a^2+4b^2}$$

leads to the equation for the reference variable π:

$$\pi_t=\frac{a}{2}\partial\pi.$$

Repeating the procedure for the second operator, we have

$$P_2\psi=\frac{a}{a^2+4b^2}\begin{pmatrix} a & 2b \\ 2b & 4\frac{b^2}{a}\end{pmatrix}\begin{pmatrix} u \\ v\end{pmatrix}=\begin{pmatrix} \lambda \\ \frac{2b}{a}\lambda\end{pmatrix}.$$

In turn, this allows to define the second reference variable

$$\lambda=a\frac{au+2bv}{a^2+4b^2}.$$

Application to the equation

$$LP_2\psi=\partial\begin{pmatrix} a & b \\ c & 0\end{pmatrix}\begin{pmatrix} \lambda \\ \frac{2b}{a}\lambda\end{pmatrix}=\begin{pmatrix} a\partial\lambda+\frac{2}{a}b^2\partial\lambda \\ c\partial\lambda\end{pmatrix}$$

gives, finally,

$$\lambda_t=a\partial\lambda+\frac{2}{a}b^2\partial\lambda=\frac{a^2+2b^2}{a}\partial\lambda.$$

To compare the results with the conventional analysis of the system (1.1) and (1.2), we rewrite it in the following manner:

$$u_t=au_x+bv_x,$$

$$v_t=cu_x+dv_x.$$

Multiplying the second equation by λ and taking a sum of equations, we obtain

$$u_t+\lambda v_t=au_x+bv_x+\lambda cu_x+\lambda dv_x,$$

$$u_t-(a+\lambda c)\,u_x=-\lambda\left(v_t-\frac{(b+\lambda d)}{\lambda}v_x\right).$$

Equalizing the directional derivatives

$$-(a+\lambda c)=-\frac{(b+\lambda d)}{\lambda},$$

yields

$$b-a\lambda+d\lambda-c\lambda^2=0.$$

The solution to the quadratic equation for nonzero c gives

$$\begin{aligned}\lambda_{+} &= -\tfrac{1}{2c}\left(a-d+\sqrt{-2ad+4bc+a^2+d^2}\right),\\ \lambda_{-} &= \tfrac{1}{2c}\left(-a+d+\sqrt{-2ad+4bc+a^2+d^2}\right),\end{aligned} \tag{1.40}$$

which proves the coincidence of both classifications.

A general hydrodynamics of a homogeneous gaseous or liquid medium is based on equations of motion, continuity, and energy. It is actually a local equilibrium level of description that neglects the relaxation connected with a finite mean free-path-length l of a fluid's molecules. There are five modes [2] that constitute a complete set of subspaces, which may be defined as eigenspaces of the linearized problem with appropriate projector operators [2,10]. The are referred to as natural forms in references [3,11]. Similar ideas may be found in [5]. Namely, the modes are acoustic (two branches), vortex (also two branches) and entropy ones. Five eigenvectors of the linear thermoviscous flow over the homogeneous background could be calculated approximately for the quasi-plane geometry of the flow [12–14]. The corresponding projectors are established and applied to derive nonlinear evolution equations for the interacting entropy, vortical, and acoustic modes.

The description of nonlinear phenomena in a fluid's flow is based on the interaction of the preceding modes. Among them, we may list heating, streaming and reflected wave. The acoustic heating and reflected wave are known as the secondary processes, which appear due to nonlinear absorption of sound. The investigation of the reflected wave (i.e., the oppositely directed wave in the field of an incident wave) began from a theoretical point of view but has been verified by some experimental data [15]. The theory concerned mainly the important case of the periodic acoustic source induced by transducer. Though the periodic perturbations are an important particular case of real processes and allow the nonlinear effects to enhance with time, nonperiodic (also impulse) acoustic waves, such as those widely used in medicine, may give rise to the heating and the reflected wave (see also [15] and the referred papers). The pulsed structure of acoustic waves could help to separate the initializing signals and the responses.

Acoustic heating of the background is a slow (mean field) phenomenon appearing due to nonlinear sound absorption. It is important in many applications of ultrasound, and in medical therapy, where an increase of temperature should be evaluated as accurately as possible. An elevation of temperature is caused by a heat producing per unit volume and followed as usual by an isobaric decrease of density. The variation of density caused by absorption has not been taken into account in many sources [16,17] since the equilibrium state was traditionally referred to purely as incompressible liquid. Variations in density proved to be important in studies of acoustic heating for the majority of fluids [15]. The secondary processes in the sound field, such as acoustic heating, change the background of acoustic wave propagation and therefore impact on the primary wave itself. Novel phenomena, such as the self-focusing of an acoustic beam, has been observed and explained by strong nonlinear ultrasound absorption [17].

If the acoustic perturbations are periodic, averaging over one period (or over an integer number of periods), it allows us to separate non-acoustic and acoustic parts of the overall perturbation. The interval of averaging must be small in comparison to the characteristic time of variations of the non-acoustic slow field. In an absorbing fluid, the total energy conservation law should be considered:

$$E_t + \vec{\nabla}\vec{J} = 0.$$

Here, $E = \rho\varepsilon + \rho(\vec{v}\cdot\vec{v})/2$ is the total energy volume density, $\vec{J} = p\vec{v} + E\vec{v}$ is the energy flux density vector, and $\varepsilon, \rho, \vec{v}, p$ are the internal energy per mass unit, mass density, velocity, and pressure, respectively. All variables are thought of as a sum of the acoustic (subscript a) and non-acoustic parts. Averaging over one period for the periodic sound wave (marked by square brackets) yields the instantaneous rate per unit volume q at which heat is produced in a fluid by ultrasound:

$$\langle q\rangle = -\vec{\nabla}\left\langle \vec{J}_a\right\rangle$$

with an acoustic source in the right side, $\left\langle \vec{J}_a\right\rangle = \langle p_a\vec{v}_a\rangle$ [18]. This way to evaluate heat generation is suitable for periodic acoustic waves. The procedure of time averaging lefts only the "slow" and deletes the "quick" part of the overall flow and fits many real applications concerning periodic ultrasound, but it has an essential shortcoming. The temporal averaging fails when pulses or other nonperiodic acoustic waves generate heating.

We propose a way to evaluate heating caused by any acoustic field, including nonperiodic ones (Section. 3.2). We continue to apply the main idea to separate modes on the level of the basic system of differential equations. Modes, like all flows that follow the local conservation laws of fluids, are thought of as eigenvectors of this system of equations. Matrix dynamic projectors allow us to separate every mode from the overall perturbation at any moment and to derive dynamic equations for the interacting modes. This idea has been developed by the authors and applied to flows over inhomogeneous and stratified media [19–21]. The steps to get an approximate solution are also pointed out there and reviewed there. The procedure does not need temporal averaging, and it allows us to account for non-isentropic initial conditions.

We do not consider here the phenomenon of three-dimensional heating where rotational fluid motion complicates the direct application of our method. Generally, after modification, the idea of the projecting is fruitful also in the case of multidimensional movement [14]. Note that rotational modes are "slow" as well as entropy mode so the temporal averaging could not separate these "slow" modes in a consistent manner. To avoid involving the heat mode, the initial point for problems relating to streaming are dynamic equations for flow over incompressible liquid. The inconsistency of this approach was emphasized in some sources [22,23]. We may add that the heating effects left by ultrasound change both the temperature and the density of the medium. Therefore, the new background of the acoustic and rotational perturbations is formed that could not be taken into account in the incompressible model of fluid.

In the following sections, we specify an equation that governs streaming caused by arbitrary acoustic waves. We trace this equation to the known results on streaming caused by a quasi-periodic source. As an important illustration, we calculate the radiation acoustic force for the mono-polar source. Traditionally, streaming is referred to as a localized mean flow generated by an acoustic source. This flow is created by the Reynolds forces, nonzero time-averaged values of the quadratic acoustic terms arising in the equation of conservation of momentum. There exist extensive reviews on this subject [18,24,25]. Steady acoustic streaming is described by the mean Eulerian velocity $\langle \vec{v} \rangle$, or the mean mass flow velocity $\vec{U}$, where $\vec{U}$ is defined by $\vec{U} = \frac{\langle \rho \vec{v} \rangle}{\rho_0} = \vec{V} + \frac{\langle \rho_a \vec{v}_a \rangle}{\rho_0}$, where $\vec{V}$ is a slow component of velocity, namely the Eulerian velocity of the streaming. The difference between $\vec{U}$ and $\vec{V}$ is approximately proportional to the acoustic intensity vector and is small compared with any of mean velocities [26]. From the continuity equation, it follows $\nabla \vec{U} = 0$ if we assume $\left\langle \frac{\partial \rho}{\partial t} \right\rangle = 0$. The usual discussion of acoustic streaming [22,23] suggests that there is an important unresolved issue concerning acoustic streaming, namely the effect of fluid incompressibility (the continuity equation is ignored). The inconsistency of this point is that fluid is supposed to be incompressible while the acoustic wave causing streaming may propagate only over a compressible medium. Not to mention that a gas is strongly compressible, and the conservation of energy in this limitation is not considered, since excluding of one variable (density) reduces an initial system. It has been proved that the effect of the heat conductivity can not be discarded in a study of temperature variation associated with streaming [23]. Actually, this approximation is well understood in a typical liquid like water, but it should be revised for other liquids [27]. This is a convincing reason to start from the full system of conservation laws, including the energy balance equation and the continuity equation.

The procedure of temporal averaging requires continuous perturbations of many periods as an acoustic source; thus, the well-known results on acoustic streaming apply only for this particular case, though they are quite realistic. There are clear inconsistencies concerning the very procedure of temporal averaging. As mentioned before, an equality $\nabla \vec{U} = 0$ is valid only if $\left\langle \frac{\partial \rho}{\partial t} \right\rangle = 0$. In fact, the overall field includes not only acoustic and vortical modes, but also the entropy mode. The entropy mode caused by acoustic waves is known as a slow decrease of density and an isobaric increase of temperature of the background of acoustic wave propagation. The importance of variations in the density of the surroundings (not only in temperature) has been emphasized in the review [15]. If nonlinear interactions are accounted in a proper way, the density of the entropy mode is described by the following equation :

$$\frac{\partial \rho_e}{\partial t} + \frac{\eta}{\rho_0} \Delta \rho_e = S_a,$$

where η is shear viscosity, S_a is an acoustic source proportional to a gradient of acoustic intensity, and index e concerns the entropy mode. An attenuation of an acoustic wave leads to a slow but continuous decrease of density, so that $\langle \frac{\partial \rho_e}{\partial t} \rangle = 0$

becomes incorrect as well as $\left\langle \frac{\partial \rho}{\partial t} \right\rangle = 0$, since $\rho = \rho_0 + \rho_a + \rho_e$. Therefore, if the acoustic heating caused by a periodic acoustic wave is accounted for $\nabla \vec{U} = 0$ becomes incorrect.

These inconsistencies call for new ways to conduct theoretical investigations of acoustic streaming, including streaming caused by nonperiodic acoustic waves. We plan to start with all conservation equations for compressible thermoviscous flow, then to define modes by establishing the relations between perturbations (Sections 3.1,4.1), to demonstrate how the well-known linear and nonlinear evolution equations may be obtained by projecting (Section 4.2), and, finally, to apply projectors to calculations of the streaming caused by pulse acoustic waves (Section 4.3).

The nonlinear phenomena in flows over inhomogeneous media are of great importance. Boundaries in the viscous flow introduce inhomogeneity in the equilibrium hydrodynamic parameters, and, in particular, in the mean velocity of a flow. The total hydro-thermodynamic perturbations with account for a boundary layer are split into eigen modes: a Tollmien–Schlichting (TS) wave and two acoustic ones. A mode definition is realized via local relation equations, which are extracted from the linearizion of the general system over an equilibrium boundary layer flow. Again, such relations define the subspace and the appropriate projector. So, the subdivision may be performed locally in this essentially inhomogenous problem [28].

In general, nonlinearity specifies the form of interactions between acoustic and TS fields. By means of projecting into a subspace of an Orr–Sommerfeld equation solution for the TS mode and the corresponding procedure for acoustics, the equations rearrange to a one-dimensional system that describes evolution along the mean stream. A new mechanism of nonlinear resonance excitation of the TS wave by sound is proposed and modeled via four-wave interaction.

The idea to decompose the linear flow into specific modes is not novel and has been exploited for a long time, see again [2], where a homogeneous background with sources of heat and mass are considered. The principle advance is an expansion of the ideas into areas of nonlinear flow: to get nonlinear coupled evolution equations for interacting modes and to solve the system approximately, for asymptotic solutions see [7]. Deriving the final nonlinear system is algorithmic; one just applies projectors at the initial system of nonlinear equations. Some ideas concerning the separation and interaction of acoustic-gravity disturbances [10] and all waves in geophysical hydrodynamics may be found in [4].

A general electrodynamics naturally starts from a self-consistent Maxwell system in differential form for a continuous medium that is supplemented by material (thermodynamic) relations between a polarization vector and an electric field as well as a magnetization vector and a magnetic field. Both fields represent conventionally the electromagnetic field vector of state. The material relations generally are integral if dispersion is taken into account and can include nonlinear terms [30]. More general background contains the dependencies of the polarization and magnetization vectors with both fields $\vec{E}, \vec{B}$ [31].

The system of Maxwell equations including the charge density and the density of current may be closed in minimal approach by the local form of Ohm's Law $j_k =$

$\sigma_{ki}E_i$ and, more generally, by the appropriate integral relation; see also Chapter 10. The additional components of the state vector may appear if one accounts for the motion of a charged particle that enters the system via charges and current densities. In this platform we go up to plasma physics, which may include multicomponent a description for each ion density/velocity/pressure dynamics.

The complete system of Maxwell equations splits into independent subsystems by means of a special dynamic projecting technique presented in this book. The technique relies upon a direct link between the field components that determine appropriate modal subspaces. The explicit form of links and their corresponding subspaces evolution equations are obtained in conditions of some symmetry; it is illustrated by examples of plane waves (Chapters 7 and 8), rectangular (Chapter 9) and cylindrical waveguides (Chapter 9), as well as spherical and quasi-one-dimensional ones (Chapter 10). The principal part makes use of a linearized statement of the problem, and the nonlinear terms are included as extra terms, stipulating the modes interaction.

Describing the unidirectional propagation of light pulses in a medium with cubic nonlinearity, we start from the nonlinear Schrödinger equation (NLSE), which was derived by Zakharov [32]. This derivation is based on an envelope wave approximation, its validity is restricted by pulses of a duration larger than the carrier wave period by order. The equation is proved to be so-called integrable, i.e. to belong to a class of solitonic ones as the pioneering KdV [33]. To go down to shorter pulses, we need another approach. Our interest is for ultra-short pulses whose length is shorter than a few cycles of the center frequency. Such pulses were investigated by other studies, which explored many alternative ideas [34–38]. In the derivation of equations describing the propagation of ultra-short pulses, there are two basic approximations for an electromagnetic field, which are described in the review by Maimistov [39], namely the slowly varying envelope approximation and the unidirectional wave approximation. For ultra-short pulses, whose length does not exceed a few cycles, the first approximation leads to inaccurate results. Therefore, the validity of the application of NLSE to describe the propagation of ultra-short pulses of light is questionable by different groups [40–42]. The approach to the problem makes use of a slowly evolving wave approximation SEWA [42] or an application of Bloch equations [43]. Another approach is to introduce new field variables in the form

$$\psi^{\pm} = \varepsilon\frac{1}{2}E_i \pm \mu\frac{1}{2}H_j,$$

as did Fleck [34] and Kinsler [38,44,45] in their works. It is a method similar partially to the projection operator method, which was described in the Leble's book [4], with many examples in different fields of physics and geophysics. In accordance to this method, the specification of fields is accompanied by a corresponding combination of basic equations. Being completed by the projection operator technique, it is a powerful tool, which has been initially set for the Cauchy problem and which allows for detailed analysis of the total field and individual modes of field perturbations. Applying the projection operators to the one-dimensional model of the light pulse propagation has led directly to the generalized short pulse equation (GSPE) [9], in which the physical form of cofactors has been shown and the interactions between waves propagating in opposite directions has been taken into consideration. The case of the interaction of two counter-propagating waves was the topic of experimental studies [46], where authors presented the experimental proof of interaction of counter-propagating pulses with different polarizations. The theory for such case was proposed and discussed in [47]. In the other case, the authors of the book [48] have proven the existence of the wave propagating backward in optical resonators. The most fitting (short pulse) case is a paper that concerns backward pulse generation in a transmission line [49], which lacks a theoretical description of the phenomenon.

A mode decomposition of a problem for the general electrodynamics in a medium may be considered in a manner similar to hydrodynamics [4,9]. We should take into account that the Maxwell equations already contain two projecting equations that do not include time derivatives and hence define differential links between fields components. The simplest nontrivial version of the electrodynamics reduces to a planar one-dimensional wave equation with a fixed polarization. Such a version has important applications in such physically specified circumstances as laser generation [34] or optical fibers [50,51]. The coexistence and, in some amplitude ranges, interaction, of oppositely directed waves in a similar situation is an important feature of a phenomenon. It is of essential reason to study equations of directed short [4,50] and ultrashort [37] directed waves, as well as generalizations with the simultaneous presence of waves in both directions of propagation [47] and, next, with an account for polarization [51]. We also study a one-dimensional model, similar to Schäfer, Wayne [37] and Kuszner, Leble [9], in which the x-axis was chosen as the direction of a pulse propagation. Like the mentioned authors, we assume $D_x = 0$ and $B = 0$, taking into account only one polarization of electromagnetic waves. We establish and apply projecting operators to a Maxwell system that allows us to split the total field into directed waves. In the case of weak nonlinearity, it is possible to generalize the linear result for arbitrary dispersion and to derive the system of interacting rightward/leftward waves with combined (hybrid) magnitude and polarizations. We also discuss and investigate solutions to the system related to some boundary regimes. The significance of this approach to the problem of self-action and the interaction of the directed waves is also supported by some developing directions of mathematical physics. They concern both integrable models in this context [33,52,53] and their embedding to general electrodynamics problems [4,54].

An interest to the simultaneous study of dynamics of individual directed waves arises in different contexts (see, e.g., [4,46]), for example, naturally, in nonlinear resonator theory. The mentioned book introduces a general tool of theoretical physics to derive evolution equations of the first order in time that naturally include unidirectional ones corresponding to elementary roots of a dispersion equation. It is based on a complete set of projecting operators, each one for a dispersion relation that fixes the corresponding subspace of a linearized fundamental system such as Maxwell equations. The method allows us to combine equations of the complex basic system in the most rational way with account for dispersion, dissipation, and, after some development—nonlinearity, introducing also combined (hybrid) fields as basic modes. It allows us therefore to formulate a corresponding mathematical problem effectively in appropriate physical language. This concerns initial or boundary conditions problems.

Continuing on the mentioned investigations and especially their nonlinear development, we go to the general method of integrable evolution equations, whose implementation was adjusted to the problems to be considered. The theory links a unidirectional pulse propagation approach with derivations of Nonlinear Schrödinger (NS), Coupled NS, Manakov, and generalizations in conditions of optical waveguides [51,55]. One of examples of this theory, which is based on the second-order wave equation, is given in Ref. [38]. The author shows that a unidirectional approach for hybrid electromagnetic fields allows us to write down the first-order wave equation for pulse propagation. Moreover, the author of the article reveals a generalization of the unidirectional [44] and bidirectional [45] approaches presented in previous works. The results, however, do not account for dispersion effectively and hence require generalization. The propagation of a few cycle optical pulses in a nonlinear regime was also intensely studied from different points of view [36,56].

Recently, the trend toward a detailed description of fairly short pulses was successfully advanced in a wave range from approximately 680 nm to infrared [37]. The resulting equation (named the short pulse equation - [SPE]) was properly investigated numerically [57] and was proven to be integrable [52,53]. The derivation of the SPE was based on a few approximations, including the dispersion relation adjusted to the mentioned wave range, accounting for the third-order nonlinearity and unidirectionality of a pulse and making use of a slow time variable in the reference frame following the pulse [58] with developments in [56,57]. We start from the formulation of a Cauchy problem for the 1D electrodynamics modeling of a pulse propagation in a medium with a conventional material relation and with application to short and ultrashort pulses with a derivation of systems for leftward/rightward progressive wave modes that may be generalized as celebrated NLS and SPE integrable equations.

Further, we continue to study a boundary regime problem for 1D electrodynamics modeling, which considers propagation of a pulse in a medium with a material relation to be general enough to include metamaterials. It relates also to a problem of wave excitation in a metameterial layer boundary with reflection so the nonlinearity serves as an additional reflecting boundary. More general, and hence more complicated, 3D or 2D problems in electrodynamics rely upon a combination of two types of projecting [4,54]. The first is transversal mode projecting, in essence a generalized Fourier series expansion with respect to special functions basis, established by symmetry of the problem. The second is the dynamical matrix projecting operating at a subspace of a transversal mode, which is the main topic of this book.

A metamaterial is conventionally an artificial material that has dielectric permittivity and magnetic permeability that may not exist in nature. Perhaps the first research on metamaterials was conducted by J.C. Bose in 1898 [59]. He studied the rotation of a polarization plane of electromagnetic waves by a twisted structure created by himself. In 1946, W.E. Kock suggested to use a mixture of metal spheres in a matrix as an artificial material [60,61]. He was probably the first to use the term *artificial dielectric*, which later was widely accepted in microwave literature [62,63].

Problems involving metamaterials with simultaneously negative dielectric permittivity and magnetic permeability are most relevant to this book's method. Malyuzhinets proposed backward-wave transmission lines in 1951 [64], which hint at the use of directed waves notion implementation. Conducting stick composites [65] and bi-helix media [66] were studied in microwave range with the accent on negative values of dielectric permittivity and magnetic permeability.

A visible gap in the theory relates to the textbook formula for energy density, $W = \varepsilon E^2 + \mu H^2$, which appears in the Pointing Theorem, derived in conditions of dispersionless media, and which obviously does not allow negative material constants. The first theoretician who studied the general properties of dispersive materials was Victor Veselago [67]. He showed that for a medium with frequency dispersion, the energy density of electromagnetic wave packet is approximately described by the expression:

$$w = \frac{\partial \omega \varepsilon(\omega)}{\partial \omega} E^2 + \frac{\partial \omega \mu(\omega)}{\partial \omega} H^2,$$

in which connection $w > 0$, if

$$\frac{\partial \omega \varepsilon(\omega)}{\partial \omega} > 0, \; \frac{\partial \omega \mu(\omega)}{\partial \omega} > 0,$$

for details, see for example [68]. Such formalism admits simultaneously negative medium parameters $\varepsilon < 0$ and $\mu < 0$. To achieve negative values of the constitutive parameters ε and μ, metamaterials must be dispersive in some range, i.e., their permittivity and permeability should be frequency dependent; otherwise, they would not be causal [69]. The two-time derivative Lorentz material (2TDLM) model incorporates the most commonly discussed metamaterial models; they have the frequency domain susceptibility as follows [70]:

$$\chi = \frac{\omega_p^2 \chi_\alpha + i\omega_p \chi_\beta \omega - \chi_\gamma \omega^2}{\omega_0^2 + i\omega\Gamma - \omega^2}. \tag{1.41}$$

There would be independent models for the permittivity and permeability $\varepsilon(\omega) = \omega_0(1+\chi_e)$ and $\mu(\omega) = \mu_0(1+\chi_m)$. This 2TDLM model produces a resonant response at $\omega = \omega_0$ when Γ is small. It recovers the Drude model when this resonant frequency goes to zero, and the constants $\chi_\alpha = 1$, $\chi_\beta = \chi_\gamma = 0$, so that, for instance,

$$\varepsilon = \varepsilon_0 \left(1 - \frac{\omega_p^2}{\omega(\omega - i\Gamma)}\right).$$

The real part of this permittivity is clearly negative for all $\omega < \sqrt{\omega_p^2 + \Gamma^2}$. The Drude model is obtained with the bed-of-nails medium [71,72]. In the reference [73], a metallic mesostructure split-ring resonator (SRR) medium was created, which allowed David Smith and his group to investigate such types of structures and to establish that their permittivity and permeability were negative [74].

The applications for metamaterials are broad and vary from the celebrated electromagnetic cloaking [75,76] to new imaging capabilities [77] such as superlensing data storage [78]. An important advance in nonlinear Kerr metamaterials was made in laser technologies [79]. As for directed waves applications in waveguides, Alu and Engheta [80], and Grbic and Eleftheriades [81] studied the concept of cavity resonators and the transmission of line-based metamaterials to be applied to waveguides and generally guided-wave structures, which sub-wavelength ranges admit to overcome the diffraction limit. The directional couplers were described in [82,83]. A high-efficiency directional coupler is described in [84]. In 2010, physicists from China demonstrated experimentally the first broadband illusion-optics device [85].

Metamaterials are also applicable in acoustics. We may list some directions of application: controlling the path of waves leads to incredibly efficient lenses [86] and invisibility cloaking [87,88], and controlling their transmission and reflection leads to highly efficient filters [89], diodes [90–92] or superabsorbers [93]. It was demonstrated that an acoustic metadevice can impact on the acoustic control any of the ways in which waves travel. It opens a possibility to apply electromagnetic and acoustic directed modes division altering dynamically a geometry of a three-dimensional colloidal crystal in real time [94,95].

In most cases, metamaterials are fabricated using planar lithography techniques that can not produce large, volumetric metamaterials. This is valid in particular at optical frequencies, but even at THz and microwave frequencies, achieving volumetric rather than two-dimensional samples requires typically the manual assembly of two-dimensional metamaterial boards or sheets. The simplest example of such metamaterials is the fibre array, with an effective permittivity along the fibers having a plasmonic response that can be adjusted through the diameter and separation of the wires [73].

The nonlinear interaction of ultrashort pulses in a metamaterials is well understood in the nonlinear optics [96]. Structures with simultaneously negative dielectric permittivity and magnetic permeability have been called by many names: Veselago media, negative-index media, negative-refraction media, backward wave media, double-negative media, negative phase-velocity media, and even left-handed media (LHM) [97].

Projecting operators in electrodynamics have specific properties due to the structure of Maxwell equations and the physicall nature of the basic field components. Many approaches can help simplify a general electrodynamics statement of problem. Many are based on an anzatz for a solution such as a space- and time-dependent amplitude factor with a small parameter to cut terms with higher derivatives. An exemplary approach to the problem makes use of a slowly evolving wave approximation (SEWA) [108]. Another approach is to combine basic variables, as has been already mentioned, in a dispersionless medium

$$\psi^{\pm} = \varepsilon\frac{1}{2}E_i \pm \mu\frac{1}{2}H_j,$$

as Fleck [34], Kinsler [38,45] and Amiranashvili [98] did in their works. It is a trick, a change of field variables that is an ingredient of the projection operator method, which is the main tool of this book. In Chapters 7 and 8, we establish and apply the projecting operators in electrodynamics, as in the previous chapters; see again the publications [4,54], where some aspects of the method were initiated.

That is a fairly general tool in theoretical physics, to derive evolution equations of the first order in time, which naturally include unidirectional waves corresponding to elementary roots of a dispersion equation. The procedure is based on a complete set of projecting operators, each for a dispersion relation that fixes the corresponding subspace of a linearized fundamental system such as Maxwell equations. The method, compared to this one used in [34,38,99], allows us to combine simultaneously *equations* of the complex basic system in an algorithmic way, taking into account polarization, dispersion, dissipation and, after some development, nonlinearity. It introduces combined (hybrid) fields as basic modes. The method therefore gives a possibility to formulate equivalent mathematical problems of lower order, including initial or boundary conditions in an appropriate physical language in mathematically convenient form.

In Chapter 8, we apply the method of projecting to the problem of 1D-metamaterial wave propagation with the specific dispersion of both ε and μ. The main aim of this application is similar to the recent study [38,47]: we want to derive an evolution equation for the mentioned conditions with the minimal simplifications. We highlight and discuss methodical differences between the papers and this book's presentation and results. To conclude, let us mention applications of the method in quantum mechanics. There may be found some applications in connection with the theory of relativistic particles in textbooks. The equation of Dirac may be solved as the system of equations similar to the hydrodynamic and electrodynamic ones [100].

Our starting point of Chapter 8 is the Maxwell equations for a simple case of linear isotropic but dispersive dielectric media in the SI unit system:

$$\mathrm{div}\vec{D}(\vec{r},t) = 0, \tag{1.42}$$

$$\mathrm{div}\vec{B}(\vec{r},t) = 0, \tag{1.43}$$

$$\frac{\partial\vec{B}(\vec{r},t)}{\partial t} = -\mathrm{rot}\vec{E}(\vec{r},t), \tag{1.44}$$

$$\frac{\partial\vec{D}(\vec{r},t)}{\partial t} = \mathrm{rot}\vec{H}(\vec{r},t). \tag{1.45}$$

We also do not include space charges into the system but fix attention on waves propagation. Note also that the units of the Chapter 7 are in the Gauss system for a reader alternative.

The system is completed by material equations of a very similar structure:

$$\vec{D}(\vec{r},t) = \frac{\varepsilon_0}{2\pi}\int_{-\infty}^{\infty}\int_{-\infty}^{\infty}\varepsilon(\omega)\exp(i\omega(t-s))d\omega\vec{E}(\vec{r},s)ds,$$
$$\vec{B}(\vec{r},t) = \frac{\mu_0}{2\pi}\int_{-\infty}^{\infty}\int_{-\infty}^{\infty}\mu(\omega)\exp(i\omega(t-s))d\omega\vec{H}(\vec{r},s)ds. \quad (1.46)$$

The form of the system (1.42)–(1.46) is the starting point for the three-dimensional (3D) problems of Chapter 10. More complicated 3D problems require more advanced constructions and geometry (such as, for example, cylindrical or spherical constructions in geophysical applications [101]), and may lead to very nontrivial generalizations of the technique and algorithm as well as the norm of the appropriate construction of subspaces. We begin from a rectangular waveguide problem, making use of a basis of transverse modes, the simplest elementary functions that guarantee zero-boundary conditions for physically reasonable fields. In cylindrical waveguides such as optical fibers, we use the Bessel functions, which, in the linear case, lead to Hondros–Debye theory [102], and in the nonlinear case end in multimode interaction [108]. The open 3D space implies some geometrical restrictions on the conditions of wave generation (quasi-one-dimensional or spherical), which will be considered in the last chapter.

A problem of a wave mode identification has interdisciplinary character; it is important, for example, in physics of atmosphere, where the superposition of acoustical, gravity, and planetary waves occurs [103]. For planetary Rossby and Poincare waves, the spatial scales are of the same order; hence their separation, and an estimation of contributions is complicated. Taking an account of their periods does not lead to high precision because of the presence of other atmospheric phenomena.

A similar problem relates to the distinction between acoustic and internal gravity waves in planetary atmospheres [104]. The situation is even more complicated in plasma physics because of additional specific branches of waves coexisting with ones for a neutral gas. In geophysics, the wave field diagnostics by means of wavelengths generally needs many-point observations that cover an extant space domain. This is fairly expensive and hence sometimes does not look feasible. Next, if an important problem of a specific wave mode source localization is studied, it is formulated as the typical inverse problem that is generally ill-posed [105]. To solve such a problem, it requires the so-called *a priori* information that may include the *generalized polarization relations* that fix a mode (e.g., directed wave) which is determined in this book.

Thinking about a novel alternative approach [101], we propose to use measurements in the vicinity of an only point, but this method implies many component observations with the aid of a special mathematical tool, namely the projecting operators technique, which was developed in this study for the purpose of the wave's identification. There are lot of important problems in theoretical physics of the same level of description: in electrodynamics, see [9,38,106], in acoustics, [12,13], in hydrodynamics, including the boundary layer Tollmien–Schlichting waves, [28]. All mentioned problems may be directly formulated as a system of equations, so such a problem implies the vector description (in general electrodynamics there are six reference field variables, components of electric and magnetic fields). Such components have direct physical sense; therefore, they are convenient as basic variables of a problem study. The projecting gives the link between the components that represents the already-mentioned "generalized polarization relation" and, hence, allows us to extract and identify the contribution of the specific mode in the total perturbations [107].

Let us describe the content of chapters very briefly for readers' convenience.

Chapter 2 contains mainly mathematical description of the algorithms with accent to difference in formalism of boundary regime and initial problems. Some hints for nonlinearity, inhomogeneity, and spectral properties account are made. **Chapter 3** may be divided into two parts. The first one is about heating of a medium induced by acoustic waves. An effectivity of the process depends on both attenuation coefficients: thermoconductivity and viscosity. Namely the projecting *approach* allows to build instant action transmission, hence the possibility of pulsed acoustics usage for the effects generation. The second part describes problems of planetary atmospheres. Its stratification (nonuniformity) gives a good possibility to demonstrate a power of the method, separating such nontrivial waves as Rossby, internal and acoustic ones. **Chapter 4** contains information about such interesting phenomenon as a transition of momentum and energy from acoustic oscillation of a medium to a rotational mode. Such possibility may realize itself only in the presence of two effects: viscosity and nonlinearity. **Chapter 5** relates to media with various kinds of relaxations. It includes such important issues as vibration relaxations of molecules and chemical reactions in gas and relaxation in plasma. This leads to peculiarities *in the field's behavior* such as energy and momentum losses and conductivity presence. Then, the entropy mode definition and proper account for its energy becomes significant. There is rather a general difficulty when we apply the projecting method to problems in which inhomogeneity of propagation medium leads to equations with coefficients dependent on coordinates. One way to generate an universal tool for such cases, we study in **Chapter 6** the subdivision of acoustic and Tollmien-Schlichting waves in a boundary layer. We also build and apply dynamic projecting, but do not use the Fourier transformation. Nevertheless we achieve the result: we derive the system that describes modes interaction and even link it with cases of the known integrable N-wave systems. The application of the dynamic projecting method to electrodynamics begins in **Chapter 7** from a conventional example of 1D wave processes *in a matter* with dispersion and nonlinearity account. It leads to so-called hybrid fields (modes), this name was coined to such E-B combination by Kinsler. The consideration is focused on short pulses in conditions of Shafer-Wayne dispersion and nonlinearity, but taking into account both directions of propagations and both polarizations. So, the four-component equations are derived, investigated, and, after some simplifications, solved. **Chapter 8** contains 1D theory of electromagnetic wave propagation in metamaterials. The celebrated Drude dispersion for both magnetic and electric material relations is studied in details. Solutions for initial or boundary excitations by short pulses are based on two-component Shafer-Wayne equations, its construction in terms of elliptic functions is built and illustrated. **Chapter 9** is devoted to waveguide propagation in two geometries: rectangular metal waveguides and cylindrical dielectric (fibers). The main issue is the separation of directed waves with polarization account. It is also done when both waveguide and material dispersions are superposed. The Kerr nonlinearity is also taken into account. **Chapter 10** relates to general electrodynamics, it contains the dynamic projecting operator of maximum dimensions of this book. The operators separate solution in 3D space, starting from independent six fields, describing an electromagnetic wave (or stationary solution) in matter. The inclusion of material relations and charges is studied. The cases of spherical symmetry and quasi-1D are also presented.

BIBLIOGRAPHY

1. Fock, V.A. 1936. The fundamental significance of approximate methods in theoretical physics. *Advances in the Physical Sciences* 16(8): 1070.
2. Kovasznay, L.S.G. 1953. Turbulence in supersonic flow. *Journal of Aerosol Science* 20: 657–682; Chu, B.T., and L.S.G. Kovasznay. 1958. Non-linear interactions in a viscous heat-conducting compressible gas. *Journal of Fluid Mechanics* 3: 494–514.
3. Novikov, A.A. 1976. Application of the method of coupled waves to an analysis of nonresonance interaction. *Radiophysics and Quantum Electronics* 19: 225–227.
4. Leble, S. 1988. *Nonlinear Waves in Waveguides*. Leningrad State University (extended version Springer, Heidelberg, Sankt Petersburg, 1990).
5. Belov, V.V., S.Y. Dobrokhotov, and T.Y. Tudorovskiy. 2006. Operator separation of variables for adiabatic problem in quantum and wave mechanics. *Journal of Engineering Mathematics* 55: 183–237.
6. Haberman, R. 2004. *Applied Partial Differential Equations: With Fourier Series and Boundary Value Problems*. Prentice Hall, Pearson, Upper Saddle River, NJ.
7. Maslov, V.P., and G.A. Omel'yanov. 2001. *Geometric Asymptotics for Nonlinear PDE. I*. AMS, Providence, RI.
8. Leble, S., and A. Zaitsev. 1989. *New Methods in Nonlinear Waves Theory*. Kaliningrad State University, Kaliningrad.
9. Kuszner, M., and S. Leble. 2011. Directed electromagnetic pulse dynamics: Projecting operators method. *Journal of the Physical Society of Japan* 80: 024002.
10. Bessarab, F.S., S.P. Kshevetsky, and S. Leble. 1987. Acoustic-gravity waves interaction in the atmosphere problems of nonlinear acoustics. In: ISNA Proceedings, Novosibirsk, pp. 144–148.
11. Novikov, A.A. 1981. The perturbation method and natural forms in nonlinear waves theory. *Waves and Diffraction* 2: 66–69.
12. Perelomova, A. 2003. Interaction of modes in nonlinear acoustics: Theory and applications to pulse dynamics. *Acta Acustica united with Acustica* 89: 86–94.
13. Perelomova, A. 2006. Development of linear projecting in studies of nonlinear flow. Acoustic heating induced by non-periodic sound. *Physics Letters A* 357: 42–47.
14. Perelomova, A., and A. Perclomova. 2004. Acoustic streaming caused by modulated sound and wavepackets. *Archives of Acoustics* 29(4): 647–654.
15. Makarov, S., and M. Ochmann. 1996. Nonlinear and thermoviscous phenomena in acoustics, part I. *Acustica* 82: 579–606.
16. Bakhvalov, N.S., Y.M. Zhileikin, and E.A. Zabolotskaya. 1987. *Nonlinear Theory of Sound Beams*. American Institute of Physics, New York.
17. Karabutov, A.A., O.V. Rudenko, and O.A. Sapozhnikov. 1989. Thermal self-focusing of weak shock waves. *Soviet Physics Acoustics* 34: 371–374.

18. Rudenko, O.V., and S.I. Soluyan. 1977. *Theoretical Foundations of Nonlinear Acoustics*. Plenum, New York.
19. Perelomova, A.A. 1998. Nonlinear dynamics of vertically propagating acoustic waves in stratified atmosphere. *Acta Acustica* 84: 1002–1006.
20. Leble S., and A. Perelomova. 2013. Problem of proper decomposition and initialization of acoustic and entropy modes in a gas affected by the mass force. *Applied Mathematical Modelling* 37: 629–635.
21. Perelomova, A.A. 2000. Nonlinear dynamics of directed acoustic waves in stratified and homogeneous liquids and gases with arbitrary equation of state. *Archives of Acoustics* 25(4): 451–463.
22. Qi, Q. 1993. The effect of compressibility on acoustic streaming near a rigid boundary for a plane traveling wave. *Journal of the Acoustical Society of America* 94(2): 1090–1098.
23. Menguy, L., and J. Gilbert. 2000. Non-linear acoustic streaming accompanying a plane stationary wave in a guide. *Acta Acustica* 86: 249–259.
24. Nyborg, W.L. 1965. Acoustic streaming. In: Manson, W.P. (Ed), *Physical Acoustics, Vol. II, Part B*, Academic Press, New York, pp. 265–331.
25. Lighthill, M.J. 1978. Acoustic streaming. *Journal of Sound and Vibration* 61(3): 391–418.
26. Gusev, V.E., and O.V. Rudenko. 1980. Evolution of nonlinear two-dimensional acoustic streaming in the field of highly attenuated sound beam. *Soviet Physics Acoustics* 27: 481–484.
27. Kamakura, T., M. Kazuhisa, Y. Kumamoto, and M.A. Breazeale. 1995. Acoustic streaming induced in focused Gaussian beams. *Journal of the Acoustical Society of America* 97(5), Pt. 1: 2740–2746.
28. Leble, S., and A. Perelomova. 2002. Tollmienn-Schlichting and sound waves interacting: Nonlinear resonances. In: Rudenko, O.V., Sapozhnikov, O.A. (Eds), Nonlinear Acoustics in the Beginning of the 21st Century, vol. 1 Faculty of Physics, MSU, Moscow, pp. 203–206.
29. Perelomova, A.A. 2000. Projectors in nonlinear evolution problem: Acoustic solutions of bubbly liquid. *Applied Mathematics Letters* 13: 93–98; Perelomova, A.A. 1998. Nonlinear dynamics of vertically propagating acoustic waves in a stratified atmosphere. *Acta Acustica* 84(6): 1002–1006.
30. Boyd, R.W. 1992. *Nonlinear Optics*. Academic Press, Boston.
31. Wegener, M. 2005. *Extreme Nonlinear Optics: An Introduction. Advanced Texts in Physics*. Springer-Verlag, Berlin-Heidelberg.
32. Zakharov, V.E. 1968. Stability of periodic waves of finite amplitude on the surface of a deep fluid. *Journal of Applied Mechanics* and *Technical Physics* 9, 2, Springer, New York, 190.
33. Zakharov, V.E., and A.B. Shabat. 1974. A scheme for integrating nonlinear equations of mathematical physics by the method of the inverse scattering problem. *Functional Analysis and Its Applications* 8(3): 5456: Zakharov, V.E., and A.B. Shabat. 1979. Integration of nonlinear equations of mathematical physics by the method of inverse scattering. II. *Functional Analysis and Its Applications* 13(3): 166–174.
34. Fleck, Jr. J.A. 1970. Ultrashort pulses generation by Q-switch lasers. *Physical Review B* 1: 84.
35. Mamyshev, P.V., and S.V. Chenikov. 1990. Ultrashort-pulse propagation in optical fibers. *Optics Letters* 15: 1076.

36. Kozlov, S.A., and S.V. Sazonov. 1997. Nonlinear propagation of optical pulses of a few oscillations duration in dielectric media. *Journal of Experimental and Theoretical Physics* 84: 221.
37. Schäfer, T., and G.E. Wayne. 2004. Propagation of ultra-short optical pulses in cubic nonlinear media. *Physica D* 196: 90.
38. Kinsler, P. 2010. Unidirectional optical pulse propagation equation for materials with both electric and magnetic responses. *Physical Review A* 81: 023808.
39. Caputo, J-.G., and A.I. Maimistov. 2002. Unidirectional propagation of an ultra-short electromagnetic pulse in a resonant medium with high frequency Stark shift. *Physics Letters A* 296: 34.
40. Ranka, J.K., and A.L. Gaeta. 1998. Breakdown of the slowly varying envelope approximation in the self-focusing of ultrashort pulses. *Optics Letters* 23: 798.
41. Rothenberg, J.E. 1992. Space-time focusing: breakdown of the slowly varying envelope approximation in the self-focusing of femtosecond pulses. *Optics Letters* 17: 1340.
42. Brabec. T., and F. Krausz. 1997. Nonlinear optical pulse propagation in the single-cycle regime. *Physical Review Letters* 78: 3282.
43. Voronin, A.A., and A.M. Zheltikov. 2008. Soliton-number analysis of soliton-effect pulse compression to single-cycle pulse widths. *Physical Review A* 78: 063834.
44. Kinsler, P., B.P. Radnor, and G.H.C. New. 2005. Theory of direction pulse propagation. *Physical Review A* 72: 063807.
45. Kinsler, P. 2007. Limits of the unidirectional pulse propagation approximation. *Journal of the Optical Society of America B* 24: 2363.
46. Pitois, S., G. Millot, and S.Wabnitz. 2001. Nonlinear polarization dynamics of counter-propagating waves in an isotropic optical fiber: Theory and experiments. *Journal of the Optical Society of America B* 18(4): 432–443.
47. Kuszner, M., and S. Leble. 2014. Ultrashort opposite directed pulses dynamics with Kerr effect and polarization account. *Journal of the Physical Society of Japan* 83: 034005.
48. Montes, C., P. Aschieri, A. Picozzi, C. Durniak, and M. Taki. 2013. *Without Bounds: A Scientific Canvas of Nonlinearity and Complex Dynamics*. Springer-Verlg, Berlin-Heidelberg.
49. Crpin, T., J.F. Lampin, T. Decoopman, X. Mlique, L. Desplanque, and D. Lippens. 2005. Experimental evidence of backward waves on terahertz left-handed transmission lines. *Applied Physics Letters* 87: 104105.
50. Menyuk, C.R. 1987. Nonlinear pulse propagation in birefringent optical fibers. *IEEE Journal of Quantum Electronics* 23(2): 174.
51. Kuszner, M., and S. Leble. 2015. Waveguide electromagnetic pulse dynamics: Projecting operators method. In: Kuppuswamy, P. and Ramanathan, G. (Eds), *Odyssey of Light in Nonlinear Optical Fibers: Theory and Applications*. CRC Press, Boca Raton, FL.
52. Sakovich, S. 2008. Integrability of the vector short pulse equation. *Journal of the Physical Society of Japan* 77: 123001–123005.
53. Sakovich, A., and S. Sakovich. 2006. Solitary wave solutions of the short pulse equation. *Journal of Physics A* 39: L361–L367.
54. Leble, S. 2003. Nonlinear waves in optical waveguides and soliton theory applications. In: Porsezian, K. and Kuriakose, V.C. (Eds), *Optical Solitons, Theoretical and experimental Challenges*. Springer-Verlag, Berlin-Heidelberg, pp. 71–104.
55. Leble, S., and B. Reichel. 2008. Mode interaction in few-mode optical fibres with Kerr effect. *Journal of Modern Optics* 55: 1–11.
56. Brabec, T., and F. Krausz. 2000. Intense few-cycle laser fields: Frontiers of nonlinear optics. *Reviews of Modern Physics* 72: 545.

57. Chung, Y., and T. Schäfer. 2008. Stabilization of ultra-short pulses in cubic nonlinear media. *Physics Letters A* 361: 63.
58. Chung, Y., C.K.R.T. Jones, T. Schäfer, and C.E. Wayne. 2005. Ultra-short pulses in linear and nonlinear media. *Nonlinearity* 18: 1351.
59. Bose, J.C. 1898. On the rotation of plane of polarisation of electric waves by a twisted structure. *Proceedings of Royal Society* 63: 146–152.
60. Kock, W.E. 1948. Metallic delay lenses. *Bell System Technical Journal* 27: 58–82.
61. Kock, W.E. 1946. Metal-lens antennas. *Proceedings of the Institute of Radio Engineers and Waves and Electrons* 34: 828–836.
62. Collin, R.E. 1991. *Field Theory of Guided Waves*, 2nd ed., IEEE Press, New York.
63. Brown, J. 1960. Artificial dielectrics. In: Birks, J.B. (Ed), *Progress in Dielectrics*, vol. 2, Wiley, pp. 193–225.
64. Malyuzhinets, G.D. 1951. A note on the radiation principle. *Zhurnal Technicheskoi Fiziki* 21: 940–942 (in Russian. English translation in Sov. Phys. Technical Physics).
65. Lagarkov, A.N., A.K. Sarychev, Y.R. Smychkovich, and A.P. Vinogradov. 1992. Effective medium theory for microwave dielectric constant and magnetic permeability of conducting stick. *Journal of Electromagnetic Waves and Applications* 6: 1159–1176.
66. Semenenko, V.N., V.A. Chistyaev, and D.E. Ryabov. 1998. Microwave magnetic properties of Bi-Helix media in dependence on helix pitch. In: *Proceeding of the Bianisotropics 98 7th International Conference on Complex Media*. Braunschweig, Germany. June 3–6, pp. 313–316.
67. Veselago, V.G. FTT 8: 3571 (1966); Soviet Physics Solid State 8: 2853 (1967); Veselago, V.G. 1968. The electrodynamics of substances with simultaneously negative values of ε and μ. *Soviet Physics Uspekhi* 10: 509–514.
68. Milonni, P.W. 2005. *Fast Light, Slow Light and Left-Handed Light. Series in Optics and Optoelectronics*. IOP Publishing Ltd, Bristol.
69. Ziolkowski, R.W., and Kipple, A. 2003. Causality and double-negative metamaterials. *Physical Review E* 68: 026615.
70. Engheta, N., and R.W. Ziolkowski. 2005. A positive future for double-negative metamaterials. *IEEE Transactions on Microwave Theory and Techniques* 53(4): 1535–1556.
71. Rotman, W. 1962. Plasma simulation by artificial dielectrics and parallel plate media. *IRE Transactions on Antennas and Propagation* 10(1): 82–95.
72. Pendry, J.B., A.J. Holden, D.J. Robbins, and W.J. Stewart. 1998. Low frequency plasmons in thin-wire structures. *Journal of Physics: Condensed Matter* 10: 4785–4809.
73. Pendry, J.B., A.J. Holden, W.J. Stewart, and I. Youngs. 1996. Extremely low frequency plasmons in metallic mesostructures. *Physical Review Letters* 76: 4773–4776.
74. Smith, D.R., W.J. Padilla, D.C. Vier, S.C. Nemat-Nasser, and S. Schultz. 2000. Composite medium with simultaneously negative permeability and permittivity. *Physical Review Letters* 84: 4184–4187.
75. Shelby, R.A., D.R. Smith, and S. Schultz. 2001. Experimental verification of a negative index of refraction. *Science* 292(5514): 77–79.
76. Kundtz, N., D. Gaultney, and D.R. Smith. 2010. Scattering cross-section of a transformation optics-based metamaterial cloak. *New Journal of Physics* 12(4): 043039.
77. Lipworth, G., A. Mrozack, J. Hunt, D.L. Marks, T. Driscoll, D. Brady, and D.R. Smith. 2013. Metamaterial apertures for coherent computational imaging on the physical layer. *Journal of the Optical Society of America A* 30: 1603–1612.
78. Wuttig, M., and N. Yamada. 2007. Phase-change materials for rewriteable data storage. *Nature Materials* 6(11): 824–832.

79. Lee, J., M. Tymchenko, C. Argyropoulos, P.-Y. Chen, F. Lu, F. Demmerle, G. Boehm, M.-C. Amann, A. Alù, and M.A. Belkin. 2014. Giant non-linear response from plasmonic metasurfaces coupled to intersubband transitions. *Nature* 511: 65–69.
80. Alu, A., and N. Engheta. 2004. Guided modes in a waveguide filled with a pair of single-negative (SNG), double-negative (DNG), and/or double-positive (DPS) layers. *IEEE Transactions on Microwave Theory and Techniques* 52(1): 199–210.
81. Grbic, A., and G.V. Eleftheriades. 2004. Overcoming the diffraction limit with a planar left-handed transmission lines. *Physical Review Letters* 92(11): 117–403.
82. Caloz, C., and T. Itoh. 2004. A novel mixed conventional microstrip and composite right/lefthanded backward wave directional coupler with broadband and tight coupling characteristics, *IEEE Microwave and Wireless Components Letters* 14(1): 31–33.
83. Islam, R., F. Eleck, and G.V. Eleftheriades. 2004. Coupled-line metamaterial coupler having codirectional phase but contra-directional power flow. *Electronics Letters* 40(5): 315–317.
84. Killen, W.D., and R.T. Pike. 2004. High efficiency directional coupler. US Patent 6,731,244.
85. Li, C., X. Meng, X. Liu, F. Li, G. Fang, H. Chen, and C.T. Chan. 2010. Experimental realization of a circuit-based broadband illusion-optics analogue. *Physical Review Letters* 105: 233906.
86. Zhang, X., and Z. Liu. 2008. Superlenses to overcome the diffraction limit. *Nature Materials* 7(6): 435–441.
87. Zhang, X., C. Xia, and N. Fang. 2011. Broadband acoustic cloak for ultrasound waves. *Physical Review Letters* 106(2): 024301.
88. Garca-Meca, C. et al. 2013. Analogue transformations in physics and their application to acoustics. *Scientific Reports* 3, Article number: 2009.
89. Zhu, H., and F. Semperlotti. 2013. Metamaterial based embedded acoustic filters for structural applications. *AIP Advances* 3(9): 092121.
90. Liang, B., X.S. Guo, J. Tu, D. Zhang, and J.C. Cheng. 2010. An acoustic rectifier. *Nature Materials* 9(12): 989–992.
91. Liang, B., B. Yuan, and J.C. Cheng. 2009. Acoustic diode: Rectification of acoustic energy flux in one-dimensional systems. *Physical Review Letters* 103(10): 104301.
92. Li, Y., B. Liang, Z. Gu, X. Zou, and J. Cheng. 2013. Unidirectional acoustic transmission through a prism with near-zero refractive index. *Applied Physics Letters* 103: 053505.
93. Mei, J. et al. 2012. Dark acoustic metamaterials as super absorbers for low-frequency sound. *Nature Communications* 3: 756.
94. Caleap, M., and B.W. Drinkwater. 2014. Acoustically trapped colloidal crystals that are reconfigurable in real time. *Proceedings of the National Academy of Sciences of USA* 111(17): 6226–6230.
95. Caleap, M., and B.W. Drinkwater. 2014. Reconfigurable acoustic metamaterials: From random to periodic in a split second. http://www.2physics.com/2014/05/reconfigurable-acoustic-metamaterials.html.
96. Wen, S., Y. Xiang, X. Dai, Z. Tang, W. Su, and D. Fan. 2007. Theoretical models for ultrashort electromagnetic pulse propagation in nonlinear metamaterials. *Physical Review A* 75: 033815.
97. Sihvola, A. 2007. Metamaterials in electromagnetics. *Metamaterials* 1: 2–11.
98. Amiranashvili, S., and A. Demircan. 2011. Ultrashort optical pulse propagation in terms of analytic signal. *Advances in Optical Technologies* 2011: 989515.
99. New, G.H.C., P. Kinsler, and B.P. Radnor. 2005. Theory of direction pulse propagation. *Physical Review A*, 72: 063807.

100. Biaynicki-Birula, I., M. Cieplak, and J. Kamiski. 1992. *Theory of Quanta*. Oxford University Press.
101. Karpov, I., and S. Leble. 2017. On problem of atmosphere Rossby and Poincare waves separation and spectra. *TASK Quarterly* 221: 85–96.
102. Hondros, D., and P. Debye. 1910. ElektromagnetischeWellen an dielektrischen Drhten. *Annals of Physics* 32: 465.
103. Pedlosky, J. 2003. *Waves in the Ocean and Atmosphere*. Springer-Verlag, Berlin-Heidelberg.
104. Gossard, E.E. 1975. *W.H. Hooke Waves in the Atmosphere: Atmospheric Infrasound and Gravity Waves, Their Generation and Propagation*. Developments in Atmospheric Science, New York.
105. Lavrentiev, M.M. 1967. *Some Improperly Posed Problems in Mathematical Physics*. Springer, Berlin, 1967; Lavrentév, M.M., and Savelév, L.Y. 1995. *Linear Operators and Ill-Posed Problems*. Consultants Bureau, New York, Division of Plenum Publishing Corporation.
106. Lekner, J. 2000. Omnidirectional reflection by multilayer dielectric mirrors. *Journal of Optics A: Pure and Applied Optics* 2: 349352.
107. Leble, S., and I. Vereshchagina. 2016. Problem of disturbance identification by measurement in the vicinity of a point. *TASK Quarterly* 20(2): 131141.
108. Kuszner, M., S. Leble, and B. 2011. Reichel. Multimode systems of nonlinear equations: derivation, integrability, and numerical solutions. *Theoretical and Mathematical Physics* 168(1): 977.

2 General Technique

2.1 GENERAL PROPER SPACE DEFINITION – EIGENVECTOR PROBLEM FOR PERTURBATIONS OVER A HOMOGENEOUS GROUND STATE

This chapter content is partially published in [1].

2.1.1 GENERAL 1+1 D PROBLEM. LINEAR EVOLUTION IN HOMOGENEOUS CASE

We start from mathematical statement of a linear evolution problem in one-dimensional position coordinate space.

Let the independent variables ranges $t \geq 0, x \in (-\infty, \infty)$ define a half-space and $L(\partial)$ is a $n \times n$ matrix differential operator with constant coefficients.

Consider the evolution equation

$$\psi_t = L(\partial)\psi, \tag{2.1}$$

for $\psi(x,t) \in A^n$, this is n-dimensional vector on the 1+1 half space, while

$$\partial = \partial / \partial x.$$

The initial condition for the vector

$$\psi(x,0) = \phi(x) \tag{2.2}$$

defines the Cauchy problem to be solved.

Let the Fourier transformation operator be denoted as F

$$F\psi(x,t) = \frac{1}{\sqrt{2\pi}} \int_{-\infty}^{\infty} \exp[-ikx]\psi(x)dx = \tilde{\psi}(k,t), \tag{2.3}$$

that acts on each component, and its inverse is marked as F^{-1},

$$F^{-1}\tilde{\psi}(k) = \frac{1}{\sqrt{2\pi}} \int_{-\infty}^{\infty} \exp[ikx]\tilde{\psi}(k)dk. \tag{2.4}$$

Note that for physical variables the components are real, so that

$$\tilde{\psi}^*(-k) = \tilde{\psi}(k). \tag{2.5}$$

Hence, the identity holds

$$F\psi_t = FL(\partial)F^{-1}F\psi,$$

that leads to the system of ordinary differential equations

$$L(\tilde{k})\tilde{\psi} = \tilde{\psi}_t, \tag{2.6}$$

where $L(\tilde{k}) = FL(\partial)F^{-1}$, $F\psi = \tilde{\psi}$.

Consider the matrix $n \times n$ eigenvalue problem

$$\tilde{L}\phi = \lambda\phi, \tag{2.7}$$

that introduces n subspaces, which we represent by the matrix of solutions Ψ

$$\tilde{L}\Psi = \Psi\Lambda, \tag{2.8}$$

with the diagonal matrix $\Lambda = \text{diag}\{\lambda_1, \ldots, \lambda_n\}$. If some eigenvalue is degenerate, we choose the eigenvectors in this subspace as linear independent. We also fix the normalization of all eigenvectors such that the first component is unit

$$\Psi_{1j} = 1, \tag{2.9}$$

if it does not contradict a physical sense. By the choice of linear independent eigenvectors, the inverse matrix Ψ^{-1} exists and

$$\Psi^{-1}\tilde{L} = \Lambda\Psi^{-1}. \tag{2.10}$$

Multiplying (2.8) from the right side by Ψ^{-1} gives

$$\tilde{L} = \Psi\Lambda\Psi^{-1}, \tag{2.11}$$

and this gives the spectral decomposition of the matrix operator $\tilde{L}$, or in components,

$$\tilde{L}_{ij} = \Psi_{ik}\Lambda_{kl}\Psi^{-1}_{lj} = \Psi_{ik}\lambda_k\Psi^{-1}_{kj} = \sum_k \lambda_k\Psi_{ik}\Psi^{-1}_{kj} = \sum_s \lambda_s(\tilde{P}^s)_{ij}, \tag{2.12}$$

where the matrices

$$(\tilde{P}^s)_{ij} = \Psi_{is}\Psi^{-1}_{sj},\ s = 1,\ldots,n, \tag{2.13}$$

are projecting operators in k-representation.

As a corollary, we get basic properties of projectors (summation by all repeated indices, excluding s, is implied)

$$\sum_{s=1}^{n}(\tilde{P}^s)_{ij} = \sum_{s=1}^{n}\Psi_{is}\Psi^{-1}_{sj} = \delta_{ij},\ \Psi_{is}\Psi^{-1}_{sj}\Psi_{jt}\Psi^{-1}_{tk} = \Psi_{jt}\Psi^{-1}_{tk}, \tag{2.14}$$

which reads as

$$\sum_{s=1}^{n}\tilde{P}^s = I,\ \tilde{P}^s\tilde{P}^s = \tilde{P}^s. \tag{2.15}$$

Now we arrive at the evolution operator diagonalization theorem.

Theorem 2.1

For the matrix $\tilde{L} = FLF^{-1}$, parameterized by the wavenumber k, it is necessary and sufficient that $\tilde{L}$ is equivalent to

$$\tilde{L} = \sum_s \lambda_s(k)\tilde{P}^s(k), \tag{2.16}$$

where $\lambda_i(k)$ is found as a root of the algebraic (dispersion) equation

$$det(\tilde{L}(k) - \lambda I) = 0, \tag{2.17}$$

and $P^iP^k = \delta_{ik}\tilde{P}^k$. The system (2.6) splits as

$$\tilde{L}(\tilde{P}^i\tilde{\psi}) = (\tilde{P}^i\tilde{\psi})_t. \tag{2.18}$$

■

For a proof it is enough to commute

$$\tilde{P}^i\tilde{L} = \tilde{L}\tilde{P}^i, \tag{2.19}$$

which follows from the spectral decomposition (2.16)

$$\tilde{P}^i\tilde{L} = \sum_s \lambda_s\tilde{P}^i\tilde{P}^s = \lambda_i\tilde{P}^i = \sum_s \lambda_s\tilde{P}^s\tilde{P}^i. \tag{2.20}$$

Then, Eq. (2.18) reads as

$$\tilde{L}(\tilde{P}^i\tilde{\psi}) = (\tilde{P}^i\tilde{\psi})_t = \lambda_i\tilde{P}^i\tilde{\psi}, \tag{2.21}$$

with the solution

$$\tilde{P}^i(k)\tilde{\psi}(k,t) = C_i(k)\exp[\lambda_i(k)t], \tag{2.22}$$

where $C_i(k)$ is defined by the initial condition

$$\tilde{P}^i(k)\tilde{\psi}(k,0) = C_i(k). \tag{2.23}$$

It is clear that the functions $C_i(k)$ are defined by the initial conditions in x-representation (2.2). The general solution of (2.6) is the sum of the particular solutions (2.22)

$$\sum_{i=1}^{n} \tilde{P}^i(k)\tilde{\psi}(k,t) = \tilde{\psi}(k,t) = \sum_{i=1}^{n} C_i(k)\exp[\lambda_i(k)t], \tag{2.24}$$

if $\lambda_i \neq \lambda_k$ due to the completeness property of projectors (2.15). Having in mind the condition (2.5), we write

$$\tilde{\psi}^*(-k,t) = \sum_{i=1}^{n} C_i^*(-k)\exp[\lambda_i^*(-k)t] = \tilde{\psi}(k,t), \tag{2.25}$$

so the reality of a solution implies conditions on the particular solutions and evolution operator spectrum (dispersion relation). In the case of the choice

$$\lambda_i^*(-k) = \lambda_i(k), \tag{2.26}$$

one has

$$C_i^*(-k) = C_i(k), \tag{2.27}$$

which is necessary to take into account constructing the directed or other fields in x- or t-domains.

Next, let us introduce new variables

$$\tilde{P}_{1j}^s(k)\tilde{\psi}_j(k,t) = \Psi_{1s}\Psi_{sj}^{-1}\tilde{\psi}_j(k,t) = \tilde{\Pi}^s(k,t), \tag{2.28}$$

These explain a convenience of the matrix Ψ normalization choice (2.9). The explicit form of the matrix $\tilde{P}^i$ allows to write

$$\Psi_{2s}\Psi_{sj}^{-1}\tilde{\psi}_j(k,t) = \Psi_{2s}\Psi_{1s}^{-1}\Psi_{1s}\Psi_{sj}^{-1}\tilde{\psi}_j(k,t) = \Psi_{2s}\Psi_{1s}^{-1}\tilde{\Pi}^s(k,t), \tag{2.29}$$

and similar for the indexes $3,...,n$. Summation by s of all such equations starting from (2.28) gives the system of n algebraic equations for $r = 1,...,n$

$$\sum_{s=1}^{n} \tilde{P}_{rj}^s(k)\tilde{\psi}_j(k,t) = \delta_{rj}\tilde{\psi}_j(k,t) = \tilde{\psi}_r = \sum_{s=1}^{n} \Psi_{rs}\Psi_{1s}^{-1}\tilde{\Pi}^s(k,t). \tag{2.30}$$

as a corollary of the completeness of the set of P^s (2.14). The result gives the explicit form of the Fourier transformed components of the vector $\tilde{\psi}_j(k,t)$, for all physical variables. The simplest expression is obtained for $\tilde{\psi}_1$:

$$\tilde{\psi}_1 = \sum_{s=1}^{n} \tilde{\Pi}^s(k,t). \tag{2.31}$$

Solving the system (2.30) with respect to $\tilde{\Pi}^s(k,t)$, one obtains the inverse transformation, expressing $\tilde{\Pi}^s$ as a linear function of the principle components transforms $\tilde{\psi}_j(k,t)$.

2.1.2 TRANSITION TO X-REPRESENTATION

After the inverse Fourier transform, the original system (2.1) splits as

$$(P^i\psi)_t = LP^i\psi, \tag{2.32}$$

where the formula $P^i(x) = F^{-1}\tilde{P}^iF$ represents the matrix integral operator of the convolution type, so that

$$(F^{-1}\tilde{P}_{jk}^iF)\phi(x) = P_{jk}^i\phi(x) = \int_{-\infty}^{\infty} K_{jk}^i(x-y)\phi(y)dy, \tag{2.33}$$

with the kernel

$$K^i_{jk}(x-y) = \frac{1}{2\pi}\int_{-\infty}^{\infty} \tilde{P}^i_{jk}(k)\exp[ik(x-y)]. \tag{2.34}$$

The operators $P^i(x)$ are named here as the **dynamic projecting operators**, as the basic notion of this book.

Let us check the statement. The transformations sequence, starting from (2.18),

$$(\tilde{P}^i\tilde{\psi})_t = (\tilde{P}^i F\psi)_t = FLF^{-1}\tilde{P}^i\tilde{\psi} = FLF^{-1}(\tilde{P}^i F\psi), \tag{2.35}$$

and action by F^{-1} from the left, proves the resulting formula (2.32). The splitting is more obvious if one performs the substitutions based on the identity

$$\sum_{i=1}^{n} P^i\psi = \psi, \tag{2.36}$$

which is the transform of (2.14); it reads as the system of equations for the components transition to new variables

$$\sum_{j=1}^{n} P^i_{kj}\psi_j = \Pi^i_k,\ k = 1,\ldots,n. \tag{2.37}$$

In fact, we have n copies of such a system, which differ only by an operator factor. Therefore, we can fix the choice for the new variables in most compact form as $\Pi^i = P^i_{1j}\psi_j$ (here and after summation by repeating indices is implied) from the relation (2.37), or

$$\Pi^i(x,t) = P^i_{1k}\psi_k(x,t) = \int_{-\infty}^{\infty} K^i_{1k}(x-y)\psi_k(y,t)dy, \tag{2.38}$$

which we shall name as **mode variables**. The transform (2.38) is a one-to-one map.

So, we check the statement by means of the projectors definition in k-space (2.13)

$$\sum_{s=1}^{n}\tilde{\Pi}^s = \sum_{s=1}^{n}\Psi_{1s}\Psi^{-1}_{sj} = \delta_{1j} = \tilde{\psi}_j \tag{2.39}$$

other relations are transforms of (2.30) with k = 2,3,..., n, that gives equivalent description in x-representation.

$$\sum_{s=1}^{n} F^{-1}\tilde{P}^s_{rj}(k)FF^{-1}\tilde{\psi}_j(k,t) = F^{-1}\tilde{\psi}_r = \sum_{s=1}^{n} F^{-1}\Psi_{rs}\Psi^{-1}_{1s}FF^{-1}\tilde{\Pi}^s(k,t). \tag{2.40}$$

In integral form, we have the inverse transform

$$\psi_r(x,t) = \sum_{s=1}^{n} \int_{-\infty}^{\infty}\int_{-\infty}^{\infty} \Psi_{rs}(k,t)\Psi_{1s}^{-1}(k,t)\exp[ik(x-y)]dk\Pi^s(y,t)dy, \tag{2.41}$$

which completes the proof of the statement.

Initial conditions for the mode variables $\Pi^s(y,0)$ are extracted from the identity $\Sigma_1^n P^i\phi = \phi$ (the system (2.42) at $t=0$), for a given vector ϕ from (2.2) or, taking (2.42), at $t=0$

$$\psi_r(x,0) = \phi_r(x) = \sum_{s=1}^{n} \int_{-\infty}^{\infty}\int_{-\infty}^{\infty} \Psi_{rs}(k,t)\Psi_{1s}^{-1}(k,t)\exp[ik(x-y)]dk\Pi^s(y,0)dy, \tag{2.42}$$

which is solved with respect to $\Pi^s(y,0)$ by (2.38)

Theorem 2.2

$$\Pi^i(x,0) = P_{1k}^i\psi_k(x,0) = \int_{-\infty}^{\infty} K_{1k}^i(x-y)\phi_k(y)dy, \tag{2.43}$$

which accomplishes the problem formulation for the mode variables Π^s together with the x-representation (2.44) for the equation (2.18):

$$\Pi_t^s = \lambda_s(\partial)\Pi^s. \tag{2.44}$$

The action of the operator in R.H.S. is defined by

$$\lambda_s(\partial)\Pi^s = F^{-1}\lambda_s(k)F\Pi^s = \int_{-\infty}^{\infty}\int_{-\infty}^{\infty} \lambda_s(k)\exp[ik(x-y)]dk\Pi^s(t,y)dy. \tag{2.45}$$

■

The problem is, in fact, solved by the sequence of formulas (2.22) and (2.23) after the inverse Fourier transform, but its nonlinear version (see Section 2.1.4) is a non-trivial generalization that, generally, cannot be solved by the Fourier transformation method. It also gives us the possibility to use approximations via $\lambda_s(\partial)$ expansions as a pseudodifferential operator in a realm of analiticity of $\lambda_s(k)$.

2.1.3 BOUNDARY REGIME PROPAGATION

A geometry of an experiment often needs a transition from a Cauchy problem formulation to a boundary regime one by the given functions of time for each variable at some fixed point usually chosen at $x = 0$. Such a situation is realized if we have a plane wave that falls at a plane interface between two media. Similarly, a waveguide mode is excited by some perturbation at an end of the waveguide. Formally, such a $1+1$ problem may be written for an equation quite similar to (2.1), in the form

$$\partial_x \psi = Ł(\partial_t)\psi, \tag{2.46}$$

for a vector $\psi(x,t)$ generally of the same or higher number of components with the boundary condition

$$\psi(0,t) = \phi(t). \tag{2.47}$$

A reformulation of the equation (2.1) in the case when we have in $L(\partial_x)$ only x-derivatives of the first order is simple: we shift the terms with a time derivative to the R.H.S. and vice versa for x-derivatives. The result has the R.H.S. in the form of $Ł\psi$ automatically. If the operator L contains second or higher derivatives, sometimes one can manipulate the equation with differentiation and rearrangement of the terms until the form of (2.46) is achieved. Generally, the known trick is applied: when we have the second derivative, say a_{xx}, we denote $a_x = b$, having

$$\partial_x b = f(a,\ b,\ a_t, \dots)$$

and, consider the result of new notation introduction as a set of new equations:

$$\partial_x a = b.$$

As you see, the order of the system grows from n to $n+1$. For example, if we start from the diffusion equation for a concentration $u(x,t)$ with given boundary values $u(0,t) = \mu(t)$ and mass flow proportional to $u_x(0,t) = \nu(t)$,

$$u_t = Du_{xx}. \tag{2.48}$$

Considering it as the system of the order 1, we apply the preceding trick, arriving at

$$\begin{aligned} v_x &= D^{-1}u_t, \\ u_x &= v. \end{aligned} \tag{2.49}$$

Going to the matrix form of the Eq. (2.46), we write

$$\partial_x \begin{pmatrix} u \\ v \end{pmatrix} = \begin{pmatrix} 0 & 1 \\ D^{-1}\partial_t & 0 \end{pmatrix} \begin{pmatrix} u \\ v \end{pmatrix}, \tag{2.50}$$

with the boundary conditions (regime)

$$\begin{pmatrix} u(0,t) \\ v(0,t) \end{pmatrix} = \begin{pmatrix} \mu(t) \\ \nu(t) \end{pmatrix}, \tag{2.51}$$

if the conditions are physically compatible. Now we should deliver a Fourier transformation in the t-variable. The natural obstacle is the quarter plane $t, x \leq 0$ domain under consideration. The continuation of the function u, v to the negative time values may be realized by taking the boundary conditions (2.60) structure into account. For physics reasons, the switching type conditions may be chosen so that u and v have zero values for $t = 0$, which is guaranteed by means of a choice of odd functions with respect to the reflection $t \to -t$. Hence, the Fourier transformation

$$u(x,t) = \frac{1}{\sqrt{2\pi}} \int_{-\infty}^{\infty} \tilde{u}(x,\omega) \exp[i\omega t] d\omega. \tag{2.52}$$

implies the function $\tilde{u}(x,\omega)$ symmetry that follows from

$$\begin{aligned} u(x,-t) &= \frac{1}{\sqrt{2\pi}} \int_{-\infty}^{\infty} \tilde{u}(x,\omega) \exp[-i\omega t] d\omega \\ &= \frac{1}{\sqrt{2\pi}} \int_{-\infty}^{\infty} \tilde{u}(x,-\omega') \exp[i\omega' t] d\omega' = -u(x,t). \end{aligned} \tag{2.53}$$

after the substitution $\omega = -\omega'$. It yields

$$\tilde{u}(x,-\omega) = -\tilde{u}(x,\omega). \tag{2.54}$$

Similarly, we transform $v \to \tilde{v}$, arriving at the system of ODE

$$\partial_x \begin{pmatrix} \tilde{u} \\ \tilde{v} \end{pmatrix} = \begin{pmatrix} 0 & 1 \\ i\omega D^{-1} & 0 \end{pmatrix} \begin{pmatrix} \tilde{u} \\ \tilde{v} \end{pmatrix}. \tag{2.55}$$

The boundary values at each point $\omega \in (-\infty, \infty)$ are

$$\begin{aligned} \begin{pmatrix} \tilde{u}(0,\omega) \\ \tilde{v}(0,\omega) \end{pmatrix} &= \begin{pmatrix} \frac{1}{\sqrt{2\pi}} \int_{-\infty}^{\infty} u(0,t) \exp[-i\omega t] dt. \\ \frac{1}{\sqrt{2\pi}} \int_{-\infty}^{\infty} v(0,t) \exp[-i\omega t] dt \end{pmatrix} \\ &= \begin{pmatrix} \frac{1}{\sqrt{2\pi}} \int_{-\infty}^{\infty} \mu(t) \exp[-i\omega t] dt, \\ \frac{1}{\sqrt{2\pi}} \int_{-\infty}^{\infty} \nu(t) \exp[-i\omega t] dt \end{pmatrix}. \end{aligned} \tag{2.56}$$

The reality condition reads

$$u(x,t)^* = \frac{1}{\sqrt{2\pi}} \int_{-\infty}^{\infty} \tilde{u}^*(x,\omega) \exp[-i\omega t] d\omega = \frac{1}{\sqrt{2\pi}} \int_{-\infty}^{\infty} \tilde{u}^*(x,-\omega') \exp[i\omega' t] d\omega', \tag{2.57}$$

for $\omega = -\omega'$. Then

$$\tilde{u}^*(x,-\omega) = \tilde{u}(x,\omega). \tag{2.58}$$

The same as with t dependence symmetry, which gives (2.54). So we should account for both (2.58) and (2.54).

Now we can apply the projecting technique in the spirit of this book. Consider the eigenproblem

$$\begin{pmatrix} 0 & 1 \\ i\omega D^{-1} & 0 \end{pmatrix} \begin{pmatrix} \phi_1 \\ \phi_2 \end{pmatrix} = \lambda \begin{pmatrix} \phi_1 \\ \phi_2 \end{pmatrix}, \tag{2.59}$$

which gives eigenvalues

$$\lambda_\pm = \pm\sqrt{i\omega D^{-1}} \tag{2.60}$$

and eigenvectors

$$\phi_+ = \begin{pmatrix} 1 \\ \lambda_+ \end{pmatrix}, \phi_- = \begin{pmatrix} 1 \\ \lambda_- \end{pmatrix}. \tag{2.61}$$

Then

$$\Psi = \begin{pmatrix} 1 & 1 \\ \lambda_+ & \lambda_- \end{pmatrix}. \tag{2.62}$$

Hence, the inverse matrix is

$$\Psi^{-1} = \frac{1}{\lambda_- - \lambda_+} \begin{pmatrix} \lambda_- & -1 \\ -\lambda_+ & 1 \end{pmatrix}, \tag{2.63}$$

which yields in the projecting matrices in ω-representation applying the Eq. (2.13).

In our case,

$$\lambda_- = -\lambda_+;$$

therefore, the expression (2.63) simplifies as

$$\Psi^{-1} = \frac{1}{2} \begin{pmatrix} 1 & \lambda_+^{-1} \\ 1 & -\lambda_+^{-1} \end{pmatrix}, \tag{2.64}$$

which results in the projectors

$$P^\pm = \frac{1}{2} \begin{pmatrix} 1 & \pm\lambda_+^{-1} \\ \pm\lambda_+ & 1 \end{pmatrix}. \tag{2.65}$$

Acting on the Eq. (2.55) by the projectors (2.65)

$$\begin{aligned} &\partial_x \begin{pmatrix} 1 & \pm\lambda_+^{-1} \\ \pm\lambda_+ & 1 \end{pmatrix} \begin{pmatrix} \tilde{u} \\ \tilde{v} \end{pmatrix} \\ &= \begin{pmatrix} 1 & \pm\lambda_+^{-1} \\ \pm\lambda_+ & 1 \end{pmatrix} \begin{pmatrix} 0 & 1 \\ i\omega D^{-1} & 0 \end{pmatrix} \begin{pmatrix} \tilde{u} \\ \tilde{v} \end{pmatrix} \end{aligned} \tag{2.66}$$

The first line of the result reads

$$\partial_x \left(\tilde{u} \pm \lambda_+^{-1} \tilde{v} \right) = \pm\lambda_+ \tilde{u} + \tilde{v}. \tag{2.67}$$

The second line gives as usual the same (proportional) result. Multiplying (2.67) by λ_+, one has

$$\partial_x(\lambda_+\tilde{u} \pm \tilde{v}) = \pm\lambda_+(\lambda_+\tilde{u} \pm \tilde{v}). \tag{2.68}$$

Denoting

$$\tilde{\Pi}_\pm = \lambda_+\tilde{u} \pm \tilde{v}, \tag{2.69}$$

we arrive at the x-evolution equations in the ω-domain

$$\partial_x\tilde{\Pi}_\pm = \pm\lambda_+\tilde{\Pi}_\pm \tag{2.70}$$

and the boundary conditions

$$\tilde{\Pi}_\pm(0,\omega) = \lambda_+\tilde{u}(0,\omega) \pm \tilde{v}(0,\omega). \tag{2.71}$$

These conditions are originated from (2.56)

$$\tilde{\Pi}_\pm(0,\omega) = \frac{1}{\sqrt{2\pi}}\int_{-\infty}^{\infty}[\lambda_+\mu(t) \pm \nu(t)]\exp[-i\omega t]dt \tag{2.72}$$

The solutions of the (2.70) are parametrized by ω and given by

$$\tilde{\Pi}_\pm(x,\omega) = \exp[\pm\lambda_+(\omega)x]\tilde{\Pi}_\pm(0,\omega). \tag{2.73}$$

Now we should return in (2.70) to the Fourier original via

$$\tilde{u}(x,\omega) = \frac{1}{\sqrt{2\pi}}\int_{-\infty}^{\infty}u(x,t)\exp[-i\omega t]d\omega = Fu. \tag{2.74}$$

Acting on Eq. (2.70) by F^{-1} from the left side, we write

$$\partial_x F^{-1}\tilde{\Pi}_\pm = \pm F^{-1}\lambda_+ F F^{-1}\tilde{\Pi}_\pm, \tag{2.75}$$

or, denoting $\Pi_\pm = F^{-1}\tilde{\Pi}_\pm$, we obtain the x-evolution equations for the mode amplitudes $\Pi_\pm$,

$$\partial_x\Pi_\pm = \pm F^{-1}\lambda_+(\omega)F\Pi_\pm. \tag{2.76}$$

Next, in terms of the time derivative operator ∂_t, we have

$$\partial_x\Pi_\pm = \pm\lambda_+(i\partial_t)\Pi_\pm. \tag{2.77}$$

The solution may be presented either in pseudodifferential form

$$\Pi_\pm = \exp[\pm\lambda_+(i\partial_t)x]\Pi_\pm(0,t), \tag{2.78}$$

or in integral one via (2.73)

$$\Pi_\pm(x,t) = F\tilde{\Pi}_\pm(x,\omega) = F\exp[\pm\lambda_+(\omega)x]\tilde{\Pi}_\pm(0,\omega). \tag{2.79}$$

Finally, by means of (2.72) we write

$$\Pi_{\pm}(x,t) = \frac{1}{2\pi}\int_{-\infty}^{\infty}\int_{-\infty}^{\infty} \exp[\pm\lambda_{+}(\omega)x]\exp[i\omega(t-\tau)][\mu(\tau)\lambda_{+}(\omega)\pm v(\tau)]d\tau d\omega. \tag{2.80}$$

or plugging (2.60) we yield

$$\Pi_{\pm}(x,t) = \frac{1}{2\pi}\int_{-\infty}^{\infty}\int_{-\infty}^{\infty} \exp[\pm\sqrt{i\omega D^{-1}}x]\exp[i\omega(t-\tau)][\mu(\tau)\sqrt{i\omega D^{-1}}\pm v(\tau)]d\tau d\omega. \tag{2.81}$$

It solves the boundary problem (2.50, 2.51) by means of an inverse to the t-version of (2.71)

$$\Pi_{\pm} = \lambda_{+}(i\partial_t)u \pm v \tag{2.82}$$

resulting in the original variable (2.60)

$$u = \frac{1}{2}\lambda_{+}(i\partial_t)^{-1}(\Pi_{+}+\Pi_{-}). \tag{2.83}$$

The variable v is expressed as

$$v = \frac{1}{2}(\Pi_{+}-\Pi_{-}). \tag{2.84}$$

Next, we recall that the variable v originated from $v = u_x$, so there is a link between (2.83) and (2.85). Namely

$$(\Pi_{+}-\Pi_{-}) = \lambda_{+}(i\partial_t)^{-1}(\Pi_{+}+\Pi_{-})_x, \tag{2.85}$$

which automatically holds because of (2.77).

There are celebrated results on the boundary regime problem, going up to Fourier's seminal paper devoted to the processes mentioned as heat waves. In the textbook [2] the heat conduction problem

$$\begin{aligned} u_t &= k u_{xx}, \\ u(0,t) &= \mu(t) \end{aligned} \tag{2.86}$$

and a theorem of uniqueness are considered for a class of bounded solutions. The coefficient k denotes the thermo-conductivity. The equation has the form equivalent to the diffusion equation (2.48).

2.1.4 ON WEAK NONLINEARITY ACCOUNT PROBLEMS

Let again the independent variables ranges $t \geq 0, x \in (-\infty,\infty)$ define a half-space and the evolution equation

$$\psi_t - L(\partial)\psi = \varepsilon N(\psi) \tag{2.87}$$

For $\psi(x,t) \in A^n$, it contains now so-called weak nonlinear terms $N(\psi)$, the small parameter ε appears after rescaling of the wave vector $\psi \to \varepsilon\psi$.

The initial condition for the vector

$$\psi(x,0) = \phi(x) \tag{2.88}$$

defines the Cauchy problem to be solved.

The formal idea to make a first step to reformulate the problem in terms of the linear modes introduced in the previous sections is based on the substitution (2.42). In such terms, the L.H.S. splits into independent mode evolution subequations and allows us to apply the procedure of perturbation expansion of the equation operator in power series with respect to the amplitude parameter ε [1]. In nonlinearity terms, we would be left with only quadratic powers here. The procedure is performed by action of the projecting operators on the Eq. (2.87):

$$P^s(\psi_t - L(\partial)\psi) = \varepsilon P^s N\left(\sum_{j=1}^{n} P^j\psi\right), \tag{2.89}$$

The unit operator is inserted into the nonlinear part of the equation. It was proved that the projecting operators commute with the linear evolution operator L by construction (see Section 2.1.2); therefore,

$$(P^s\psi)_t - L(\partial)P^s\psi = \varepsilon P^s N\left(\sum_{j=1}^{n} P^j\psi\right). \tag{2.90}$$

The result of the operator P^s application to ψ gives (2.38), so the first line of the Eq. (2.90) reads as

$$\Pi^s_t - \lambda_s(\partial)\Pi^s = \varepsilon \sum_{j=1}^{n} P^s_{1j}N_j\left(\sum_{j=1}^{n} P^j\psi\right). \tag{2.91}$$

The components of the nonlinear R.H.S of the system may be understood as

$$N_j\left(\sum_{j=1}^{n} P^j\psi\right) = N_j(\psi_1,\ldots,\psi_n), \tag{2.92}$$

where the components of the basic vector ψ_i are expressed in terms of the mode amplitudes Π^s via the Eq. (2.42) having

$$N_j\left(\sum_{j=1}^{n} P^j\psi\right) = \tilde{N}_j(\Pi^1,\ldots,\Pi^n) \tag{2.93}$$

Plugging Eq. (2.93) into (2.91), we obtain the system of equations that describe interactions of the principal modes, having

$$\Pi^s_t - \lambda_s(\partial)\Pi^s = \varepsilon \sum_{j=1}^{n} P^s_{1j}\tilde{N}_j(\Pi^1,\ldots,\Pi^n). \tag{2.94}$$

The L.H.S. of the system may be considered a result of the so-called linearization of a basic system like general hydrodynamic or electrodynamics; see (2.45) for the operator $\lambda_s(\partial)$ action definition.

The initial conditions for $\Pi^s(x,t)$ are determined by projecting procedure (2.43), or

$$\Pi^i(x,0) = \int_{-\infty}^{\infty} K^i_{1k}(x-y)\phi_k(y)dy = \chi^i(x). \tag{2.95}$$

If a set of initial conditions (2.88) is chosen as

$$\chi^i(x) = \delta_{i1}u(x), \tag{2.96}$$

and only one mode is excited, we can neglect the other modes generation and a feedback action for a time T and operate by the equation for $U(x,t) = \Pi^1(x,t)$:

$$U_t - \lambda_1(\partial)U = \varepsilon P^1_{1j}N_j(\psi_1,\ldots,\psi_n). \tag{2.97}$$

The expressions for ψ_r, $r = 1,\ldots,n$, in the R.H.S as the function of U are given by (2.42)

$$\psi_r(x,t) = \int_{-\infty}^{\infty}\int_{-\infty}^{\infty} \Psi_{r1}(k,t)\Psi^{-1}_{11}(k,t)\exp[ik(x-y)]dkU(y,t)dy. \tag{2.98}$$

Such an equation describes a self-action of the first mode, valid at $t \in [0,T]$. Similar equations may be written for each mode function $\Pi^2,\ldots,\Pi^n$.

The integral operators that appear in the theory may be considered as pseudodifferential ones if we expand the kernels in Taylor series with respect to k. Such an expansion is used to build an approximation if we restrict ourselves to only a few terms of it, which may be used for a wave process whose spectrum (wavelength values) is located by some range. Such a location is controlled by a class C of initial conditions $\phi(x) \in C$ and a corresponding small parameter that appears after the rescaling of basic variables. The scales for a Cauchy problem may be introduced as follows. If we have the only variable $u(x,t)$, then, if

$$max|u(x,0)| = max_C|\phi(x)| = u_0, \tag{2.99}$$

and

$$max_C|u_x(x,0)| = max|\phi_x(x)| = k_0u_0, \tag{2.100}$$

the space scale may be chosen as $\lambda = 2\pi k_0^{-1}$. The time scale choice is defined by the evolution itself as $\tau = 2\pi\omega^{-1}(k_0)$.

In the case of multicomponent state description, that is described by a vector (2.88) from a normed space, the amplitude parameter is introduced by

$$max_C||\phi(x,0)|| = u_0, \tag{2.101}$$

and the derivatives are estimated as

$$max_C||\phi_x(x)|| = k_0 u_0. \tag{2.102}$$

The evolution of a mode s is described by (2.44), which comes from (2.7). For a conventional relation $\lambda_s = i\omega_s$, the expansions of the kernels may be cut at some powers dependent on the time evolution scale

$$\tau_s = 2\pi\omega_s^{-1}(k_0) \tag{2.103}$$

for a given mode Π^s, because of a different small parameter definition.

The formal solution of the Eq. (2.44)

$$\Pi^s(x,t) = \exp[\lambda_s(\partial)t]\Pi(x,0), \tag{2.104}$$

yields an estimation of the L.H.S.

$$|\Pi^s(x,t)| \leq ||\exp[\lambda_s(\partial)t]|||\Pi(x,0)|, \tag{2.105}$$

where $||A(\partial)||$ is the pseudodifferential operator $A(\partial)$ norm [3].

In such cases, we approximate the expansion by a finite number of terms — a polynomial in k of $\lambda_s(k)$ in the kernel of the integral operator (2.45), which allows us to change the integral operators by differential ones.

The famous nonlinear equations for long directed waves such as integrable Korteweg-de Vries and Burgers equations are derived namely by the presented method. These approximate equations are obtained from general hydrodynamic or other basic system with account of minimal nonlinear (quadratic) terms.

As for iterations of the procedure leading to higher powers of amplitude parameter ε, it is rather sophisticated. Let us mention only one article [4], where the dynamical projecting accounts for nonlinear terms.

2.1.5 WEAKLY INHOMOGENEOUS GROUND STATE. HYPERBOLIC EQUATION

Consider the system

$$\frac{\partial u(x,t)}{\partial t} - \varepsilon b(x)\frac{\partial v(x,t)}{\partial x} = 0, \tag{2.106}$$

$$\frac{\partial v(x,t)}{\partial t} - \varepsilon c(x)\frac{\partial u(x,t)}{\partial x} = 0, \tag{2.107}$$

where the small parameter ε defines the strong condition

$$max\frac{\partial b(x)}{\partial x} = \varepsilon << 1, \tag{2.108}$$

with similar restrictions for the coefficient $c(x)$, which we name as a condition of a *weak inhomogeneity*.

The formal evolution operator may be written as

$$\exp[\hat{\lambda}(\varepsilon D)]. \tag{2.109}$$

As the **principle step**, introduce a system:

$$\varepsilon \cdot b(x)D\tilde{v}(x) = \hat{\lambda}(\varepsilon D) \cdot \tilde{u}(x), \tag{2.110}$$

$$\varepsilon \cdot c(x)D\tilde{u}(x) = \hat{\lambda}(\varepsilon D) \cdot \tilde{v}(x), \tag{2.111}$$

which defines **pseudodifferential operators** $\hat{\lambda}_i(D)$, $i = 1,2$. Such an operator system opens a way to a pseudodifferential representation of projecting and separates the evolution to directed waves subspaces. Expanding the operator $\hat{\lambda}_i$ as

$$\hat{\lambda}_i(\varepsilon D) = \sum_{n=0}^{\infty} s_n^{(i)}(x)(\varepsilon D)^n, \tag{2.112}$$

we go to the algorithm of the coefficients $s_n^{(i)}$ evaluation, that arises from substitution of (2.112) to the corollary of (2.110).

This corollary gives the link

$$\varepsilon(\hat{\lambda}(D))^{-1} \cdot c(x)D\tilde{u}(x) = \tilde{v}(x), \tag{2.113}$$

which yields the equation for $\hat{\lambda}$ in the form of (2.112)

$$[\varepsilon^2 \cdot b(x)D(\hat{\lambda}(D))^{-1}c(x)D - \hat{\lambda}(D)]\tilde{u}(x) = 0. \tag{2.114}$$

2.1.6 WEAK INHOMOGENEITY. DIRECTED WAVES

Operator matrix idempotents. For the operator $L(\varepsilon D)$ to be diagonalized, the projectors P^i are constructed via the link (2.113) as the function of the pseudodifferential operator. Consider a 2×2 matrix with operator-valued, generally non-commutative elements;

$$P = \begin{pmatrix} p & \pi \\ \xi & \eta \end{pmatrix} \tag{2.115}$$

with basic

$$P^2 = P$$

$$\begin{aligned} p^2 + \pi\xi = p \quad & p\pi + \pi\eta = \pi \\ \xi p + \eta\xi = \xi \quad & \xi\pi + \eta^2 = \eta \end{aligned} \tag{2.116}$$

algebraic corollaries. Further, if $\exists \pi^{-1}, \xi^{-1}$,

$$\eta = 1 - \pi^{-1} p\pi = 1 - \xi p \xi^{-1}, \tag{2.117}$$

and

$$\pi^{-1} p\pi = \xi p \xi^{-1}, \tag{2.118}$$

or

$$p\pi\xi = \pi\xi p.$$

The problem is reformulated in terms of directed waves. The projecting operators in this case are calculated via the basic relation for the projection subspaces that are derived directly from Eqs (2.110, 2.111), in which evolution is fixed by the pseudodifferential spectral operators as expansion in ε [5]. In the first order, arriving at a supermatrix:

$$P_{1,2} = \frac{1}{2}\begin{pmatrix} 1 & \pm M^{-1} \\ \pm M & 1 \end{pmatrix} \tag{2.119}$$

Now the evolution operator L, in the same notations, is also a supermatrix:

$$L = \begin{pmatrix} 0 & \varepsilon b(x)D \\ \varepsilon c(x)D & 0 \end{pmatrix}. \tag{2.120}$$

where $f = \sqrt{\frac{c}{b}}, D = \frac{\partial}{\partial x}$, and the operator valued matrix elements are expressed in terms of $M = D^{-1}fD$.

A desired splitting of the system, as it is formulated in the Theorem 2.1 is approved by (2.19), therefore we base on the commutation relation $[P_{1,2}, L] = 0$ that is valid automatically in the case of constant coefficients b, c. For the x-dependent case of $b(x), c(x)$, the commutator L and P_1 is equal to

$$[P_1, L] = \frac{\varepsilon}{2}\begin{pmatrix} M^{-1}cD - bDM & 0 \\ 0 & MbD - cDM^{-1} \end{pmatrix}. \tag{2.121}$$

The condition that the commutator is zero can be written as

$$Dc - fDbf = 0 \tag{2.122}$$

or from the expression for f:

$$c'b - b'c = 0. \tag{2.123}$$

It fixes the case of the complete reduction (diagonalization) of the evolution operator.

As the further development of the method, we suggest an approximate procedure (see e.g. [6]), generally treating the condition $[P_{1,2}, L] = O(\varepsilon)$.

Using the projecting operators, we reduce (2.110, 2.111) to a couple of equations that, as in the previous section, we traditionally name for the left and right waves, splitting the problem of evolution. The approximate splitting could be achieved if one could neglect the commutators of $P_{1,2}$ and L. This is possible if the coefficients b, c are of the zero order ($\cong O(1)$), whereas the order of the derivative $(\frac{c}{b})'$ is of a higher order, e.g. $\cong O(\varepsilon)$. This is guaranteed by the evolution operator's dependence on ε and the conditions of the spectral operators expansion domain. Acting by P_1 to the system (2.110), (2.111) written in matrix operator form, one proceeds as

$$(P_{1,2}\Psi)_t = P_{1,2}L\Psi, \tag{2.124}$$

or approximately

$$(P_{1,2}\Psi)_t = L(P_{1,2}\Psi), \tag{2.125}$$

where

$$P_1\Psi == \frac{1}{2}\begin{pmatrix} \Pi \\ M\Pi \end{pmatrix}, \quad (2.126)$$

$$P_2\Psi == \frac{1}{2}\begin{pmatrix} \Lambda \\ -M\Lambda \end{pmatrix}. \quad (2.127)$$

Reading the first lines of these relations yields

$$\Pi = \frac{1}{2}(u + M^{-1}v), \quad (2.128)$$

and

$$\Lambda = \frac{1}{2}(u - M^{-1}v). \quad (2.129)$$

This relation allows us to establish the Cauchy problems for directed waves.

$$u(x,0) = \phi(x), v(x,0) = \psi(x)$$
$$\Pi(x,0) = \frac{1}{2}(\phi + M^{-1}\psi), \quad (2.130)$$

and

$$\Lambda(x,0) = \frac{1}{2}(\phi - M^{-1}\psi). \quad (2.131)$$

From the equations (2.128), (2.129) one extracts

$$\begin{aligned} u &= \Pi + \Lambda, \\ v &= M(\Pi - \Lambda). \end{aligned} \quad (2.132)$$

Considering equations (6.44), (2.129) and the relation for the commutator $P_1L = LP_1 - [P_1, L]$, one obtains approximately

$$\Pi_t = -\sqrt{bc}\Pi_x. \quad (2.133)$$

This could be interpreted as the equation for the right wave. Similarly, the equation for the left wave variable Λ looks as follows:

$$\Lambda_t = \sqrt{bc}\Lambda_x. \quad (2.134)$$

Solving the first-order equations by the method of characteristics gives u, v by the relation (2.132); formally, the system coincides with the case that includes constant coefficients (1.35) from the preceding chapter, but the velocity of propagation and coefficients in (2.132) are functions depending on coordinate.

2.1.7 LINK TO SPECTRAL THEOREM

It is known that there is a general statement concerning normal operators (the property $A^+A = AA^+$ is satisfied), defined on a Hilbert space [7]. It reads as an expansion of the operator in terms of the Stiltjes integral

$$L = \int \lambda d\Pi(\lambda). \quad (2.135)$$

for a spectral problem related to the equation

$$L\psi = \lambda\psi, \tag{2.136}$$

and projecting operator $\Pi(\lambda)$ is a function that grows only in points of spectrum. For an alternative theory with many important details, see [8].

Let the spectrum be continuous at $\lambda \in [\lambda_0, \infty)$, and let there exist the operator $P(\lambda)$ such that

$$\Pi(\lambda) = \int_{\lambda}^{\infty} P(\lambda')d\lambda', \, d\Pi(\lambda) = P(\lambda)d\lambda. \tag{2.137}$$

allow us to factorize $d\Pi$ as in (2.137); an important example of such an operator is the operator of multiplication by a coordinate which is described, for example, in the excellent textbook on quantum mechanics [9], Chapter II.

The spectral theorem for a given normal $n \times n$ matrix L is given by (2.166), or, more succinctly by

$$L = \sum_{s=1}^{n} \lambda_s P^s. \tag{2.138}$$

The statement is directly generalized for operators with point spectrum, including the case of $n = \infty$.

In the case of the simplest differential operator

$$L = \frac{d}{dx}, \tag{2.139}$$

the Eq. (6.31) has an explicit solution in elementary functions; namely, the eigen functions are proportional to

$$\chi(x,k) = \exp ikx. \tag{2.140}$$

Its spectrum is continuous and lies in an imaginary axis of the parameter $\lambda = ik$, $\Re k = 0$ [9]. Let us look for the projecting operators in the form of the integral operator

$$P(k)f(x) = \int_{-\infty}^{\infty} a_k(x,y)f(y)dy \tag{2.141}$$

By definition, its action gives

$$\int_{-\infty}^{\infty} a_k(x,y)f(y)dy = C\exp ikx, \tag{2.142}$$

which gives

$$a_k(x,y) = C_k \exp[ik(x-y)]. \tag{2.143}$$

This statement may be proved applying the Fourier transformation as

$$f(y) = \int_{-\infty}^{\infty} \exp[imy]\hat{f}(m)dm, \tag{2.144}$$

and plugging it into (2.142) results in

$$P(k)f(x) = \int_{-\infty}^{\infty} C_k \exp ik(x-y) \int_{-\infty}^{\infty} \exp[imy]\hat{f}(m)dmdy. \tag{2.145}$$

Using the Fubini theorem yields

$$\begin{aligned} P(k)f(x) &= C_k \exp[ikx] \int_{-\infty}^{\infty}\int_{-\infty}^{\infty} \exp[-iky]\exp[imy]dy\hat{f}(m)dm \\ &= 2\pi C_k \exp[ikx] \int_{-\infty}^{\infty} \delta(k-m)\hat{f}(m)dm = 2\pi C_k \exp[ikx]\hat{f}(k). \end{aligned} \tag{2.146}$$

Checking the main properties of projecting operators, we start with the expansion of unit

$$\int_{-\infty}^{\infty} P(k)dkf(x) = \int_{-\infty}^{\infty}\int_{-\infty}^{\infty} C_k \exp[ik(x-y)]dkf(y)dy, \tag{2.147}$$

which gives

$$\int_{-\infty}^{\infty} P(k)dkf(x) = 2\pi \int_{-\infty}^{\infty} C_k\delta(x-y)f(y)dy = f(y), \tag{2.148}$$

if $2\pi C_k = 1$. The orthogonality property is established by

$$\begin{aligned} P(m)P(k)f(x) &= P(m)\int_{-\infty}^{\infty} a_k(x,y)f(y)dy \\ &= \int_{-\infty}^{\infty} a_m(x,z)\int_{-\infty}^{\infty} a_k(z,y)f(y)dydz \\ &= C_mC_k \int_{-\infty}^{\infty}\int_{-\infty}^{\infty} \exp[im(x-z)]\exp[ik(z-y)]f(y)dydz \\ &= 2\pi C_mC_k\delta(m-k)\int_{-\infty}^{\infty} \exp[imx]\exp[-iky]f(y)dy. \end{aligned} \tag{2.149}$$

It coincides with the projector $P(k)$ that acts as (2.141) if $C_k = 1/2\pi$, having

$$P(m)P(k) = \delta(m-k)P(k). \tag{2.150}$$

An action of the operator

$$\int_{-\infty}^{\infty} ikP(k)dkf(x) = \int_{-\infty}^{\infty} ik \int_{-\infty}^{\infty} \frac{1}{2\pi} \exp[ik(x-y)]dkf(y)dy$$
$$= \frac{d}{dx}\int_{-\infty}^{\infty}\int_{-\infty}^{\infty} \frac{1}{2\pi} \exp[ik(x-y)]dkf(y)dy$$
$$= \frac{d}{dx}\int_{-\infty}^{\infty} \delta(x-y)f(y)dy = \frac{df}{dx}, \tag{2.151}$$

reproduces the spectral theorem statement.

Let us take the square of $\frac{d}{dx}$,

$$\frac{d^2f}{dx^2} = \int_{-\infty}^{\infty} imP(m)dm \int_{-\infty}^{\infty} ikP(k)dkf(x)$$
$$= \frac{1}{(2\pi)^2}\int_{-\infty}^{\infty} im \int_{-\infty}^{\infty} \exp[im(x-z)] \int_{-\infty}^{\infty} ik \int_{-\infty}^{\infty} \exp[ik(z-y)]dzdkf(y)dydm$$
$$= \frac{1}{2\pi}\int_{-\infty}^{\infty} im\exp[im(x)] \int_{-\infty}^{\infty} ik \int_{-\infty}^{\infty} \delta(k-m)\exp[ik(-y)]dkf(y)dydm$$
$$= \frac{1}{2\pi}\int_{-\infty}^{\infty} (ik)^2 \int_{-\infty}^{\infty} \exp[ik(x-y)]dkf(y)dy. \tag{2.152}$$

Denote $k^2 = -\lambda, 2kdk = -d\lambda, k = \kappa > 0, k = -\kappa < 0$, then

$$Lf = -\frac{d^2f}{dx^2} = \frac{1}{2\pi}\int_{0}^{\infty} (i\kappa)^2 \int_{-\infty}^{\infty} \exp[i\kappa(x-y)]dkf(y)dy$$
$$+\frac{1}{2\pi}\int_{0}^{\infty} (-i\kappa)^2 \int_{-\infty}^{\infty} \exp[-i\kappa(x-y)]d\kappa f(y)dy$$
$$= \frac{1}{2\pi}\int_{-\infty}^{\infty}\int_{0}^{\infty} \lambda(exp[i\sqrt{\lambda}(x-y)] + \exp[-i\sqrt{\lambda}(x-y)])\frac{d\lambda}{\sqrt{\lambda}}f(y)dy$$
$$= \int_{0}^{\infty} \lambda(P_+(\lambda) + P_-(\lambda))d\lambda f. \tag{2.153}$$

The spectral expansion within the two subspaces of positive and negative square roots of λ is easily recognized:

$$P_+(\lambda)f = \frac{1}{2\pi\sqrt{\lambda}} \int_{-\infty}^{\infty} \exp[i\sqrt{\lambda}(x-y)]f(y)dy,$$
$$P_-(\lambda)f = \frac{1}{2\pi\sqrt{\lambda}} \int_{-\infty}^{\infty} \exp[-i\sqrt{\lambda}(x-y)]f(y)dy \tag{2.154}$$

The second-order derivative operator

$$L = -\frac{d^2}{dx^2}, \tag{2.155}$$

is directly linked to the matrix operators of the preceding chapter used in the case of a wave equation. Its spectrum may be expressed as $\lambda = -\kappa^2, \kappa > 0$, each point in eigenspace is two-dimensional:

$$\psi = c_1 \exp[i\kappa x] + c_2 \exp[-i\kappa x]. \tag{2.156}$$

The projectors to the corresponding solutions' subspaces are built formally identical to (2.141), but for $\kappa > 0$, so that

$$P_\pm(\kappa)f(x) = \int_{-\infty}^{\infty} a_{\kappa\pm}(x,y)f(y)dy = C_\pm \exp[\pm\kappa x], \tag{2.157}$$

with

$$a_{\kappa\pm}(x,y) = \frac{1}{2\pi}\exp[\pm i\kappa(x-y)]. \tag{2.158}$$

so that

$$P_+(\kappa) + P_-(\kappa) = P(\kappa). \tag{2.159}$$

It also define the projectors to the eigenspaces (2.156)

$$\int_0^\infty P_\pm(\kappa)d\kappa = P^\pm, \tag{2.160}$$

that are direct links with dynamic projectors with completeness

$$P^- + P^+ = I. \tag{2.161}$$

It implies that for the projecting operator with the kernel

$$\int_0^\infty a_{\kappa+}(x,y)d\kappa = C_\kappa \int_0^\infty \exp[i\kappa(x-y)]d\kappa. \tag{2.162}$$

the identity

$$\frac{1}{\sqrt{2\pi}}\int_0^\infty \exp[i\kappa(x-y)]d\kappa = \frac{1}{\sqrt{2\pi}}\int_{-\infty}^\infty \theta(k)\exp[ik(x-y)]dk$$
$$= \frac{\sqrt{\pi}}{2}\left(\delta(x-y)+\frac{1}{i\pi(y-x)}\right). \quad (2.163)$$

holds, and also

$$\int_0^\infty P_-(\kappa)d\kappa = P^-. \quad (2.164)$$

So,

$$\int_0^\infty a_{\kappa-}(x,y)d\kappa = C_k\int_0^\infty \exp[-i\kappa(x-y)]d\kappa. \quad (2.165)$$

The integral is transformed as

$$\frac{1}{\sqrt{2\pi}}\int_0^\infty \exp[-ik(x-y)]dk = \frac{1}{\sqrt{2\pi}}\int_{-\infty}^\infty \theta(k)\exp[-ik(x-y)]dk$$
$$= \frac{\sqrt{\pi}}{2}\left(\delta(x-y)-\frac{1}{i\pi(y-x)}\right). \quad (2.166)$$

Adding the relations gives the expansion of unit

$$\int_0^\infty (P_+(\kappa)+P_-(\kappa))d\kappa f(x) = \int_{-\infty}^\infty\int_{-\infty}^\infty C_k\exp[ik(x-y)]dkf(y)dy \quad (2.167)$$

which gives

$$\int_{-\infty}^\infty P(k)dkf(x) = 2\pi\int_{-\infty}^\infty C_k\delta(x-y)f(y)dy. \quad (2.168)$$

A direct correspondence to the approach of this book, named by "dynamic projecting" may be demonstrated by example. It is established by means of a matrix operator problem for

$$\mathcal{L} = \begin{pmatrix} 0 & \frac{d}{dx} \\ \frac{d}{dx} & 0 \end{pmatrix}. \quad (2.169)$$

The spectral problem

$$\mathcal{L}\Psi = \lambda\Psi, \Psi = \begin{pmatrix} \psi_1 \\ \psi_2 \end{pmatrix}. \quad (2.170)$$

is equivalent to one of (2.155). The link is established by

$$\frac{d}{dx}\psi_2 = ik\psi_1, \frac{d}{dx}\psi_1 = ik\psi_2. \quad (2.171)$$

Both ψ_i satisfy

$$\frac{d^2}{dx^2}\psi_i = -k^2\psi_i, \tag{2.172}$$

which means the connection with the problem (2.155) by

$$\lambda = k^2, \tag{2.173}$$

for two linear independent solutions for $k > 0$

$$\psi_{1,\pm} = \exp[\pm ikx], \tag{2.174}$$

and the corresponding eigen vectors

$$\Psi_\pm = \begin{pmatrix} \psi_{1\pm} \\ \frac{d}{dx}\psi_{1\pm} \end{pmatrix} = \begin{pmatrix} 1 \\ \pm ik \end{pmatrix} \exp[\pm ikx]. \tag{2.175}$$

The subspaces are fixed by matrix projectors. The projectors

$$P^\pm(k) = \frac{1}{2}\begin{pmatrix} 1 & \pm(ik)^{-1} \\ \pm ik & 1 \end{pmatrix} P_\pm(k), \tag{2.176}$$

see also (2.176), act as

$$P^\pm(k)\Psi = \frac{1}{2}\begin{pmatrix} 1 & \pm(ik)^{-1} \\ \pm ik & 1 \end{pmatrix} P_\pm(k)\begin{pmatrix} \psi_1 \\ \psi_2 \end{pmatrix}. \tag{2.177}$$

The dynamical projectors then are obtained as

$$\int_0^\infty P^\pm(k)2kdk = P^\pm = \frac{1}{2}\begin{pmatrix} 1 & \pm 1 \\ \pm 1 & 1 \end{pmatrix}; \tag{2.178}$$

compare with (1.20) of the preceding chapter with $v_2 = -v_1 = 1$.

REFERENCES

1. Leble, S. 2016. General remarks on dynamic projection method. *Task Quarterly* 20(2): 113–130.
2. Tikhonov, A.N., and A.A. Samarskii. 2013. *Equations of Mathematical Physics*. Dover Publications Inc., New York.
3. Shubin, M.A. 2001. *Pseudodifferential Operators and Spectral Theory*. Springer, Berlin.
4. Perelomova, A. 2006. Development of linear projecting in studies of non-linear flow. Acoustic heating induced by non-periodic sound. *Physics Letters A* 357: 42–47.
5. Leble, S., and I. Vereshchagina. 2017. Dynamic projection operator method in the theory of hyperbolic systems of partial differential equations with variable coefficients. *Task Quarterly* 21(2): 705716.
6. Perelomova, A., and S. Leble. 2005. Vortical and acoustic waves interaction. From general equations to integrable cases. *Theoretical and Mathematical Physics* 144: 1030–1039.

7. Dunford, N., and J.T. Schwartz. 1988. *Linear Operators, Part 1: General Theory*, Vol. 1. Wiley-Interscience, Hoboken.
8. Ljusternik, L.A., and V.A. Sobolev. 1965. Elements of functional analysis, Nauka, Moskva (in Russian). Elemente der Funktionalanalysis, volume 8 of Mathematische Lehrbücher und Monographien. I. Abteilung: Mathematische Lehrbücher. Akademie-Verlag, Berlin (in Deutche).
9. Fock, V.A. 1978. *Fundamentals of Quantum Mechanics*. Mir Publishing, Moscow.

3 One-Dimensional Problem in Hydrodynamics

3.1 ON THE HYDRO-THERMODYNAMIC RELATIONS FOR QUASI-ISENTROPIC PROCESSES

The important issue of projecting in dynamic problems, and, in particular, in hydrodynamics, is to establish individual modes as relations of specific perturbations of a medium. As with the first approximation, these links may be established as they follow from the linearized equations, that is, in the flow of infinitely-small magnitudes of perturbations. These links may be corrected with any desired accuracy by involving the nonlinear terms. Sound is usually determined as the isentropic or nearly isentropic mode. Hence, the links of perturbations in sound should support its adiabaticity with some accuracy. The modes propagate independently in a linear flow exclusively. The nonlinear coupling of modes may be described by the means of linear projecting.

To illustrate the efficiency of projecting, let us briefly discuss other more common approaches in nonlinear acoustics. We mean the usual scheme of deriving a thermo-hydrodynamic relation that serves as a modified equation of state (following the review by Makarov and Ochmann [1]). The basic postulates and a detailed discussion may be found in the paper by Perelomova [2]. The Taylor series of pressure as function of entropy and density $p(s,\rho)$ in the vicinity of the thermodynamic equilibrium state (s_0,ρ_0) reads:

$$\begin{aligned} p =& p_0 + A\frac{(\rho-\rho_0)}{\rho_0} + \frac{B}{2}\frac{(\rho-\rho_0)^2}{\rho_0^2} \\ &+ \left(\frac{\partial p}{\partial s}\right)\bigg|_{\rho=\rho_0,s=s_0}(s-s_0)\left(\frac{\partial^2 p}{\partial\rho\partial s}\right)\bigg|_{\rho=\rho_0,s=s_0}(s-s_0)(\rho-\rho_0), \end{aligned} \tag{3.1}$$

where

$$A=\rho_0\left(\frac{\partial p}{\partial\rho}\right)\bigg|_{\rho=\rho_0,s=s_0}=\rho_0 c_0^2,\quad \frac{B}{A}=2\rho_0 c_0\,\frac{\partial c}{\partial p}\bigg|_{s=s_0,p=p_0}. \tag{3.2}$$

c denotes the sound speed, and c_0 its equilibrium vaue evaluated at equilibrium state (p_0,ρ_0). The ratio B/A is the first parameter of nonlinearity of a fluid. As a rule, nonlinear corrections of an order higher than second are neglected as minor. Recent research accounts for the cubic and higher order nonlinear corrections [3,4]. When the effect of linear thermal conductivity is included, an excess entropy takes the form

$$s-s_0=-\frac{\kappa}{T_0}\left(\frac{\partial T}{\partial p}\right)\bigg|_{s=s_0,p=p_0}\vec{\nabla}\cdot\vec{v}=-\kappa\frac{b}{\rho_0 C_p}\vec{\nabla}\cdot\vec{v}. \tag{3.3}$$

This follows from the equation of entropy balance under the assumption irrotational nature of a pressure wave, where T designates temperature, C_p is the heat capacity under the constant pressure, and κ and b are compressibility and thermal expansion, respectively:

$$\kappa = -\frac{1}{V}\left(\frac{\partial V}{\partial p}\right)_T = \frac{1}{\rho}\left(\frac{\partial \rho}{\partial p}\right)_T, b = \frac{1}{V}\left(\frac{\partial V}{\partial T}\right)_p = -\frac{1}{\rho}\left(\frac{\partial \rho}{\partial T}\right)_p. \tag{3.4}$$

An Eq. (3.1) with account of (3.3) and the thermodynamic relation (C_V denotes the heat capacity under the constant volume V)

$$\left(\frac{\partial p}{\partial s}\right)\bigg|_{\rho=\rho_0, s=s_0} = \left(\frac{1}{C_V} - \frac{1}{C_p}\right)\frac{\rho_0 C_p}{b},$$

may be thus rewritten as follows:

$$p = p_0 + A\frac{(\rho-\rho_0)}{\rho_0} + \frac{B}{2}\frac{(\rho-\rho_0)^2}{\rho_0^2} - \kappa\left(\frac{1}{C_V} - \frac{1}{C_p}\right)v + \left(\frac{\partial^2 p}{\partial\rho\partial s}\right)\bigg|_{\rho=\rho_0, s=s_0} \times (s-s_0)(\rho-\rho_0). \tag{3.5}$$

It is claimed at this point that $(s-s_0)$ at the end of (3.5) means an irreversible increase in entropy that does not include the effect of thermal conductivity that relates to the fluid temperature increment $\Delta T(x,t)$ in the following way:

$$s - s_0 = \frac{C_p}{T_0}\Delta T. \tag{3.6}$$

The last two terms in (3.5) express corrections due to variations in ambient density and temperature. After some algebra,

$$\begin{aligned} \frac{C_p}{T_0}\left(\frac{\partial p}{\partial s}\right)\bigg|_{\rho=\rho_0, s=s_0} &= Ab = \tau_1, \\ \frac{\rho_0 C_p}{T_0}\left(\frac{\partial^2 p}{\partial\rho\partial s}\right)\bigg|_{\rho=\rho_0, s=s_0} &= 2\rho_0 c_0 \left(\frac{\partial c}{\partial T}\right)\bigg|_{\rho=\rho_0, T=T_0} = \tau_2, \end{aligned} \tag{3.7}$$

the relation (3.5) rearranges as

$$p = p_0 + A\frac{(\rho-\rho_0)}{\rho_0} + \frac{B}{2}\frac{(\rho-\rho_0)^2}{\rho_0^2} - \kappa\left(\frac{1}{C_V} - \frac{1}{C_p}\right)v + \tau_1\Delta T + \tau_2\Delta T\frac{\rho-\rho_0}{\rho_0}. \tag{3.8}$$

Assuming that the total quantities consist of the unperturbed ones, acoustic ones (denoted by lower index a), and, in addition, density, involves the perturbation, which associates with the entropy mode, $\Delta\rho$:

$$\begin{aligned} p &= p_0 + p_a \\ \rho &= \rho_0 + \Delta\rho + \rho_a, \end{aligned} \tag{3.9}$$

Equation (3.8) takes the form

$$p_a = A\frac{\rho_a}{\rho_0} + \frac{B}{2}\frac{\rho_a^2}{\rho_0^2} - k\left(\frac{1}{C_V} - \frac{1}{C_p}\right) v_a + \left(\tau_2 - \frac{B}{A}\tau_1\right)\Delta T\frac{\rho_a}{\rho_0}. \tag{3.10}$$

Finally, Eq. (3.10) accounts for corrections to acoustic pressure due to an increase in entropy caused by the irreversible processes that occur in viscous and/or heat-conducting flow. In turn, the irreversible mean temperature increment $\Delta T(x,t)$ is not an acoustic quantity and relates to a fluid density decrease as

$$\frac{\Delta\rho}{\rho_0} = -\frac{\tau_1}{A}\Delta T. \tag{3.11}$$

Suppose that $\Delta T(x,t)$ could be found additionally from the heat-transfer equation.

We have repeated a common scheme of deriving the hydrodynamic relation $p_a = p(\rho_a, v_a)$ (3.10) by following the ideas of Ref. [1]. Equation (3.10) considers the nonlinear coupling of sound and the entropy mode by means of a term that includes ΔT. Let us discuss this scheme from the initial form (3.1) to the final one (3.10).

1. The terms including $(s - s_0)$ are evaluated with the additional assumption of the irrotational character of acoustic mode. The acoustic mode is determined as an isentropic irrotational flow except of the effect of linear thermal conductivity in the equation of state (3.8). This looks somewhat artificial and should follow immediately from the system of conservaton equation. This requires additional explanation. An acoustic mode is also related as a pressure wave. This imposes that other modes that may exist in a fluid are isobaric. This requires justification, and, although it is true for flows over uniform media, it may be broken in the case of more complex flows. Subdividing excess entropy into reversible and irreversible parts looks doubtful since both thermal conductivity and viscosity lead to an irreversible increase in entropy. To evaluate $\Delta T(x,t)$, one should extract information from the dynamical equations such as a balance equation for entropy. To avoid this indefiniteness and the complex interpretation of intermediate steps, one needs a unique definition of the acoustic (and other) modes in any problem involving fluid's flow, which should be a starting point for studies. This concerns also multi-dimensional flows, flows with an additional degree of the thermodynamic state, those affected by internal forces, etc.
2. The very form of (3.1), (3.10) implies explication of all terms. This may lead to different interpretations. In some papers [5,6] an effect of ambient density perturbation is neglected in problems of nonlinear coupling of acoustic and non-acoustic modes. Just recently, it has been justified that ambient density perturbation should be considered [1].
3. The interaction of modes is understood as a variation in the background forming the new conditions for pressure wave propagation. The physical

conditions of a flow often look more complex. In inhomogeneous media (affected by an external force, for instance), the entropy mode is specified by variations both in temperature and pressure [7]. Other modes (vortex, relaxation) may contribute to features of sound propagation. On the other hand, sound may affect all non-wave modes. In a number of problems, the non-acoustic modes couldn't be treated as secondary. The thermohydrodynamical relation (3.10) is just a starting issue; it should be again involved in the system. So, the most convenient way to derive nonlinear evolution equations is rather thorny.
In contrast, the projecting gives the following possibilities:

a. To determine modes as relations between specific perturbations. This follows from the linearized system of dynamic equations that describe a flow. The linear definition of modes leads to the application at the total vector of perturbations of projectors that separate every mode uniquely.
b. To fix these leading-order relations going to investigations of the weakly nonlinear dynamics of a fluid and to correct them with the required accuracy. This also may be done by making use of projecting.
c. To derive a system of nonlinear coupling equations in terms of specific perturbations by means of the application of the projectors on the nonlinear dynamic system of equations. This may be solved according to the geometry of the problem, the type of initial/boundary conditions, and the initial magnitudes of modes.

Comment on a):
The idea to define modes in accordance to the dispersion relations of a llinear problem is long known; see, for example, Ref. [1]. We develop it, determining relations of specific perturbations and making use of them in analysis of the weakly nonlinear flows. Studies of the fluid motion of infinitely small amplitudes start usually with representing all perturbations as a sum of planar waves, where $\tilde{q}(\vec{k})\exp(i\omega(\vec{k})t)$ is the Fourier-transform of any perturbation q':

$$q'(x,t) = \int_{-\infty}^{\infty}\int_{-\infty}^{\infty}\int_{-\infty}^{\infty} \tilde{q}(\vec{k}) \cdot \exp i(\omega t - \vec{k}\cdot\vec{r}) dk_x dk_y dk_z + cc$$

where $\vec{r}$ is the radius vector, and $\vec{k}$ is the wave vector with Carthesian components k_x, k_y, k_z. This determines the dispersion equation. Its roots, that is, dispersion relations, in turn define all possible modes in a fluid as links of perturbations specifying every mode.
The standard issue in hydrodynamics is to determine modes due to rotational character as vorticity modes, potential (acoustic) modes, and the entropy mode. The detailed study by Chu and Kovásznay [8] is probably the first attemt to consider weak interactions of different modes by means of this kind of projecting. The arguments to define modes due to the branches of the dispersion relation and to apply projecting

are as follows: the procedure is algorithmic, guarantees that all possible movements are taken into account (with subdivision of sound into branches). This leads to a system of nonlinear coupling evolution equations. Projecting is useful not only in the simplest case of homogeneous media without mean flows but also in much more complicated problems, such as motion in the field of external force [7], dispersive flows in a bubbly liquid [9], or flows with different kinds of thermodynamic relaxation. The external heating or cooling of a media also affects definition of modes and, as a result, the peculiarities of nonlinear interaction of modes in a flow.

Comment on b):
Making use of linear links in specific perturbations results in a system for coupling nonlinear evolution equations that describe the nonlinear dynamics of a fluid. A general procedure is algorithmic: we apply projectors on the dynamic system of equations. The nonlinear interaction in a planar flow leads to resonant generation of other modes by the dominant wave, a non-dominant mode consisting of two parts of different velocities: one resonant and the second moving with its own linear velocity. In order to classify the overall field due to the direction of propagation, all resonant additions should be picked up. The detailed analysis, explanation, and examples may be found in the Ref. [2]. An accurate account of resonant inputs leads to nonlinear corrections in the relations between the specific components for pure progressive wave, which, in the case of ideal gas, is described by the famous Earnshaw equation. These corrections support adiabaticity of sound and may be evaluated with any desired accuracy in the different flows.

To comment c):
Arbitrary initial or boundary conditions may be readily considered. The final system consists of equations in the first-order partial derivatives with respect to time, which contain specific perturbations of different modes. As for descriptions of sound propagation, it is simpler for an approximate solution than the nonlinear wave equation reported by Kuznetsov and widely used till now [10]. The Kuznetsov equation contains the second-order time derivative since the acoustic modes of the different directions of propagation are not subdivided there. The interaction with non-waves modes are not considered by the Kuznetsov equation at all. In our approach, the entropy mode is involved as well as acoustic modes; it is not obviously secondary. This concerns also other non-wave modes that may exist in a fluid. The approximate solution of the system of coupling equations depends on the type of initial/boundary problem. Two- or three-dimensional problems are treated in the same way. The linear viscous terms of different origin affect both projectors and relations between the specific perturbations of every mode. They affect also the form of coupling equations. Moreover, the

coupling between wave and non-wave modes takes place due to the attenuation and relative irreversible transfer of energy and momentum.

3.2 THERMOCONDUCTING FLOW OF AN UNIFORM NEWTONIAN GAS. MODES, PROJECTORS AND DYNAMIC EQUATIONS. ACOUSTIC HEATING

3.2.1 AN IDEAL GAS

The starting point is a set of hydrodynamic equations of the plane flow in the differential form:

$$\rho\left(\frac{\partial v}{\partial t}+v\frac{\partial v}{\partial x}\right)+\frac{\partial p}{\partial x}=\frac{4\eta}{3}\frac{\partial^2 v}{\partial x^2},$$

$$\rho\left(\frac{\partial e}{\partial t}+v\frac{\partial e}{\partial x}\right)=\chi\frac{\partial^2 T}{\partial x^2}+\frac{4\eta}{3}\left(\frac{\partial v}{\partial x}\right)^2, \tag{3.12}$$

$$\frac{\partial \rho}{\partial t}+\frac{\partial(\rho v)}{\partial x}=0.$$

where e denotes internal energy per unit mass, and χ,η mark thermal conductivity and viscosity, both supposed to be constants. The system (3.12) should be completed by the equations of state for an ideal gas,

$$e=C_V T=\frac{p}{(\gamma-1)\rho}. \tag{3.13}$$

$\gamma=C_p/C_V$ denotes the ratio of specific heats. In the excess quantities, Eq. (3.12) rearranges in the leading order as

$$\frac{\partial v}{\partial t}+\frac{1}{\rho_0}\frac{\partial p'}{\partial x}-\delta_1\frac{\partial^2 v}{\partial x^2}=-v\frac{\partial v}{\partial x}+\frac{\rho'}{\rho_0^2}\frac{\partial p'}{\partial x}-\delta_1\frac{\rho'}{\rho_0}\frac{\partial^2 v}{\partial x^2},$$

$$\frac{\partial p'}{\partial t}+\rho_0 c_0^2\frac{\partial v}{\partial x}-\frac{\delta_2}{\gamma-1}\frac{\partial^2(\gamma p'-c_0^2\rho')}{\partial x^2}=-v\frac{\partial p'}{\partial x}-\gamma p'\frac{\partial v}{\partial x}+\delta_1\rho_0(\gamma-1)\left(\frac{\partial v}{\partial x}\right)^2$$
$$-\frac{\delta_2}{\rho_0(\gamma-1)}\frac{\partial^2(\gamma p'\rho'-c_0^2\rho'^2)}{\partial x^2}, \tag{3.14}$$

$$\frac{\partial \rho'}{\partial t}+\rho_0\frac{\partial v}{\partial x}=-v\frac{\partial \rho'}{\partial x}-\rho'\frac{\partial v}{\partial x}.$$

where

$$c_0=\sqrt{\gamma p_0/\rho_0},\quad \delta_1=\frac{4\eta}{3\rho_0},\quad \delta_2=\frac{\chi}{\rho_0}\left(\frac{1}{C_V}-\frac{1}{C_p}\right).$$

We consider an initially unperturbed medium with zero mean flow: $v=v'$. The system (3.14) is equivalent to the initial one (3.12) with accuracy up to quadratic nonlinear terms (including those standing by the dissipative coefficients), which are of major importance in weakly nonlinear acoustics. Three eigenvectors of the corresponding matrix operator take the form as follows [4,5]. The first two are acoustic, and the third is the entropy:

$$\psi_1 = \begin{pmatrix} \frac{c_0}{\rho_0} - \frac{\beta}{2\rho_0}\partial/\partial x \\ c_0^2 - \delta_2 c_0 \partial/\partial x \\ 1 \end{pmatrix} \rho_1, \psi_2 = \begin{pmatrix} -\frac{c_0}{\rho_0} - \frac{\beta}{2\rho_0}\partial/\partial x \\ c_0^2 + \delta_2 c_0 \partial/\partial x \\ 1 \end{pmatrix} \rho_2,$$

$$\psi_3 = \begin{pmatrix} -\frac{\delta_2}{\rho_0(\gamma-1)}\partial/\partial x \\ 0 \\ 1 \end{pmatrix} \rho_3, \tag{3.15}$$

where

$$\psi = \begin{pmatrix} v & p' & \rho' \end{pmatrix}^T = \psi_1 + \psi_2 + \psi_3$$

is the vector of total perturbations, ρ_1, ρ_2, ρ_3 designate excess densities specifying the correspondent mode, and $\beta = \delta_1 + \delta_2$ designates the total attenuation. These three eigenvectors manifest existence of three basic types of the planar flow. They relate to three dispersion relations:

$$\omega_1 = c_0 k + i\frac{\beta}{2}k^2, \quad \omega_2 = -c_0 k + i\frac{\beta}{2}k^2, \quad \omega_3 = i\frac{\delta_2}{(\gamma-1)}k^2. \tag{3.16}$$

The projecting rows which subdivide the individual specific excess densities from the total perturbation,

$$d_1\psi = \rho_1, \quad d_2\psi = \rho_2, \quad d_3\psi = \rho_3,$$

take the form:

$$d_1 = \left(\frac{\rho_0}{2c_0} + \frac{\delta_2\rho_0}{2c_0^2}\frac{\partial}{\partial x}, \frac{1}{2c_0^2} + \left(\frac{\beta}{4c_0^3} - \frac{\delta_2}{2(\gamma-1)c_0^3}\right)\frac{\partial}{\partial x}, \frac{\delta_2\rho_0}{2c_0(\gamma-1)}\frac{\partial}{\partial x}\right),$$

$$d_2 = \left(-\frac{\rho_0}{2c_0} + \frac{\delta_2\rho_0}{2c_0^2}\frac{\partial}{\partial x}, \frac{1}{2c_0^2} - \left(\frac{\beta}{4c_0^3} - \frac{\delta_2}{2(\gamma-1)c_0^3}\right)\frac{\partial}{\partial x}, -\frac{\delta_2\rho_0}{2c_0(\gamma-1)}\frac{\partial}{\partial x}\right), \tag{3.17}$$

$$d_3 = \left(-\frac{\delta_2\rho_0}{c_0^2}\frac{\partial}{\partial x}, -\frac{1}{c_0^2}, 1\right).$$

Applying at the linearized system (3.14), they distinquish linear dynamic equations of everyone from specific excess densities, and applying at Eq. (3.14), they lead to coupling equations which reflect interaction of modes. Without loss of generality, we will consider coupling of the first acoustic mode, which propagates in the positive direction of the x-axis, and the entropy mode, which is imposed to be a secondary mode with magnitude of perturbations much less than that of sound [12]. Links for the rightwards progressive mode should be corrected by the involvement of quadratic nonlinear terms specific for the Riemann wave [13,14]. These terms make sound quasi-isentropic within the corresponding accuracy (as usual, up to quadratic nonlinear terms, but corrrections of the higher order may be considered in accordance to the physically meaning of a problem). They may be readily established in the way of resonant corrections of the acoustic mode by means of linear projecting [2]. These links must result in the equivalent dynamic equations for sound while substituted in every one from the set of conservation equations, if all other modes are ignored. This may also serve as a foundation in establishing these corrections. The corrected links contribute to the total perturbations:

$$v(x,t) = \frac{c_0}{\rho_0}\rho_1(x,t) + \frac{c_0}{\rho_0^2}\frac{\gamma-3}{4}\rho_1^2(x,t) - \frac{\beta}{2\rho_0}\frac{\partial \rho_1(x,t)}{\partial x} - \frac{\delta_2}{\rho_0(\gamma-1)}\frac{\partial \rho_3(x,t)}{\partial x},$$

$$p(x,t) = c_0^2\rho_1(x,t) + \frac{(\gamma-1)c_0^2}{2\rho_0}\rho_1^2(x,t) - \delta_2 c_0\frac{\partial \rho_1(x,t)}{\partial x}, \qquad (3.18)$$

$$\rho(x,t) = \rho_1(x,t) + \rho_3(x,t).$$

The acoustic modes are correctd by the terms referring to the excess density, but they may take different form in dependence on choice of the reference variable. It should differ from zero. As an example, perturbation in pressure can not be the reference variable for the entropy mode. By use of (3.18), two dynamic equations follow from system (3.14). The first one governs excess density in acoustic mode, which is the famous Burgers equation:

$$\frac{\partial \rho_1}{\partial t} + c_0\frac{\partial \rho_1}{\partial x} + \frac{\gamma+1}{2\rho_0}\rho_1\frac{\partial \rho_1}{\partial x} - \frac{\beta}{2}\frac{\partial^2 \rho_1}{\partial x^2} = 0. \qquad (3.19)$$

The second equation describes the excitation of the entropy mode in the field of sound:

$$\frac{\partial \rho_3}{\partial t} - \frac{\delta_2}{\gamma-1}\frac{\partial^2 \rho_3}{\partial x^2} = -\delta_2\frac{\rho_1}{\rho_0}\frac{\partial^2 \rho_1}{\partial x^2} - \frac{((\gamma-1)\delta_1+\gamma\delta_2)}{\rho_0}\left(\frac{\partial \rho_1}{\partial x}\right)^2 - \frac{(\gamma-5)\delta_2}{4\rho_0}\frac{\partial^2 \rho_1^2}{\partial x^2}. \qquad (3.20)$$

Equations (3.19) and (3.20) are valid over the time domain, when the magnitude of excess density in sound ρ_1 is much larger than that of the entropy mode, ρ_3. Both equations are instanteneous. We do not make use of temporal averaging which in fact is a special kind of projecting.

The irreversible enlargement of entropy, which is associated with dissipation of sound energy, takes the form:

$$\Delta s = \frac{c_0^2}{\rho_0^2}\frac{4/3\eta}{Pr}\left(\frac{\gamma-3}{2}\rho_1\frac{\partial^2 \rho_1}{\partial x^2} + \frac{3\gamma+2Pr-5}{2}\left(\frac{\partial \rho_1}{\partial x}\right)^2\right), \qquad (3.21)$$

where Pr is the Prandtl number:

$$Pr = \frac{(\gamma-1)\delta_1}{\delta_2}.$$

Equation (3.20) may be solved numerically. An excess acoustic density in its right-hand side, in turn, is a solution of the Burgers equation (3.19), which reduces to the linear diffusity equation by the Hopf-Cole transformation [14]. Among solutions to Eq. (3.19), the self-similar impulses are of importance [14]:

$$\rho_1 = -\frac{4\rho_0}{(\gamma+1)}\sqrt{\frac{\beta}{2\pi c_0(x+x_0)}}\frac{\exp\left(-\frac{c_0(c_0(t+t_0)-(x+x_0))^2}{2\beta(x+x_0)}\right)}{\left(C-Erf\left(-\frac{c_0(c_0(t+t_0)-(x+x_0))^2}{2\beta(x+x_0)}\right)\right)}, \tag{3.22}$$

where x_0, t_0 and C are some constants. Impulses with smaller $|C|$ reveal larger deviation from the symmetry. The main conclusions are that the efficiency of heating is considerably larger if produced by the impulses of negative polarity. The difference between positive $|C|$ and negative impulses with the same modul of C increases for smaller values of C [12]. The relative efficiency tends to 1 when t tends to infinity. Impulses of smaller $|C|$ among groups of positive and negative impulses are more effective.

3.2.2 FLUIDS DIFFERENT FROM IDEAL GASES

The fluids that behave differently from ideal gas equations of state may be readily considered. An excess internal energy and temperature in caloric and thermal equations of state should be expanded in series with accuracy up to quadratic nonlinear terms. In the general case, the leading-order series for the smooth-enough thermodynamic functions look like [11]:

$$\rho_0 e' = E_1 p' + \frac{E_2 p_0}{\rho_0}\rho' + \frac{E_3}{p_0}p'^2 + \frac{E_4 p_0}{\rho_0^2}\rho'^2 + \frac{E_5}{\rho_0}p'\rho', \tag{3.23}$$

$$T' = \frac{\theta_1}{\rho_0 C_V}p' + \frac{\theta_2 p_0}{\rho_0^2 C_V}\rho' + \frac{\theta_3}{\rho_0 p_0 C_V}p'^2 + \frac{\theta_4 p_0}{\rho_0^3 C_V}\rho'^2 + \frac{\theta_5}{\rho_0^2 C_V}p'\rho'.$$

All coefficients in series are evaluated in equilibrium state. The first coefficient in the set may be readily expressed in terms of compresiibility and thermal expansion:

$$E_1 = \frac{\rho_0 C_V \kappa}{b},\ E_2 = -\frac{C_p \rho_0}{b p_0} + 1.$$

An excess entropy is a total differential that imposes a relation of coefficients in both series in the form of equality:

$$\theta_2 = \frac{C_V \rho_0 T_0}{E_1 p_0} - \frac{(1-E_2)\theta_1}{E_1}.$$

The first series yields a small-signal sound velocity:

$$c_0 = \sqrt{\frac{(1-E_2)p_0}{E_1 \rho_0}}.$$

It is convenient to introduce viscous coefficients (involving the bulk viscosity ζ):

$$\delta_1^1 = \frac{(\zeta + \eta/3)}{\rho_0}, \delta_1^2 = \frac{\eta}{\rho_0}, \delta_2^1 = \frac{\chi\theta_1}{\rho_0 C_V E_1}, \delta_2^2 = \frac{\chi\theta_2}{\rho_0 C_V (1-E_2)}. \tag{3.24}$$

The total attenuation β is a sum of viscous and thermal parts, δ_1 and δ_2:

$$\beta = \delta_1^1 + \delta_1^2 + \delta_2^1 + \delta_2^2,$$

$$\delta_1 = \delta_1^1 + \delta_1^2, \delta_2 = \delta_2^1 + \delta_2^2 = \frac{\chi T_0}{c_0^2 E_1^2 \rho_0}.$$

The first and third equations in (3.14) remain unchanged, but ensure that δ_1 denote the summary mechanical attenuation due to shear and bulk viscosity. The analog of the second equation from the set follows from the energy balance equation, which in fact is similar to that in Eq. (3.12) where $4/3\eta$ is replaced by the total mechanical attenuation $4/3\eta + \zeta$. It takes the form:

$$\frac{\partial p'}{\partial t} + \rho_0 c_0^2 \frac{\partial v}{\partial x} - \delta_2^1 \frac{\partial^2 p'}{\partial x^2} - \delta_2^2 c_0^2 \frac{\partial^2 \rho'}{\partial x^2} = -v \frac{\partial p'}{\partial x} + (D_1 p' + D_2 \rho') \frac{\partial v}{\partial x} + \frac{\delta_1}{E_1} \left(\frac{\partial v}{\partial x}\right)^2$$
$$+ \frac{\delta_2^3}{c_0^2 \rho_0} \frac{\partial^2 p'^2}{\partial x^2} + \frac{\delta_2^4 c_0^2}{\rho_0} \frac{\partial^2 \rho'^2}{\partial x^2} + \frac{\delta_2^5}{\rho_0} \frac{\partial^2 (p'\rho')}{\partial x^2}, \tag{3.25}$$

where

$$D_1 = \frac{1}{E_1}\left(-1 + 2\frac{1-E_2}{E_1} E_3 + E_5\right), D_2 = \frac{1}{1-E_2}\left(1 + E_2 + 2E_4 + \frac{1-E_2}{E_1} E_5\right),$$

$$\delta_2^3 = \frac{\theta_3 \chi}{E_1 \rho_0 C_V} \frac{1-E_2}{E_1}, \delta_2^4 = \frac{\theta_4 \chi}{(1-E_2)\rho_0 C_V}, \delta_2^5 = \frac{\theta_5 \chi}{E_1 \rho_0 C_V}. \tag{3.26}$$

The acoustic dispersion relations are identical to those established by Eq. (3.16), and the third root is

$$\omega_3 = -i\delta_2^2 k^2.$$

The corrected links specifying acoustic first mode and the third mode are

$$\psi_1 = \begin{pmatrix} \frac{c_0}{\rho_0} - \frac{\beta}{2\rho_0}\partial/\partial x \\ c_0^2 - \delta_2 c_0 \partial/\partial x \\ 1 \end{pmatrix} \rho_1 - \begin{pmatrix} \frac{c_0}{4\rho_0^2}(3 + D_1 + D_2) \\ \frac{c_0^2}{2\rho_0}(1 + D_1 + D_2) \\ 0 \end{pmatrix} \rho_1^2, \psi_3 = \begin{pmatrix} -\frac{\delta_2^2}{\rho_0}\frac{\partial}{\partial x} \\ 0 \\ 1 \end{pmatrix} \rho_3. \tag{3.27}$$

The quadratic corrections to acoustic modes make them isentropic in the flow without mechanical and thermal damping. The corresponding projecting rows, which extract specific excess densities for acoustic modes, take the form

$$d_1 = \left(\frac{\rho_0}{2c_0} + \frac{\delta_2 \rho_0}{2c_0^2}\frac{\partial}{\partial x}, \frac{1}{2c_0^2} + \left(\frac{\beta}{4c_0^3} + \frac{\delta_2^2}{2c_0^3}\right)\frac{\partial}{\partial x}, -\frac{\delta_2^2 \rho_0}{2c_0}\frac{\partial}{\partial x}\right),$$

$$d_2 = \left(-\frac{\rho_0}{2c_0} + \frac{\delta_2\rho_0}{2c_0^2}\frac{\partial}{\partial x}, \frac{1}{2c_0^2} - \left(\frac{\beta}{4c_0^3} + \frac{\delta_2^2}{2c_0^3}\right)\frac{\partial}{\partial x}, \frac{\delta_2^2\rho_0}{2c_0}\frac{\partial}{\partial x}\right). \tag{3.28}$$

The projecting row for entropy excess density remains unchanged and can still be described by the last formula in Eq. (3.17). Making use of projecting, we readily derive equations that govern excess densities of sound and the entropy mode. For definiteness, we consider the dominative first acoustic mode. An excess density for the first acoustic mode is described by the equation in partial derivatives:

$$\frac{\partial\rho_1}{\partial t} + c_0\frac{\partial\rho_1}{\partial x} + \frac{1-D_1-D_2}{2\rho_0}\rho_1\frac{\partial\rho_1}{\partial x} - \frac{\beta}{2}\frac{\partial^2\rho_1}{\partial x^2} = 0. \tag{3.29}$$

The second equation describes excitation of the entropy mode in the field of sound:

$$\frac{\partial\rho_3}{\partial t} + \delta_2^2\frac{\partial^2\rho_3}{\partial x^2} = -\delta_2\frac{\rho_1}{\rho_0}\frac{\partial^2\rho_1}{\partial x^2} + \frac{1}{\rho_0}\left(D_1\delta_2 - \frac{\delta_1}{E_1}\right)\left(\frac{\partial\rho_1}{\partial x}\right)^2$$
$$+\frac{1}{\rho_0}\left(-\frac{(D_1+D_2-1)\delta_2}{4} + \delta_2^1\frac{D_1+D_2+1}{2} - \delta_2^3 - \delta_2^4 - \delta_2^5\right)\frac{\partial^2\rho_1^2}{\partial x^2}. \tag{3.30}$$

The caloric equation of state of an ideal gas establishes coefficients as follows:

$$E_1 = E_4 = \theta_1 = \frac{1}{\gamma-1}, E_2 = E_5 = \theta_2 = \theta_5 = -\frac{1}{\gamma-1}, E_3 = \theta_3 = D_2 = 0, D_1 = -\gamma,$$

which ensures the results of the previous subsection in the case of an ideal gas. Equation (3.30) is instantaneous. It is valid for any acoustic perturbations in the role of the source of acoustic heating. The important case is the periodic or nearly periodic in time perturbations. In this particular case, averaging Eq. (3.30) over the sound period (designated by angle brackets), one arrives at:

$$\left\langle\frac{\partial\rho_3}{\partial t} + \delta_2^2\frac{\partial^2\rho_3}{\partial x^2}\right\rangle = \frac{1}{\rho_0}\left(\frac{\delta_1}{E_1} - \delta_2(1+D_1)\right)\left\langle\rho_1\frac{\partial^2\rho_1}{\partial x^2}\right\rangle.$$

That is, the coefficient standing on the right-hand side is not in general proportional to the total attenuation β. This is the case of exact equality

$$1+D_1 = -\frac{1}{E_1},$$

which is valid for an ideal gas but may be broken in general. Examples of van der Waals gases and semi-ideal carbon dioxide has been considered in Ref. [11]. For the van der Waals gas with equations of state

$$\frac{p}{\rho} = \frac{RT}{m}\left(1+\frac{\rho}{m}\left(b-\frac{a}{RT}\right)\right), \quad e = \frac{f}{2}\frac{RT}{m} - \frac{a\rho}{m^2},$$

where a and b are some constants, f is the number of degrees of a molecule, m is a molar mass of gas, R denotes the universal gas constant, the quantity $1+D_1+\frac{1}{E_1} = -\frac{b\rho}{m}$ differs from zero. Liquids, which are very poorly described by the van der Waals equations, are expected to reveal much more discrepancy as for proportionality of the acoustic force to the total attenuation.

3.3 NON-NEWTONIAN FLUIDS

The Newtonian fluids are in fact a special case of viscous fluids with the Newtonian form of stress tensor. In general, a flow of any viscous thermoconducting fluid is described by the momentum and energy equations and the mass conservation equation as follows:

$$\frac{\partial \rho}{\partial t} + \vec{\nabla} \cdot (\rho \vec{v}) = 0, \tag{3.31}$$

$$\frac{\partial \vec{v}}{\partial t} + (\vec{v} \cdot \vec{\nabla})\vec{v} = \frac{1}{\rho}\left(-\vec{\nabla} p + Div\,\mathbf{P}\right),$$

$$\frac{\partial e}{\partial t} + (\vec{v} \cdot \vec{\nabla})e = \frac{1}{\rho}\left(-p(\vec{\nabla} \cdot \vec{v}) + \chi \Delta T + \mathbf{P} : Grad\,\vec{v}\right).$$

The huge variety of stress tensors yield, respectively, the huge variety of dispersion relations, definition of modes and sets of coupling equations. The important case among all varieties is the Maxwellian fluid with the viscous tensor in the reological form

$$\mathbf{P}_{i,k} = \mu \int\limits_{-\infty}^{t} \left(\frac{\partial v_i}{\partial x_k} + \frac{\partial v_k}{\partial x_i}\right) e^{-(t-t')/\tau_R} dt', \tag{3.32}$$

where $\mu = \rho_0(c_\infty^2 - c_0^2)$ is dispersion coefficient (c_∞ is a frozen sound speed, that is, the sound speed of infinitively large frequency of perturbations). In the case of one-dimensional planar flow along axis x, recalling equations of state (3.23), and Eqs (3.26), and (3.31) rearrange as [15]

$$\frac{\partial v}{\partial t} + \frac{1}{\rho_0}\frac{\partial p'}{\partial x} - 2c_0\widehat{A}\frac{\partial^2 v}{\partial x^2} = -v\frac{\partial v}{\partial x} + \frac{\rho'}{\rho_0^2}\frac{\partial p'}{\partial x} - 2\frac{c_0}{\rho_0}\rho'\widehat{A}\frac{\partial^2 v}{\partial x^2},$$

$$\begin{aligned}\frac{\partial p'}{\partial t} + \rho_0 c_0^2\frac{\partial v}{\partial x} - \delta_2^1\frac{\partial^2 p'}{\partial x^2} - \delta_2^2 c_0^2\frac{\partial^2 \rho'}{\partial x^2} = & -v\frac{\partial p'}{\partial x} + \left(D_1 p' + c_0^2 D_2 \rho'\right)\frac{\partial v}{\partial x} \\ & + \frac{2\rho_0 c_0}{E_1}\frac{\partial v}{\partial x}\widehat{A}\frac{\partial v}{\partial x} + \frac{\delta_2^3}{\rho_0 c_0^2}\frac{\partial^2 p'^2}{\partial x^2} \\ & + \frac{\delta_2^4 c_0^2}{\rho_0}\frac{\partial^2 \rho'^2}{\partial x^2} + \frac{\delta_2^5}{\rho_0}\frac{\partial^2(\rho p')}{\partial x^2},\end{aligned}$$

$$\frac{\partial \rho'}{\partial t} + \rho_0\frac{\partial v}{\partial x} = -v\frac{\partial \rho'}{\partial x} - \rho'\frac{\partial v}{\partial x}. \tag{3.33}$$

Its linear form, this corresponds to the modes

$$\psi_1 = \begin{pmatrix} \frac{c_0}{\rho_0} - \frac{\delta_2}{2\rho_0}\frac{\partial}{\partial x} - \frac{\widehat{A}}{\rho_0}\frac{\partial}{\partial x} \\ c_0^2 - \delta_2 c_0 \partial/\partial x \\ 1 \end{pmatrix}\rho_1, \quad \psi_2 = \begin{pmatrix} -\frac{c_0}{\rho_0} - \frac{\delta_2}{2\rho_0}\frac{\partial}{\partial x} - \frac{\widehat{A}}{\rho_0}\frac{\partial}{\partial x} \\ c_0^2 + \delta_2 c_0 \partial/\partial x \\ 1 \end{pmatrix}\rho_2,$$

$$\psi_3 = \begin{pmatrix} -\frac{\delta_2^2}{\rho_0}\frac{\partial}{\partial x} \\ 0 \\ 1 \end{pmatrix}\rho_3 \tag{3.34}$$

where

$$\widehat{A}\phi = \frac{\mu}{\rho_0 c_0} \int_{-\infty}^{t} \phi e^{-(t-t')/\tau_R} dt'. \tag{3.35}$$

Because only the momentum equation involves linear terms associated with dispersion, it exclusively contributes to corresponding corrections in acoustic modes. The corrections that make sound isentropic in the leading order coincide with that in Eq. (3.27). The dynamic equation that describes evolution of the first acoustic mode, is

$$\frac{\partial \rho_1}{\partial t} + c_0 \frac{\partial \rho_1}{\partial x} - \frac{\delta_2}{2} \frac{\partial^2 \rho_1}{\partial x^2} - c_0 \widehat{A} \frac{\partial^2 \rho_1}{\partial x^2} + \frac{1 - D_1 - D_2}{2\rho_0} \rho_1 \frac{\partial \rho_1}{\partial x} = 0. \tag{3.36}$$

The row d_3, which projects the vector of total perturbations into ρ_3, coincides with that from Eq. (3.17). Applying it and considering only acoustic terms among all variety of nonlinear ones, one readly arrives at

$$\begin{aligned}\frac{\partial \rho_3}{\partial t} + \delta_2^2 \frac{\partial^2 \rho_3}{\partial x^2} &= -\frac{2c_0}{E_1 \rho_0} \frac{\partial \rho_1}{\partial x} \widehat{A} \frac{\partial \rho_1}{\partial x} - \delta_2 \frac{\rho_1}{\rho_0} \frac{\partial^2 \rho_1}{\partial x^2} + \frac{1}{\rho_0} D_1 \delta_2 \left(\frac{\partial \rho_1}{\partial x} \right)^2 \\ &+ \frac{1}{\rho_0} \left(-\frac{(D_1 + D_2 - 1)\delta_2}{4} + \delta_2^1 \frac{D_1 + D_2 + 1}{2} - \delta_2^3 - \delta_2^4 - \delta_2^5 \right) \frac{\partial^2 \rho_1^2}{\partial x^2}. \end{aligned} \tag{3.37}$$

We consider the acoustic field represented by the rightward progressive sound, though it may be easily expanded leftwards by one or any mixture of two acoustic branches. It is remarkable that the dynamic equation for acoustic heating is a result of linear combining of energy and continuity equations in the absence of thermal conduction. This concerns flows with any kind of mechanical attenuation that relates to the Newtonian or to a non-Newtonian stress tensor. In the case of thermoconducting fluids, it is a result of combining of the energy, continuity, and momentum equations. The Newtonian fluids correspond to limiting case of operator $\widehat{A}$, which tends to $\frac{1}{2c_0}\delta_1$ at small relaxation times as compared to the characteritic duration of an impulse. Eq. (3.37) is actually the diffusity equation with the source in the right-hand side, which associates with the thermal and mechanical damping in a flow.

The dynamic equation that governs excess temperature associated with the entropy mode in weakly thermoconducting flows, is in fact a result of projecting, it takes the form

$$\begin{aligned}\frac{\partial T_3}{\partial t} &= \frac{2\mu}{bE_1\rho_0^3} \frac{\partial \rho_1}{\partial x} \int_{-\infty}^{t} \frac{\partial \rho_1}{\partial x} exp(-(t-t')/\tau_R) dt' \\ &= \frac{2\mu}{C_V \kappa \rho_0^4} \frac{\partial \rho_1}{\partial x} \int_{-\infty}^{t} \frac{\partial \rho_1}{\partial x} exp(-(t-t')/\tau_R) dt'. \end{aligned} \tag{3.38}$$

Equation (3.38) is instantaneous, valid for periodic and aperiodic sound including impulses. The total increase in temperature of the background after a pulse has gone away, associates with an excess temperature of the entropy mode, equals $\int_{-\infty}^{\infty} \frac{\partial T_3}{\partial t'} dt'$.

Acoustic heating produced by three exemplary impulses has been studied in Ref. [15] In fact, acoustic excess density satisfies nonlinear Eq. (3.36). In view of the difficulty in establishing its analytical solutions, the exemplary acoustic impulses were taken in the form of the travelling waves proportional to $\exp\left(-\eta^2\right)$, $\exp\left(-\eta^2/4\right)$, or $\eta \exp\left(-\eta^2\right)$, where $\eta = \omega(t - x/c_0)$. Their magnitudes were selected in order to provide equal energies of impulses. The conclusion is that the third impulse is much more effective in producing heating, and the largest heating is achieved at the characteristic duration of impulse, which approximately equals the relaxation time, $\omega\tau_R = 1$.

3.4 ACOUSTICS OF A FLUID WHICH IS AFFECTED BY CONSTANT MASS FORCE

3.4.1 ISOTHERMAL ATMOSPHERE 1D DYNAMICS

The dynamics of fluids affected by external forces is a fairly complex problem that is of great importance in many applications in the physics of atmosphere and ocean and technical applications. External forces make the background of waves propagation dependent on spatial coordinates, and, probably, on time. This essentially complicates even linear analysis of flows in such media. Types of wave motion correspond to the roots of the dispersion equation. In the simplest case of flow in one dimension, the dispersion relations may be introduced over the entire wave-length range only if the background pressure and density depend exponentially on the coordinate. This is the case of constant mass external force [16]. In one dimension, there are three types of motion: two acoustic branches and the entropy mode. In the flows going out of one dimension, the buoyancy waves may contribute to the total flow. The possibility of distinguishing modes analytically and describing their dynamics is of importance also in Earth meteorology and the sun's atmosphere dynamics applications [17,18]. It may be resolved by means of projecting linear operators. Exponentially stratified volumes of an ideal gas with initially constant temperature, affected by constant mass force or moving with constant acceleration will be considered. We concentrate on the evaluation of contributions of every mode in the total energy. They are valid at any instant: it has been discovered that the proportions of different modes in the total energy do not depend on time.

The nonlinear dynamics of sound in an exponentially stratified medium have been considered in Refs [7,19]. In these studies, the dispersion properties of a flow were accounted for exactly. We start from the conservation equations of momentum, mass, and energy in a one-dimensional gas without mechanical and thermal losses, which are affected by an external constant mass force g:

$$\rho\left(\frac{\partial v}{\partial t} + v\frac{\partial v}{\partial z}\right) = -\frac{\partial p}{\partial z} - g,$$

$$\rho\left(\frac{\partial e}{\partial t}+v\frac{\partial e}{\partial z}\right)+p\frac{\partial v}{\partial z}=0, \tag{3.39}$$

$$\frac{\partial \rho}{\partial t}+\frac{\partial(\rho v)}{\partial z}=0,$$

where the mass force is directed oppositely to the axis OZ (it may be readily associated with constant gravity acceleration or with other constant mass force), and z designates a distance from a boundary (which may correspond to the Earth's surface). The unperturbed pressure and density are functions on coordinate: $\rho_0=\rho_{00}\exp(-z/H), p_0=p_{00}\exp(-z/H)=\rho_{00}gH\exp(-z/H)$, where $\rho_{00}=\rho_0(0)$, $p_{00}=p_0(0)$ denote values of density and pressure at $z=0$, and $H=(C_p-C_V)T_0/g$, where T_0 is unperturbed temperature of a gas. We consider ideal gases with caloric equation of state, Eq. (3.13). The new variables

$$R=\rho'\cdot\exp(z/2H),\ P=p'\cdot\exp(z/2H),\ V=v\cdot\exp(-z/2H)$$

make it possible to eliminate exponential factors in linear equations and to apply the Fourier-analysis in the studies of infinitely-small signal flows [16,19]. The weakly nonlinear dynamics of a gas is described by the following system of equations:

$$\frac{\partial V}{\partial t}+\frac{1}{\rho_{00}}\left(\frac{\partial}{\partial z}-\frac{1}{2H}\right)P+\frac{gR}{\rho_{00}}=\varphi_1,$$

$$\frac{\partial P}{\partial t}+\gamma gH\rho_{00}\left(\frac{\partial V}{\partial z}+\frac{1}{\gamma H}(\gamma/2-1)V\right)=\varphi_2, \tag{3.40}$$

$$\frac{\partial R}{\partial t}+\rho_{00}\left(\frac{\partial}{\partial z}-\frac{1}{2H}\right)V=\varphi_3,$$

where φ_1, φ_2, φ_3 take the forms

$$\varphi_1=-\exp(z/2H)\left(V\left(\frac{\partial}{\partial z}+\frac{1}{2H}\right)V-\frac{R}{\rho_{00}^2}\left(\frac{\partial}{\partial z}-\frac{1}{2H}\right)P-\frac{g}{\rho_{00}^2}R^2\right), \tag{3.41}$$

$$\varphi_2=-\exp(z/2H)\left(V\left(\frac{\partial}{\partial z}-\frac{1}{2H}\right)P+\gamma P\left(\frac{\partial}{\partial z}+\frac{1}{2H}\right)V\right),$$

$$\varphi_3=-\exp(z/2H)\left(R\frac{\partial V}{\partial z}+V\frac{\partial R}{\partial z}\right).$$

It accounts for only quadratic nonlinear terms among all other nonlinear ones in Eq. (3.39), which are represented by the right-hand parts of equations φ_1, φ_2, φ_3. Applying the Fourier analysis, one arrives at the well-known dispersion relations of the flow of infinitely small magnitudes. We remind them:

$$\omega_1=c_0\sqrt{k^2+\frac{1}{4H^2}},\ \omega_2=-c_0\sqrt{k^2+\frac{1}{4H^2}},\ \omega_3=0, \tag{3.42}$$

where

$$c_0 = \sqrt{\gamma g H} = \sqrt{\gamma (C_P - C_V) T_0}$$

is the speed of sound of infinitely-small scale as compared to H. The two first roots correspond to the acoustic modes, progressive in different directions of axis OZ, and the last determines the entropy mode.

It is convenient to introduce variables φ, Φ instead of perturbation in density,

$$\varphi' = p' - \gamma g H \rho', \; \Phi = \varphi' \cdot \exp(z/2H),$$

and to make use of equations

$$\frac{\partial V}{\partial t} + \frac{1}{\rho_{00}}\left(\frac{\partial}{\partial z} - \frac{1}{2H}\right) P + \frac{gR}{\rho_{00}} - \frac{\Phi}{\gamma H \rho_0} = 0,$$

$$\frac{\partial \Phi}{\partial t} + (\gamma - 1)\,\rho_{00} g V = 0$$

instead of the first and third equations in the system (3.40), (3.41). The completeness of the set of eigenvectors makes it possible to represent the total vector of perturbations as a sum of acoustic and entropy vectors at any instant,

$$\Psi(z,t) = \begin{pmatrix} V \\ P \\ \Phi \end{pmatrix} = \Psi_1(z,t) + \Psi_2(z,t) + \Psi_3(z,t)$$

$$= \begin{pmatrix} V_a \\ \frac{1}{\gamma - 1}\left(\frac{\gamma - 2}{2} + \gamma H \frac{\partial}{\partial z}\right)\Phi_a \\ \Phi_a \end{pmatrix} + \begin{pmatrix} 0 \\ P_0 \\ \left(-\frac{\gamma - 2}{2} + \gamma H \frac{\partial}{\partial z}\right) P_0 \end{pmatrix}, \qquad (3.43)$$

where quantities with lower index a represent a sum of perturbations specific for both acoustic modes, and the terms associated with the entropy mode are marked by the lower index 0.

The most important condition of impermeability at the upper and lower boundaries, $z = 0$ and $z = h$, is self-adjoint: for example, $V(z = 0) = V(z = h) = 0$. The second and third kind of (homogeneous) conditions are also admissible. Vectors $\Psi_a(z,t)$ and $\Psi_0(z,t)$ are orthogonal:

$$\langle \Psi_a, \Psi_0 \rangle = \int_0^h \left(\rho_0 V_a V_0 + \frac{P_a P_0}{\gamma g H \rho_{00}} + \frac{\Phi_a \Phi_0}{\gamma(\gamma - 1) g H \rho_{00}}\right) dz = 0. \qquad (3.44)$$

The choice of boundary conditions guarantees the orthogonality of eigenvectors as a direct corollary of Hermicity of the operator L in respect to the scalar product (3.44). Taking a sum $2\Phi + \left(\gamma - 2 - 2\gamma H \frac{\partial}{\partial z}\right) P$ and making use of Eq. (3.43), one readily reduces all terms belonging to the entropy mode in the left-hand side of the differential equation of the second order,

$$\left(1-4H^2\frac{\partial^2}{\partial z^2}\right)\Phi_a(z,t)=\frac{2(\gamma-1)}{\gamma^2}\left(2\Phi(z,t)+\left(\gamma-2-2\gamma H\frac{\partial}{\partial z}\right)P(z,t)\right)$$
$$\equiv D(z,t). \tag{3.45}$$

This is in fact a kind of projection that is valid at any instant. The solution to Eq. (3.45) takes the form

$$\Phi_a(z,t)=C_1\exp(-z/2H)+C_2\exp(z/2H)+\frac{1}{4H}\left(\exp(-z/2H)\right.$$
$$\left.\times\int_0^z\exp(z'/2H)D(z',t)dz'-\exp(z/2H)\int_0^z\exp(-z'/2H)D(z',t)dz'\right),$$

where C_1,C_2 denote any real constants. It determines both $P_a(z,t)$ (in accordance to Eq. (3.43)) and $P_0(z)=P(z,t)-P_a(z,t)$, $\Phi_0(z)=\Phi(z,t)-\Phi_a(z,t)$. Hence, the projecting makes possible to conclude about composition of total perturbations in pressure and entropy, and about contribution of the specific parts in the total energy [20].

3.4.2 EXAMPLES OF PROJECTING: DECOMPOSITION OF THE TOTAL FIELD OF EXCLUSIVELY ENTROPY OR ACOUSTIC PARTS AND ENERGY RELEASE WITH MASS INJECTION

Some conclusions follow immediately from the relations Eq. (3.43) and completeness of the set of eigenvectors. If initially

$$\Phi(z,0)=\left(-\frac{\gamma-2}{2}+\gamma H\frac{d}{dz}\right)P(z,0),\quad V(z,0)=0,$$

the total field is represented exclusively by the entropy mode *at any instant*. One may establish acoustic velocity in terms of Φ_a. The relation follows from the conservation system (3.39), but takes complex integro-differential form, which involves some integro-differential operator K. The relations for different branches are asymmetric, $V_1(z,0)=K\Phi_1(z,0)$, $V_2(z,0)=-K\Phi_2(z,0)$. It may be concluded from these equalities that if $\Phi_1(z,0)=-\Phi_2(z,0)$ and $P_1(z,0)=-P_2(z,0)$, that is,

$$P(z,0)=\frac{1}{\gamma-1}\left(\frac{\gamma-2}{2}+\gamma H\frac{d}{dz}\right)\Phi(z,0),$$

the total field consists of the entropy mode and sound with non-zero initial velocity $V(z,0)=V_1(z,0)+V_2(z,0)$ (hence, the non-zero kinetic energy) and zero initial perturbations $P(z,0)$ and $\Phi(z,0)$. For the short-scale perturbations as compared to characteristic scale of inhomogeneouty H, the leading-order relations for every acoustic branch take the integral form:

$$V_1(z,0) = -\frac{g\gamma^2}{8\rho_0(\gamma-1)(\gamma g H)^{3/2}}\int_0^z \Phi_1(z',0)dz',$$

$$V_2(z,0) = \frac{g\gamma^2}{8\rho_0(\gamma-1)(\gamma g H)^{3/2}}\int_0^z \Phi_2(z',0)dz'.$$

The simple conclusion from these relations is that if $\int_0^z (\Phi_1(z',0)+\Phi_2(z',0))dz' = 0$, the contribution of the kinetic energy in the total one is zero.

Let consider the pure heating (that is, without kinetic contribution) of a gas that may occur at any instant, for definiteness at $t=0$,

$$P(z,t=0) = \Phi(z,t=0) = \Theta(z).$$

Equation (3.45) rearranges as

$$\left(1+2H\frac{d}{dz}\right)\Phi_a(z,0) = \frac{2(\gamma-1)}{\gamma}\Theta(z),$$

which has a solution

$$\Phi_a(z,0) = \frac{\gamma-1}{\gamma H}\exp(-z/2H)\int_0^z \exp(z'/2H)\Theta(z')dz'.$$

In terms of $\widetilde{\Theta}(z)$, where $\Theta(z) = \exp(-z/2H)\frac{d}{dz}(\exp(z/2H)\widetilde{\Theta}(z))$, one readily arrives at equalities

$$\Phi_0(z,0) = \Phi-\Phi_a = \frac{2-\gamma}{2\gamma}\widetilde{\Theta}(z) + H\frac{d}{dz}\widetilde{\Theta}(z),\ P_0(z,0) = P-P_a = \frac{1}{\gamma}\widetilde{\Theta}(z),$$

$$\Phi_a(z,0) = \frac{\gamma-1}{\gamma}\widetilde{\Theta}(z),\ P_a(z,0) = \frac{\gamma-2}{2\gamma}\widetilde{\Theta}(z) + H\frac{d}{dz}\widetilde{\Theta}(z).$$

The scalar product

$$\langle\Psi_a,\Psi_0\rangle = \frac{1}{\gamma g H\rho_0}\int_0^\infty \left(P_aP_0 + \frac{\Phi_a\Phi_0}{(\gamma-1)}\right)dz = \frac{1}{\gamma^2 g\rho_0}\widetilde{\Theta}(z)\Big|_0^\infty$$

equals zero if $\widetilde{\Theta}(0) = \widetilde{\Theta}(\infty) = 0$. The total energy of the sound mode, with the exception of its kinetic part Σ_a, relates to that of the entropy mode Σ_0, in the following way:

$$\Sigma_a = \frac{1}{\gamma g H\rho_0}\int_0^\infty\left(P_a^2 + \frac{\Phi_a^2}{(\gamma-1)}\right)dz = \frac{H}{\gamma g\rho_0}\int_0^\infty\left(\frac{d}{dz}\widetilde{\Theta}(z)\right)^2 dz$$

$$+\frac{1}{\gamma g H\rho_0}\int_0^\infty\left(\frac{\gamma-1}{\gamma^2} + \left(\frac{\gamma-2}{2\gamma}\right)^2\right)\widetilde{\Theta}^2(z)dz = (\gamma-1)\Sigma_0.$$

The following example considers the initial mass injection,

$$P(z,t=0) = -\frac{\Phi(z,t=0)}{\gamma-1} = \Theta(z).$$

In this case, Eq. (3.45) transforms into

$$\left(1-2H\frac{d}{dz}\right)\Phi_a(z,0) = -\frac{2(\gamma-1)}{\gamma}\Theta(z)$$

with a solution

$$\Phi_a(z,0) = \frac{\gamma-1}{\gamma H}\exp(z/2H)\int_0^z \exp(-z'/2H)\Theta(z')dz'.$$

By use of substitution $\Theta(z) = \exp(z/2H)\frac{d}{dz}(\exp(-z/2H)\widetilde{\Theta}(z))$, we obtain at $t=0$

$$\Phi_0 = \Phi - \Phi_a = \frac{(\gamma-1)(\gamma-2)}{2\gamma}\widetilde{\Theta}(z) + H(1-\gamma)\frac{d}{dz}\widetilde{\Theta}(z),\ P_0 = P - P_0 = \frac{1-\gamma}{\gamma}\widetilde{\Theta}(z),$$

$$\Phi_a(z,0) = \frac{\gamma-1}{\gamma}\widetilde{\Theta}(z),\ P_a(z,0) = \frac{\gamma-2}{2\gamma}\widetilde{\Theta}(z) + H\frac{d}{dz}\widetilde{\Theta}(z).$$

The entropy and acoustic initial energies relate as

$$\Sigma_a = \frac{H}{\gamma g\rho_0}\int_0^\infty \left(\frac{d}{dz}\widetilde{\Theta}(z)\right)^2 dz + \frac{1}{\gamma gH\rho_0}\int_0^\infty \frac{1}{4}\widetilde{\Theta}^2(z)dz = \frac{1}{\gamma-1}\Sigma_0.$$

From the mathematical point of view, decomposition is possible due to orthogonality of the correspondent subspaces with respect to the scalar product, Eq. (3.44), and because one specific perturbation determines all other perturbations in the mode at any time. From the physical standpoint, decomposition (by measurements) reflects superposition of different modes in a linear flow.

The limit $g \to 0$ and hence $H \to \infty$ may be readily traced. Note that the product gH remains constant with γgH being the squared linear sound speed in an ideal gas. In the case $g=0$, φ means the quantity proportional to perturbation in the entropy, $\varphi' = (\gamma-1)\rho_0 s'$. It is identically zero in the both acoustic branches and is no longer suitable to be a reference quantity for them. Instead, perturbation in density may be chosen. The reason that φ' was used in the case of non-zero g, is the simplicity of expression P in terms of Φ, which includes a partial derivative with respect to z but does not include integral operators. In contrast to the motion over the uniform background, an excess pressure of the stationary mode does not equal zero.

3.4.3 DYNAMICS OF THE SHORT-SCALE WAVES

The relations linking V with P and Φ (or R) in sound propagating in a fluid affected by an external force are in general integro-differential. The exact links are derived exactly with regard to one-dimensional flow by one of the authors in Refs [7,21]. For example, the link connecting acoustic pressure and velocity, which specifies the first acoustic mode, looks as:

$$P(z,t) = \frac{\rho_{00}}{\pi\sqrt{\gamma g H}} \int_{-\infty}^{\infty} dz' \left(g(1-\gamma/2)F(z-z') - \gamma g H F(z-z') \frac{\partial}{\partial z'} \right) V(z',t), \quad (3.46)$$

where $F(z)$ reflects the dispersive features of a stratified gas,

$$F(z) = \frac{2}{\pi}\left(I_0(z/2H) - L_0(z/2H)\right) = \int_0^{\infty} dk \frac{\sin(kz)}{\sqrt{k^2 + 1/4H^2}},$$

with I_0, L_0 denoting the modified Bessel function of zero order and the Struve function, respectively. The relations for the first and second acoustic modes in terms of excess density are determined by the eigenvectors

$$\overline{\psi}_1(z,t) = \begin{pmatrix} V \\ \Phi \\ R \end{pmatrix}_1 = \begin{pmatrix} \frac{2\gamma g H}{\rho_0} \int_{-\infty}^{\infty} dz' \left(\frac{\partial}{\partial z'} - \frac{1}{2H} \right) F(z'-z) \\ g(1-\gamma)\exp(z/2H) \int_z^{\infty} \exp(-z'/2H) dz' \\ 1 \end{pmatrix} R_1,$$

$$\overline{\psi}_2(z,t) = \begin{pmatrix} -\frac{2\gamma g H}{\rho_0} \int_{-\infty}^{\infty} dz' \left(\frac{\partial}{\partial z'} - \frac{1}{2H} \right) F(z'-z) \\ g(1-\gamma)\exp(z/2H) \int_z^{\infty} \exp(-z'/2H) dz' \\ 1 \end{pmatrix} R_2.$$

They are valid at any instant.

The approximate relations in the case of small-scale perturbations (as compared to H) have been derived in Ref. [22]. The dispersion relations and, following from them, the formulae may be expanded in the Taylor series in the vicinity of $1/kH = 0$. Links for acoustic upward and downward propagating modes have the forms:

$$\psi_1 = \begin{pmatrix} V \\ P \\ R \end{pmatrix}_1 = \begin{pmatrix} 1 \\ c_0\rho_{00}\left(1 + \frac{\gamma-2}{2\gamma H}\int dz\right) \\ \frac{\rho_{00}}{c_0}\left(1 - \frac{1}{2H}\int dz\right) \end{pmatrix} V_1,$$

$$\psi_2 = \begin{pmatrix} V \\ P \\ R \end{pmatrix}_2 = \begin{pmatrix} 1 \\ -c_0\rho_{00}\left(1 + \frac{\gamma-2}{2\gamma H}\int dz\right) \\ -\frac{\rho_{00}}{c_0}\left(1 - \frac{1}{2H}\int dz\right) \end{pmatrix} V_2.$$

The limits of integrating should agree with the physical meaning of the problem. The lower limit of integration is $-\infty$, and the upper one equals z for perturbations extended in the even way at the negative half-axis OZ. The entropy mode looks as follows:

$$\psi_3 = \begin{pmatrix} V \\ P \\ R \end{pmatrix}_3 = \begin{pmatrix} 0 \\ -\frac{c_0^2}{\gamma H}\int dz \\ 1 \end{pmatrix} R_3.$$

Matrix operators, projecting the vector of total perturbations into acoustic modes,

$$\Pi_n \psi = \psi_n, \quad n = 1,\ldots,3,$$

follow from the eigenvectors:

$$\Pi_1 = \begin{pmatrix} \frac{1}{2} & \frac{1}{2\rho_{00}c_0}\left(1 - \frac{1}{2H}\int dz\right) & \frac{g}{2\rho_{00}c_0}\int dz \\ \frac{\rho_{00}c_0}{2}\left(1 - \frac{2-\gamma}{2\gamma H}\int dz\right) & \frac{1}{2} - \frac{1}{2\gamma H}\int dz & \frac{g}{2}\int dz \\ \frac{\rho_{00}}{2c_0}\left(1 - \frac{1}{2H}\int dz\right) & \frac{1}{2c_0^2} - \frac{1}{2c_0^2 H}\int dz & \frac{1}{2\gamma H}\int dz \end{pmatrix},$$

$$\Pi_2 = \begin{pmatrix} \frac{1}{2} & -\frac{1}{2\rho_{00}c_0}\left(1 - \frac{1}{2H}\int dz\right) & -\frac{g}{2\rho_{00}c_0}\int dz \\ -\frac{\rho_{00}c_0}{2}\left(1 - \frac{2-\gamma}{2\gamma H}\int dz\right) & \frac{1}{2} - \frac{1}{2\gamma H}\int dz & \frac{g}{2}\int dz \\ -\frac{\rho_{00}}{2c_0}\left(1 - \frac{1}{2H}\int dz\right) & \frac{1}{2c_0^2} - \frac{1}{2c_0^2 H}\int dz & \frac{1}{2\gamma H}\int dz \end{pmatrix},$$

$$\Pi_3 = \begin{pmatrix} 0 & 0 & 0 \\ 0 & \frac{1}{\gamma H}\int dz & -g\int dz \\ 0 & \frac{1}{c_0^2}\left(-1 + \frac{1}{H}\int dz\right) & 1 - \frac{1}{\gamma H}\int dz \end{pmatrix}.$$

The links of the first acoustic mode, which makes it isentropic, recall that for the Riemann wave that propagates over the uniform initially ideal gas:

$$P_{1,R} = \rho_0 c_0^2\left(1 + \frac{\gamma-1}{2}\frac{V_{1,R}}{c_0}\right)^{\frac{2\gamma}{\gamma-1}} - \rho_0 c_0^2, \quad R_{1,R} = \rho_0\left(1 + \frac{\gamma-1}{2}\frac{V_{1,R}}{c_0}\right)^{\frac{2}{\gamma-1}} - \rho_0.$$

The Earnshaw equation is the exact dynamic equation that governs velocity in the nonlinear Riemann wave:

$$\frac{\partial V_{1,R}}{\partial t} + c_0\frac{\partial V_{1,R}}{\partial z} + \frac{\gamma+1}{2}V_{1,R}\frac{\partial V_{1,R}}{\partial z} = 0.$$

Note that links for the Riemann wave are exact along with the Earnshaw equation; that is, valid to any accuracy with respect to the powers of the Mach number, M. As for the perturbations over the stratified background, the leading order links are

$$P_1 = c_0\rho_{00}\left(1 + \frac{\gamma-2}{2\gamma H}\int dz\right)V_1 + \exp(z/2H)\rho_{00}\frac{\gamma+1}{4}V_1^2,$$

$$R_1 = \frac{\rho_{00}}{c_0}\left(1 - \frac{1}{2H}\int dz\right)V_1 + \exp(z/2H)\frac{3-\gamma}{4}\frac{\rho_{00}}{c_0^2}V_1^2.$$

The difference with the uniform case is in the factor $e^{z/2H}$, which tends to 1 if H tends to infinity. In turn, the analog of the Earnshaw equation takes the form

$$\frac{\partial V_1}{\partial t} + c_0\frac{\partial V_1}{\partial z} + \frac{\gamma+1}{2}e^{z/2H}V_1\frac{\partial V_1}{\partial z} = 0. \tag{3.47}$$

In contrast to this concerning the uniform background, this dynamic equation is approximate. It is written with an accuracy up to terms of order $max(M^3,\ M^2/kH)$.

The important issue is to establish links of acoustic perturbations that support isentropicity of sound in the leading order, with accuracy up to terms of order M^2/kH inclusively. The corrected links must yield three equivalent equations for velocity being substituded in Eqs (3.40), (3.41). For the first acoustic mode, we seek them in the form:

$$P_1 = c_0\rho_{00}\left(1+\frac{\gamma-2}{2\gamma H}\int dz\right)V_1 + \exp(z/2H)\rho_{00}\frac{\gamma+1}{4}V_1^2 + \exp(z/2H)\overline{P}_1, \tag{3.48}$$

$$R_1 = \frac{\rho_{00}}{c_0}\left(1-\frac{1}{2H}\int dz\right)V_1 + \exp(z/2H)\frac{3-\gamma}{4}\frac{\rho_{00}}{c_0^2}V_1^2 + \exp(z/2H)\overline{R}_1,$$

where $\overline{P}_1$, $\overline{R}_1$ are some quadratic functions of V_1 of order M^2/kH to be established, for example, proportional to $\frac{1}{H}V_1\int V_1dz$ or to $\frac{1}{H}\int V_1^2dz$. Links (3.48) with zero $\overline{P}_1$ and $\overline{R}_1$ make the mode isentropic up to the terms of order M^2. Eqs (3.40), (3.41) with account for Eq. (3.47), transform into the leading-order system:

$$\xi + e^{z/2H}\frac{1}{\rho_{00}}\frac{\partial \overline{P}_1}{\partial z} = -\frac{e^{z/2H}}{2H}\left(\frac{3+\gamma}{2\gamma}V_1^2 + \frac{\partial V_1}{\partial z}\int V_1dz\right),$$

$$\xi + e^{z/2H}\frac{1}{\rho_{00}c_0}\frac{\partial \overline{P}_1}{\partial t} = -\frac{e^{z/2H}}{2H}\left(\frac{3\gamma^2+\gamma-6}{4\gamma}V_1^2 - (2-\gamma)\frac{\partial V_1}{\partial z}\int V_1dz\right),$$

$$\xi + e^{z/2H}\frac{c_0}{\rho_{00}}\frac{\partial \overline{R}_1}{\partial t} = -\frac{e^{z/2H}}{2H}\left(\frac{\gamma-3}{4}V_1^2 - \frac{\partial V_1}{\partial z}\int V_1dz\right),$$

where $\xi = \frac{\partial V_1}{\partial t} + c_0\frac{\partial V_1}{\partial z} + e^{z/2H}\frac{\gamma+1}{4}\frac{\partial V_1^2}{\partial z}$. Taking a difference of the first and the second equations, one arrives at

$$c_0\frac{\partial \overline{P}_1}{\partial z} - \frac{\partial \overline{P}_1}{\partial t} \approx 2c_0\frac{\partial \overline{P}_1}{\partial z} = \frac{c_0\rho_{00}}{2H}\left(\frac{3\gamma^2-\gamma-12}{4\gamma}V_1^2 + (\gamma-3)\frac{\partial V_1}{\partial z}\int V_1dz\right),$$

and therefore, at

$$\overline{P}_1 = \frac{\rho_{00}}{4H}\int\left(\frac{3\gamma^2-\gamma-12}{4\gamma}V_1^2 + (\gamma-3)\frac{\partial V_1}{\partial z}\int V_1dz\right)dz,$$

which in turn yields the expression for $\overline{R}_1$ and the dynamic equation for velocity of sound:

$$\overline{R}_1 = -\frac{\rho_{00}}{4Hc_0^2}\int\left(\frac{\gamma+9}{4}V_1^2 + (\gamma+1)\frac{\partial V_1}{\partial z}\int V_1dz\right)dz,$$

$$\frac{\partial V_1}{\partial t}+c_0\frac{\partial V_1}{\partial z}+e^{z/2H}\left(\frac{\gamma+1}{4}\frac{\partial V_1^2}{\partial z}+\frac{3(\gamma+1)}{16H}V_1^2+\frac{\gamma-1}{4H}\frac{\partial V_1}{\partial z}\int V_1dz\right)=0. \quad (3.49)$$

The corrected links for the second mode are the following:

$$P_2=-c_0\rho_{00}\left(1+\frac{\gamma-2}{2\gamma H}\int dz\right)V_2+\exp\left(z/2H\right)\rho_{00}\frac{\gamma+1}{4}V_2^2+\exp\left(z/2H\right)\overline{P}_2,$$

$$R_2=-\frac{\rho_{00}}{c_0}\left(1-\frac{1}{2H}\int dz\right)V_2+\exp\left(z/2H\right)\frac{3-\gamma}{4}\frac{\rho_{00}}{c_0^2}V_2^2+\exp\left(z/2H\right)\overline{R}_2,$$

$$\overline{P}_2=\frac{\rho_{00}}{4H}\int\left(\frac{3\gamma^2-\gamma-12}{4\gamma}V_2^2+(\gamma-3)\frac{\partial V_2}{\partial z}\int V_2dz\right)dz,$$

$$\overline{R}_2=-\frac{\rho_{00}}{4Hc_0^2}\int\left(\frac{\gamma+9}{4}V_2^2+(\gamma+1)\frac{\partial V_2}{\partial z}\int V_2dz\right)dz,$$

$$\frac{\partial V_2}{\partial t}-c_0\frac{\partial V_2}{\partial z}+e^{z/2H}\left(\frac{\gamma+1}{4}\frac{\partial V_2^2}{\partial z}+\frac{3(\gamma+1)}{16H}V_2^2+\frac{\gamma-1}{4H}\frac{\partial V_2}{\partial z}\int V_2dz\right)=0.$$

The specific excess density and pressure of the entropy mode are not affected by the sound in the linear flow and are governed by equations:

$$\frac{\partial R_3}{\partial t}=\frac{\partial P_3}{\partial t}=0$$

in a stationary flow with $V_3=0$.

REFERENCES

1. Makarov, S., and M. Ochmann. 1996. Nonlinear and thermoviscous phenomena in acoustics, Part I. *Acta Acustica united with Acustica* 82: 579–606.
2. Perelomova, A. 2003. Interaction of modes in nonlinear acoustics: theory and applications to pulse dynamics. *Acta Acustica united with Acustica* 89: 86–94.
3. Berg, A.M., and S. Tjøtta. 1997. Higher order nonlinearity in ultrasound beam propagation. In *Proceedings of 20th Scandinavian Symposium on Physical Acoustics*. Ustaoset, 5–7.

4. Soderholm, L.H. 2000. On the Kuznetsov equation and higher order nonlinear acoustic equations. In: Lauterborn, W., and Kurz, T. (Eds.) *Nonlinear Acoustics at the Turn of the Millennium,* 133–136. *15th International Symposium on Nonlinear Acoustics (ISNA'15),* AIP Conference Proceedings, Göttingen, Germany.
5. Bakhvalov, N.S., Zhileikin, Y.M. and E.A. Zabolotskaya. 1987. *Nonlinear Theory of Sound Beams*. American Institute of Physics, New York.
6. Karabutov, A.A., O.V. Rudenko, and O.A. Sapozhnikov. 1989. Thermal self-focusing of weak shock waves. *Soviet Physics Acoustics-Ussr* 34: 371–374.
7. Perelomova, A.A. 1998. Nonlinear dynamics of vertically propagating acoustic waves in stratified atmosphere. *Acta Acustica* 84: 1002–1006.
8. Chu, B.T. and L.S.G. Kovásznay. 1958. Non-linear interactions in a viscous heat-conducting compressible gas. *Journal of Fluid Mechanics* 3: 494–514.
9. Perelomova, A. 2000. Projectors in nonlinear evolution problem: acoustic solitons of bubbly liquid. *Applied Mathematical Letters* 13:93–98.
10. Kuznetsov, V.P. 1971. Equations of nonlinear acoustics. *Soviet Physics Acoustics* 16: 467–470.
11. Perelomova, A. 2006. Development of linear projecting in studies of non-linear flow. Acoustic heating induced by non-periodic sound. *Physics Letters A* 357: 42–47.
12. Perelomova, A. 2008. Modelling of acoustic heating induced by different types of sound. *Archives of Acoustics* 33(2): 151–160.
13. Riemann, B. 1953. *The Collected Works of Bernard Riemann*. Dover, New York.
14. Rudenko, O.V., and S.I. Soluyan. 1977. *Theoretical Foundations of Nonlinear Acoustics*. Plenum, New York.
15. Perelomova, A., and W. Pelc-Garska. 2010. Efficiency of acoustic heating produced in the thermoviscous flow of a fluid with relaxation. *Central European Journal of Physics* 8(6): 855–863.
16. Brekhovskikh, L.M., and, A.O. Godin. 1990. *Acoustics of Layered Media*. Springer, Berlin.
17. Souffrin, P. 1972. Radiative relaxation of sound waves in an optically thin isothermal atmosphere. *Astronomy and Astrophysics* 17: 458.
18. Marmolino, C., and G. Severino. 1991. Phases and amplitudes of acoustic-gravity waves. I-Upward and downward solutions. *Astronomy and Astrophysics* 242: 271–278.
19. Perelomova, A.A. 2000. Nonlinear dynamics of directed acoustic waves in stratified and homogeneous liquids and gases with arbitrary equation of state. *Archives of Acoustics* 25(4): 451–463.
20. Gordin, A.V. 1987. *Mathematical problems of hydrodynamical weather prediction. Analytical aspects.* Gydrometeoizdat, Leningrad (in Russian).
21. Leble, S., and A. Perelomova. 2013. Problem of proper decomposition and initialization of acoustic and entropy modes in a gas affected by the mass force. *Applied Mathematical Modelling* 37: 629–635.
22. Perelomova, A. 2009. Weakly nonlinear dynamics of short acoustic waves in exponentially stratified gas. *Archives of Acoustics* 34(2): 127–143.

4 Coupling of Sound with Vorticity: Acoustic Streaming

4.1 3D HYDRODYNAMICS AND VORTEX MODE

The vortex modes may exist in the flows exceeding one dimension. They are generally determined by a link connecting components of velocity: $\vec{\nabla} \cdot \vec{v} = 0$. In contrast to acoustics, this mode is isobaric and possesses zero perturbation in density. Variations in entropy and in internal energy associated with this mode are also zero. The vortex flow caused by loss in sound mometum is called acoustic streaming. From the point of view of projecting, it is the secondary mode that enhances in the field of sound. This occurs due to the acoustic force of streaming, which contains quadratic acoustic terms in the leading order. Acoustic streaming is a nonlinear phenomenon that takes place exclusively in a damping medium. It is of great importance, taking into account its medical and technical applications [1].

Acoustic streaming is namely the mean motion of a fluid caused by acoustic waves. Extensive reviews on this subject exist in the Refs [2–4]. The authors of Refs [5,6] have noticed that there is an unresolved issue concerning acoustic streaming, the effect of compressibility, because the starting point in this subject, as usual, is a set of equations describing incompressible liquid. In contrast, sound propagates over compressible fluids so that there is an evident inconsistency that requires the correct interpretation and evaluation of acoustic force in the context of weakly nonlinear flow. The usual method to identify acoustic streaming consists of two successive steps: first, performing a linear combination of these modes, and second, averaging the continuity and momentum equations over one sound period or over an integer number of sound periods [3,7]. This eliminates periodic perturbations, does not account for energy balance, and hence, discards thermal conductivity. It has been well established experimentally that the velocity of streaming depends on the total attenuation, which includes also thermal conductivity. The impact of compressibility, heat conduction and nonperiodicity of sound is not fully clear in the context of acoustic streaming. As usual, streaming in compressible fluids is larger *ceteris paribus*. It is more evident in a gas than in a liquid. The weakness of the formerly applied methodology is also its inconsistency in distinction of the vorticity and entropy modes, which are both slow and isobaric, as usual. It implies a zero temporal average over the sound period of the partial derivative of total density with respect to time, $\partial\rho/\partial t$, which includes the contribution of sound and the slowly varying part belonging to the isobaric entropy mode. Hence, the average over the sound period value of $\partial\rho/\partial t$ is no longer zero. The velocity attributable to the entropy mode is also nonzero if the heat conduction of a fluid differs from zero, so it contributes to the total mean velocity. We avoid these inconsistencies by means of the immediate projection of the conservation equations onto dynamic equations governing the vortex flow. Projection results in the correct coupling of the acoustic and nonwave motions. We will consider an example of a fluid with the Maxwell relaxation will be considered, which readily reduces to the Newtonian case at small relaxation times.

The starting point is a system (3.31). Its equivalent leading-order form is

$$\frac{\partial \Psi}{\partial t} + L\Psi = \Psi_{nl}, \tag{4.1}$$

where $\Psi = \begin{pmatrix} v_x & v_y & v_z & \widetilde{p} & \widetilde{\rho} \end{pmatrix}^T$, L is a linear matrix operator including spatial derivatives, and Ψ_{nl} denotes a nonlinear vector that includes quadratic terms. All perturbations represent a sum of planar waves:

$$f(\vec{r},t) = \int_{R^3} \widetilde{f}(\vec{k},t)\exp(-i\vec{k}\cdot\vec{r})d\vec{k}, \tag{4.2}$$

$(\widetilde{f}(\vec{k},t) = \frac{1}{(2\pi)^3}\int_{R^3} f(\vec{r},t)e^{i\vec{k}\cdot\vec{r}}d\vec{r})$. Five eigenvalues of the linearized version of Eq. (4.1) (when $\Psi_{nl} = 0$) determine two wave ($n = 1$ and $n = 2$) and non-wave (the entropy mode, $n = 3$, and two vorticity branches, $n = 4$ and $n = 5$) types of motion that may exist in a fluid. The links of the Fourier transforms of perturbations in both acoustic modes in the leading order with respect to viscous parameters caused by mechanical and thermal damping are ($n = 1, 2$):

$$\begin{aligned} \widetilde{\Psi}_n &= \begin{pmatrix} \widetilde{v}_{x,n} & \widetilde{v}_{y,n} & \widetilde{v}_{z,n} & \widetilde{p}_n & \widetilde{\rho}_n \end{pmatrix}^T \\ &= \begin{pmatrix} -\frac{ic_0\widetilde{\lambda}_n k_x}{\rho_0\widetilde{\Delta}}, & -\frac{ic_0\widetilde{\lambda}_n k_y}{\rho_0\widetilde{\Delta}}, & -\frac{ic_0\widetilde{\lambda}_n k_z}{\rho_0\widetilde{\Delta}}, & c_0^2 - c_0\delta_2\lambda_n), & 1 \end{pmatrix}^T \widetilde{\rho}_n, \end{aligned} \tag{4.3}$$

$$\widetilde{\lambda}_1 = -\sqrt{\widetilde{\Delta}} - \frac{\widehat{\beta}}{2}\widetilde{\Delta}, \quad \widetilde{\lambda}_2 = \sqrt{\widetilde{\Delta}} - \frac{\widehat{\beta}}{2}\widetilde{\Delta},$$

where $\widetilde{\Delta}$ applies in the Fourier transforms space, and $\widehat{\beta}$ reflects the attenuating properties of a fluid due to relaxation and thermal conduction:

$$\widetilde{\Delta} = -k_x^2 - k_y^2 - k_z^2, \quad \sqrt{\widetilde{\Delta}} = i\sqrt{k_x^2 + k_y^2 + k_z^2}, \quad \widehat{\beta} = 2\widehat{A} + \frac{\delta_2}{c_0}.$$

The operator A is determined by Eq. (3.35). Equation (4.3) determine relations between components of sound velocity in the $(\vec{r},t)$ space, which makes the velocity of the acoustic field potential:

$$\vec{\nabla} \times \vec{v}_1 = \vec{0}, \ \vec{\nabla} \times \vec{v}_2 = \vec{0}.$$

This is the well-known property of sound velocity to be an irrotational field. The velocity of the entropy mode is also a potential field; its excess pressure equals zero, but its excess density differs from zero:

$$\vec{\nabla} \times \vec{v}_3 = \vec{0}, \quad p_3 = 0, \quad \vec{v}_3 = \delta_2^2\vec{\nabla}\rho_3.$$

The solenoidal vorticity flow is determined by relations as follows:

$$\vec{\nabla}\cdot\vec{v}_4 = 0,\ p_4 = 0,\ \rho_4 = 0,\ \vec{\nabla}\cdot\vec{v}_5 = 0,\ p_5 = 0,\ \rho_5 = 0.$$

4.2 FIVE PROJECTORS

All five projectors may be determined by the use of specific links. They distinguish any individual mode when applied at the total vector of perturbations Ψ. In particular, the projector P_{vort} decomposes a sum of vorticity modes:

$$P_{vort}\Psi = \Psi_{vort} = \Psi_4 + \Psi_5.$$

Every projector is a matrix of spatial operators that consists of five rows and five columns. The part of the vorticity projector that has the practicle value, $P_{vort,\vec{v}}$ applies at the vector of velocity, because specific perturbations in both pressure and density in the vortex mode equal zero. It consists of three rows and three columns:

$$P_{vort,\vec{v}}\begin{pmatrix} v_x \\ v_y \\ v_z \end{pmatrix} = \Delta^{-1}\left(\begin{pmatrix} \frac{\partial^2}{\partial y^2}+\frac{\partial^2}{\partial z^2} & -\frac{\partial^2}{\partial x\partial y} & -\frac{\partial^2}{\partial x\partial z} \\ -\frac{\partial^2}{\partial x\partial y} & \frac{\partial^2}{\partial x^2}+\frac{\partial^2}{\partial z^2} & -\frac{\partial^2}{\partial y\partial z} \\ -\frac{\partial^2}{\partial x\partial z} & -\frac{\partial^2}{\partial y\partial z} & \frac{\partial^2}{\partial x^2}+\frac{\partial^2}{\partial y^2} \end{pmatrix}\right)\begin{pmatrix} v_x \\ v_y \\ v_z \end{pmatrix}$$

$$= \begin{pmatrix} v_{x,4}+v_{x,5} \\ v_{y,4}+v_{y,5} \\ v_{z,4}+v_{z,5} \end{pmatrix}.$$

$P_{vort,\vec{v}}$ manifests a certain way of applying the Helmholtz vector decomposition theorem, which makes it possible to decompose irrotational and solenoidal vector fields. $P_{vort,\vec{v}}$ satisfies the equality

$$-\vec{\nabla}\times(\vec{\nabla}\times\vec{\varphi}) = \Delta P_{vort,\vec{v}}\vec{\varphi},$$

where $\vec{\varphi}$ is any three-component smooth vector.

It is important that the projecting reduces terms of all other modes in the linear part of equations completely when applied at the system of weakly nonlinear equations. That is, the dynamic equation includes the first partial derivative of the reference variable with respect to time. The projection also distributes the nonlinear terms between dynamic equations in the correct manner. The momentum equation consists of three individual equations for every component of velocity:

$$\frac{\partial v_x}{\partial t}+\frac{1}{\rho_0}\frac{\partial p'}{\partial x}-c_0\widehat{A}\left(\Delta v_x+\frac{\partial}{\partial x}(\vec{\nabla}\cdot\vec{v})\right)$$

$$=-(\vec{v}\cdot\vec{\nabla})v_x+\frac{\rho'}{\rho_0^2}\frac{\partial p'}{\partial x}-\frac{c_0\rho'}{\rho_0}\widehat{A}\left(\Delta v_x+\frac{\partial}{\partial x}(\vec{\nabla}\cdot\vec{v})\right),$$

$$\frac{\partial v_y}{\partial t}+\frac{1}{\rho_0}\frac{\partial p'}{\partial y}-c_0\widehat{A}\left(\Delta v_y+\frac{\partial}{\partial x}(\vec{\nabla}\cdot\vec{v})\right)$$

$$=-(\vec{v}\cdot\vec{\nabla})v_y+\frac{\rho'}{\rho_0^2}\frac{\partial p'}{\partial y}-\frac{c_0\rho'}{\rho_0}\widehat{A}\left(\Delta v_y+\frac{\partial}{\partial y}(\vec{\nabla}\cdot\vec{v})\right),$$

$$\frac{\partial v_z}{\partial t}+\frac{1}{\rho_0}\frac{\partial p'}{\partial z}-c_0\widehat{A}\left(\Delta v_z+\frac{\partial}{\partial x}(\vec{\nabla}\cdot\vec{v})\right)$$

$$=-(\vec{v}\cdot\vec{\nabla})v_z+\frac{\rho'}{\rho_0^2}\frac{\partial p'}{\partial z}-\frac{c_0\rho'}{\rho_0}\widehat{A}\left(\Delta v_x+\frac{\partial}{\partial z}(\vec{\nabla}\cdot\vec{v})\right). \tag{4.4}$$

Applying $P_{vort,\vec{v}}$ on Eq. (4.4), we arrive at the dynamic equation that governs velocity in the vorticity mode:

$$\frac{\partial\vec{v}_{vort}}{\partial t}-c_0\widehat{A}\Delta\vec{v}_{vort}$$

$$=P_{vort,\vec{v}}\begin{pmatrix}-(\vec{v}\cdot\vec{\nabla})v_x+\frac{\rho'}{\rho_0^2}\frac{\partial p'}{\partial x}-\frac{c_0\rho'}{\rho_0}\widehat{A}\left(\Delta v_x+\frac{\partial}{\partial x}(\vec{\nabla}\cdot\vec{v})\right)\\ -(\vec{v}\cdot\vec{\nabla})v_y+\frac{\rho'}{\rho_0^2}\frac{\partial p'}{\partial y}-\frac{c_0\rho'}{\rho_0}\widehat{A}\left(\Delta v_y+\frac{\partial}{\partial y}(\vec{\nabla}\cdot\vec{v})\right)\\ -(\vec{v}\cdot\vec{\nabla})v_z+\frac{\rho'}{\rho_0^2}\frac{\partial p'}{\partial z}-\frac{c_0\rho'}{\rho_0}\widehat{A}\left(\Delta v_x+\frac{\partial}{\partial z}(\vec{\nabla}\cdot\vec{v})\right)\end{pmatrix}_1 \tag{4.5}$$

Only the acoustic terms associated with the first acoustic mode are left in the right-hand side of Eq. (4.5). This is denoted by the lower subscript 1. It may be rearranged into the following equation:

$$\frac{\partial\vec{v}_{vort}}{\partial t}-c_0\widehat{A}\Delta\vec{v}_{vort}=\frac{1}{\rho_0}P_{vort,\vec{v}}\left(-\rho_1\frac{\partial}{\partial t}\vec{v}_1\right)=\vec{F}_{str}, \tag{4.6}$$

where $\vec{F}_{str}$ denotes the acoustic force of streaming. One may readily derive the another equivalent form of the previous equation in terms of vorticity $\vec{\Omega}=\vec{\nabla}\times\vec{v}_{vort}$:

$$\frac{\partial\vec{\Omega}}{\partial t}-c_0\widehat{A}\Delta\vec{\Omega}=\frac{1}{\rho_0}\vec{\nabla}\times\left(-\rho_1\frac{\partial}{\partial t}\vec{v}_1\right). \tag{4.7}$$

It is useful to compare the right-hand side of Eq. (4.7) with that derived by Chu and Kovasznay [8] in the Newtonian thermoviscous flows. The right-hand acoustic source by [8] takes the form

$$\vec{\nabla} \times \left(-\frac{p_1}{c_0^2 \rho_0} \frac{\partial}{\partial t} \vec{v}_1 - \frac{1}{2} \nabla \vec{v}_1^2 \right) + O(\widetilde{\delta} M^2) = -\frac{1}{2} \vec{\nabla} \times \vec{\nabla} \left(\frac{p_1^2}{c_0^2 \rho_0^2} + \vec{v}_1^2 \right) + O(\widetilde{\delta} M^2)$$
$$= O(\widetilde{\delta} M^2),$$

where $\widetilde{\delta}$ is a generic parameter reflecting thermoviscous effects (including thermodynamic relaxation), and M is the Mach number. That is, the nonzero part is quadratic nonlinear and proportional to the damping coefficients. In Ref. [8], terms of order $\widetilde{\delta} M^2$ were not considered, only those proportional to M^2. Hence, streaming can not in principle be described in frames of the accuracy accepted by Chu and Kovasznay in view of that it is a nonlinear thermoviscous phenomenon. By the use of relations (4.3), Eq. (4.7) may be rearranged into the following one with a nonzero acoustic source only if operator $\widehat{\beta}$ differs from zero:

$$\frac{\partial \vec{\Omega}}{\partial t} - c_0 \widehat{A} \Delta \vec{\Omega} = \frac{c_0}{\rho_0^2} \vec{\nabla} \rho_1 \times \vec{\nabla} \left(\widehat{\beta} \frac{\partial \rho_1}{\partial t} \right)$$
$$= -\frac{c_0}{\rho_0} \vec{\nabla} \rho_1 \times \widehat{\beta} \vec{\nabla} (\vec{\nabla} \cdot \vec{v}_1) = \frac{c_0}{\rho_0} \vec{\nabla} \rho_1 \times \widehat{\beta} \Delta \vec{v}_1.$$

An acoustic excess density for both acoustic branches satisfies an equation that includes the second-order partial derivative with respect to time in its linear part. It includes also absorption due to relaxation and the quadratic nonlinear term:

$$\frac{\partial^2 \rho_{1,2}}{\partial t^2} - c_0^2 \Delta \rho_{1,2} - \frac{\widehat{\beta}}{c_0} \frac{\partial^3 \rho_{1,2}}{\partial t^3} - \frac{1 - D_1 - D_2}{2\rho_0} \frac{\partial^2 \rho_{1,2}^2}{\partial t^2} = 0.$$

The projection of the system (3.31) results in two equations for every individual acoustic mode. They include first-order derivatives with respect to time. The governing equation for the first acoustic mode takes the form

$$\frac{\partial \rho_1}{\partial t} + c_0 \sqrt{\Delta} \rho_1 - \frac{c_0 \widehat{\beta}}{2} \Delta \rho_1 = \frac{1}{2} \left((D_1 + D_2) \rho_1 (\vec{\nabla} \cdot \vec{v}_1) - \vec{v}_1 \cdot \vec{\nabla} \rho_1 \right). \tag{4.8}$$

The general analysis of Eqs (4.8) and (4.6) is fairly difficult. It may be significantly simplified assuming the quasi-planar geometry of a flow along axis y. Let x, z be the Cartesian coordinates perpendicular to that axis, and let a flow be cylindrically symmetric. The acoustic transducer is supposed to be situated at the plane $y = 0$; it has a characteristic radius a; and it radiates at high enough frequencies satisfying inequality $\omega \gg c_0/a$. The last assumption ensures that the beam is weakly divergent in the direction perpendicular to the course of propagation. Making use of the small parameter $\varepsilon = 1/(ka)^2$, which is responsible for diffraction, allows us to derive modes and projectors in the leading order as a power series of the small parameter ε:

$$\Delta = \partial^2/\partial y^2 + \varepsilon \Delta_\perp, \quad \sqrt{\Delta} \approx \partial/\partial y + 0.5 \varepsilon \Delta_\perp \int dy,$$

where $\Delta_\perp = \frac{\partial^2}{\partial x^2} + \frac{\partial^2}{\partial z^2}$. Equation (4.6) rearranges in the leading order as

$$\frac{\partial \vec{v}_{vort}}{\partial t} - c_0 \widehat{A} \frac{\partial^2 \vec{v}_{vort}}{\partial y^2} = -\frac{c_0^2}{\rho_0^2} P_{vort,\vec{v}} \left(\rho_1 \widehat{\beta} \vec{\nabla} \frac{\partial \rho_1}{\partial y} \right),$$

or, in the other way, as

$$\frac{\partial v_{y,vort}}{\partial t} - c_0 \widehat{A} \frac{\partial^2 v_{y,vort}}{\partial y^2} = -\frac{\varepsilon c_0^2}{\rho_0^2} \begin{pmatrix} -\frac{\partial}{\partial x} \int dy \\ \Delta_\perp \int dy \int dy \\ -\frac{\partial}{\partial z} \int dy \end{pmatrix}^T \left(\rho_1 \widehat{\beta} \vec{\nabla} \frac{\partial \rho_1}{\partial y} \right)$$

$$= \frac{\varepsilon c_0^2}{\rho_0^2} \left(\frac{\partial}{\partial x} \int dy \left(\rho_1 \widehat{\beta} \frac{\partial^2 \rho_1}{\partial x \partial y} \right) - \Delta_\perp \right.$$

$$\left. \times \int dy \int dy \left(\rho_1 \widehat{\beta} \frac{\partial^2 \rho_a}{\partial y^2} \right) + \frac{\partial}{\partial z} \int dy \left(\rho_1 \widehat{\beta} \frac{\partial^2 \rho_1}{\partial y \partial z} \right) \right).$$

Introducing the transversal coordinate r, $r = \sqrt{x^2 + z^2}$, one arrives at

$$\frac{\partial v_{y,vort}}{\partial t} - c_0 \widehat{A} \frac{\partial^2 v_{y,vort}}{\partial y^2} =$$

$$= \frac{\varepsilon c_0^2}{\rho_0^2} \int dy \left(\frac{\partial \rho_1}{\partial r} \widehat{\beta} \frac{\partial^2 \rho_1}{\partial r \partial y} + \rho_a \widehat{\beta} \Delta_\perp \frac{\partial \rho_1}{\partial y} \right) - \frac{\varepsilon c_0^2}{\rho_0^2} \Delta_\perp \int dy \int dy \left(\rho_1 \widehat{\beta} \frac{\partial^2 \rho_1}{\partial y^2} \right), \tag{4.9}$$

where $\Delta_\perp = 1/r \partial/\partial r + \partial^2/\partial r^2$ is the transversal Laplacian. In turn, an excess acoustic density in the beam propagating in the positive direction of axis y satisfies the equation

$$\frac{\partial \rho_1}{\partial t} + c_0 \frac{\partial \rho_1}{\partial y} + \frac{\varepsilon}{2} \int \Delta_\perp \rho_1 dy - \frac{c_0 \widehat{\beta}}{2} \frac{\partial^2 \rho_1}{\partial y^2} + \frac{1 - D_1 - D_2}{2 c_0 \rho_0} \rho_1 \frac{\partial \rho_1}{\partial y} = 0. \tag{4.10}$$

Equation (4.10) is analogous to the celebrated KZK equation with $\widehat{\beta}$ replacing the standard Newtonian attenuation [3,9]. It may be also derived by projecting the initial system onto dynamic equations of the perturbations, which specify the first acoustic branch by means of appropriate acoustic projector [10].

4.3 EXAMPLES OF ACOUSTIC STREAMING: WEAKLY DIFRACTING BEAM AND STATIONARY WAVEFORM

An excess density of the acoustic mode that contributes to the acoustic force of streaming must satisfy Eqs (4.8) or (4.10) depending on the geometry of a flow. Equation (4.9) includes double integrating over spacial coordinate. This makes it necessary to determine the integration constants that correspond to the physical meaning of a problem. It is possible to evaluate the right-hand side of Eq. (4.6) in some simple cases analytically. Let an excess acoustic density take the form:

$$\rho_1 = R_0 \exp(-r^2/a^2) \sin(\omega(t - y/c_0)),$$

where R_0 is the magnitude of excess density at the axis of a beam. An excess acoustic density may be considered as an approximate solution of Eq. (4.10) in the case of very weak nonlinearity, attenuation, and diffraction. The longitudinal component of the acoustic mass force equals

$$F_{str,y} = \frac{R_0^2}{\rho_0^2}\left(\frac{\mu\omega^2\tau_R}{(1+\omega^2\tau_R^2)\rho_0 c_0} + \frac{\delta_2\omega^2}{2c_0}\right)\exp\left(-2r^2/a^2\right).$$

The acoustic force in this example does not depend on the longitudinal coordinate y. In the derivation of $F_{str,y}$, however, we did not use averaging over the sound period at any stage. At small $\omega\tau_R$, the longitudinal acoustic force of streaming tends to the Newtonian with the characteristic proportionality to the square frequency of the harmonic sound, ω^2.

The next example is the stationary sound waveform that propagates in the positive direction of axis x and depends exclusively on the retarded time $\xi = t - x/c_0$ in a pure relaxing fluid without heat conduction. The one-dimensional form of Eq. (4.10) sounds:

$$\frac{1-D_1-D_2}{2}\rho_1\frac{d\rho_1}{d\xi} + \frac{\mu}{c_0^2}\int_{-\infty}^{\xi}\frac{d^2\rho_1(\xi')}{d\xi'^2}\exp\left(-(\xi-\xi')/\tau_R\right)d\xi' = 0.$$

It is readily integrated with the result [1]:

$$\left(\rho_1 + \frac{\mu}{c_0^2(1-D_1-D_2)}\right)\frac{d\rho_1}{d\xi} + \frac{\rho_1^2}{2\tau_R} = const.$$

We consider a density jump $2R_0$ from $-R_0$ ahead of the wave front to R_0 behind. Applying the boundary conditions at $\xi = -\infty$: $\rho_1 = -R_0$, $d\rho_1/d\xi = 0$ and $\xi = \infty$: $\rho_1 = R_0$, $d\rho_1/d\xi = 0$, gives the integration constant $R_0^2/2\tau_R$. Finally,

$$\xi = \tau_R \ln\frac{(1+\rho_1/R_0)^{G-1}}{(1-\rho_1/R_0)^{G+1}},$$

where $G = \frac{2\mu}{(1-D_1-D_2)R_0 c_0^2}$ measures the ratio of relaxation effects to nonlinear effects. In the limit of weak nonlinearity $G \gg 1$, the solution may be written on analytically:

$$\rho_1(\xi) = R_0 \tanh(\xi/2G\tau_R).$$

The waveform

$$\rho_1(\xi) = R_0 \exp(-r^2/a^2)\tanh(\xi/2G\tau_R)$$

may be considered as an approximate formula describing weakly diffracting acoustic beam with the stationary profile. The longitudinal component of the acoustic mass force of streaming equals

$$F_{str,y} = \frac{R_0^2}{4\rho_0^2\tau_R G^2}\left(\frac{\mu}{\rho_0 c_0} + \frac{\delta_2}{2c_0\tau_R}\right)\exp(-2r^2/a^2)\cosh\left(\frac{\xi}{\tau_R G}\right)\cosh^{-4}\left(\frac{\xi}{2\tau_R G}\right).$$

The longitudinal component of the acoustic force of streaming is positive, and it decreases rapidly with enlarging of $|\xi|$.

REFERENCES

1. Hamilton, M., and D. Blackstock. 1998. *Nonlinear Acoustics*. Academic Press, New York.
2. Nyborg, W.L. 1965. Acoustic streaming. In: Manson, W.P. (Ed.), *Physical Acoustics*, vol. II, part B. Academic Press, New York, pp. 265–331.
3. Rudenko, O.V., and S.I. Soluyan. 1977. *Theoretical Foundations of Nonlinear Acoustics*. Plenum, New York.
4. Lighthill, M.J. 1978. Acoustic streaming. *Journal of Sound and Vibration* 61(3): 391–418.
5. Qi, Q. 1993. The effect of compressibility on acoustic streaming near a rigid boundary for a plane traveling wave. *The Journal of the Acoustical Society of America* 94(2): 1090–1098.
6. Menguy, L., and J. Gilbert. 2000. Non-linear acoustic streaming accompanying a plane stationary wave in a guide. *Acta Acustica United with Acustica* 86: 249–259.
7. Makarov, S., and M. Ochmann. 1996. Nonlinear and thermoviscous phenomena in acoustics, part I, *Acta Acustica United with Acustica* 82: 579–606.
8. Chu, B.-T., and L.S.G. Kovasznay. 1958. Non-linear interactions in a viscous heat-conducting compressible gas. *Journal of Fluid Mechanics* 3: 494–514.
9. Kuznetsov, V.P. 1971. Equations of nonlinear acoustics. *Soviet Physics Acoustics* 16: 467–470.
10. Perelomova, A. 2003. Acoustic radiation force and streaming caused by non-periodic acoustic source. *Acta Acustica United with Acustica* 89: 754–763.

5 Projecting in Flows with Relaxation: Effects of Sound in Acoustically Active Fluids

5.1 VIBRATIONALLY RELAXING GASES

The establishment of the basic physical features of nonequilibrium molecular physics began in the sixties, due to the laser revolution in physics and chemistry. The hydrodynamics of nonequilibrium fluids is currently passing through a stage of formulating the fundamental equations, which incorporates knowledge in fluid mechanics, statistical physics, nonequilibrium thermodynamics and experimental physics and chemistry. This concerns the comprehension and description of novel physical effects and requires the appropriate mathematical formulations and methods. Nonequilibrium flows, among others, take place in the interstellar medium, the upper atmosphere, and in the discharge of plasma. Interest in nonequilibrium phenomena in the physics of gases increases with studies of anomalous dispersion and the absorption of ultrasonic waves in acoustically active media. The reason for these anomalies is the retarded mechanism of energy exchange between the internal and translational degrees of freedom of the molecules [1–3]. A number of problems relating to the nonlinear effects and rate processes in gases with internal relaxation have been studied previously. In this context, the contributions due to Chu [4], Parker [5], Clarke, and McChesney [6] are worth mentioning.

In relation to relaxing gases, it was firstly pointed out by Molevich (a theory developed later in the papers of Molevich and her co-authors) that the nonlinear exchange of energy between sound and the entropy mode may lead to the anomalous nonlinear phenomena of sound: cooling instead of heating in the nonequilibrium gas, if its standard attenuation is small [7]. This may occur in acoustically active gases under some conditions. The possibility of an anomaly in the other nonlinear phenomenon caused by dominative sound in a vibrationally excited molecular gas, namely streaming, was also pointed out there. The flow of this induced vortex motion may occur in the opposite direction compared to that in the equilibrium gas. The analysis of Ref. [7] was undertaken in the usual way, by considering the scattering of the harmonic sound. This results in the coupling system of equations for amplitudes of the interacting harmonic waves. Its solution in turn requires averaging over the sound period in order to eliminate quickly varying sound perturbations in description of mean fields. The projecting method allows us to derive instantaneous dynamic equations for sound and non-wave modes and accounting for their coupling independently of periodicity of sound. We recall the main results of Ref. [8] in this section.

We start from the linear definition of modes as specific types of gas motion whose steady state is maintained by incoming energy into the vibrational degrees of freedom. The power of source is I, and the heat withdrawal from the translational degrees of freedom is of power Q (both I and Q refer to unit mass). The relaxation of the vibrational energy per unit mass completes the system of conservation equations:

$$\frac{d\varepsilon}{dt} = -\frac{\varepsilon - \varepsilon_{eq}(T)}{\tau} + I. \tag{5.1}$$

The equilibrium value of the vibrational energy at temperature T is denoted by $\varepsilon_{eq}(T)$, and $\tau(\rho, T)$ designates the vibrational relaxation time. The quantity $\varepsilon_{eq}(T)$ in the simplest case of a system of harmonic oscillators equals

$$\varepsilon_{eq}(T) = \frac{\hbar\Omega}{m\,(exp(\hbar\Omega/k_B T) - 1)},$$

where m is the mass of a molecule, $\hbar\Omega$ is the magnitude of the vibrational quantum, and k_B is the Boltzmann constant. It is valid below the characteristic temperatures, where one can neglect anharmonic effects. They are fairly high for most molecules [1,3]. The mass, momentum and energy equations for a thermoviscous flow of a vibrationally relaxing gas in usual notations read

$$\begin{aligned} &\frac{\partial\rho}{\partial t} + \vec{\nabla}(\rho\vec{v}) = 0, \\ &\rho\left[\frac{\partial\vec{v}}{\partial t} + (\vec{v}\cdot\vec{\nabla})\vec{v}\right] = -\vec{\nabla}p + \eta\Delta\vec{v} + \left(\zeta + \frac{\eta}{3}\right)\vec{\nabla}(\vec{\nabla}\cdot\vec{v}), \\ &\rho\left[\frac{\partial(e+\varepsilon)}{\partial t} + (\vec{v}\cdot\vec{\nabla})(e+\varepsilon)\right] + p\vec{\nabla}\cdot\vec{v} = \chi\Delta T + \rho(I - Q) + \zeta\left(\vec{\nabla}\cdot\vec{v}\right)^2 \\ &\qquad + \frac{\eta}{2}\left(\frac{\partial v_i}{\partial x_k} + \frac{\partial v_k}{\partial x_i} - \frac{2}{3}\delta_{ik}\frac{\partial v_l}{\partial x_l}\right)^2, \end{aligned} \tag{5.2}$$

Besides Eq. (5.1), two thermodynamic functions $e(p,\rho), T(p,\rho)$ complete the system (5.2). We use that of an ideal gas, (3.13). Let us consider a planar motion of infinitely small amplitude of a gas in the case $\eta = 0$, $\zeta = 0$, $\chi = 0$, $Q = const$, $I = const$ along axis Ox. At this point, we ignore the impact of thermal conductivity, viscosity, and dependence Q on temperature. Considering every quantity q as a sum of unperturbed value q_0 (in absence of the background flows, $v_0 = 0$) and its variation q', one readily rearranges the governing equations of momentum, energy balance, and continuity into the leading-order linear system:

$$\begin{aligned} &\frac{\partial v}{\partial t} + \frac{1}{\rho_0}\frac{\partial p'}{\partial x} = 0, \\ &\frac{\partial p'}{\partial t} + \gamma p_0\frac{\partial v}{\partial x} - (\gamma - 1)\rho_0\frac{\varepsilon'}{\tau} + (\gamma - 1)\rho_0 T_0 \Phi_1\left(\frac{p'}{p_0} - \frac{\rho'}{\rho_0}\right) = 0, \end{aligned}$$

$$\frac{\partial \rho'}{\partial t} + \rho_0 \frac{\partial v}{\partial x} = 0, \tag{5.3}$$

$$\frac{\partial \varepsilon'}{\partial t} + \frac{\varepsilon'}{\tau} - T_0 \Phi_1 \left(\frac{p'}{p_0} - \frac{\rho'}{\rho_0} \right) = 0,$$

where the coefficient

$$\Phi_1 = \left(\frac{C_V}{\tau} + \frac{\varepsilon - \varepsilon_{eq}}{\tau^2} \frac{d\tau}{dT} \right)_0$$

is the quantity that is evaluated at the unperturbed state p_0, T_0, and $C_V = d\varepsilon_{eq}/dT$. The expansion into a series of equations of state yields the leading-order expression of excess translation temperature and excess internal translational energy per unit mass in terms of the perturbations of pressure and density:

$$e' = \frac{p_0}{(\gamma - 1)\rho_0} \left(\frac{p'}{p_0} - \frac{\rho'}{\rho} \right) = \frac{R}{m_{mol}(\gamma - 1)} T',$$

where γ is the isentropic exponent without accounting for vibrational degrees of freedom, R is the universal gas constant, and m_{mol} is the molar mass of a gas. Also,

$$\frac{\partial \varepsilon'}{\partial t} + \frac{\varepsilon'}{\tau} = \left(\frac{C_V}{\tau} + \frac{\varepsilon - \varepsilon_{eq}}{\tau^2} \frac{d\tau}{dT} \right)_0 T' = T_0 \Phi_1 \left(\frac{p'}{p_0} - \frac{\rho'}{\rho_0} \right).$$

The Landau-Teller dependence of relaxation time provides negative values of $d\tau/dT$ [1,3]. This is the reason for acoustically active gases to exist in principle. The approximate roots of the dispersion equation for both acoustic branches, which propagate in the positive and negative directions of axis Ox, are well known in the case of the high-frequency sound $\omega\tau \gg 1$ [3,9]:

$$\omega_1 = ck + \frac{i}{2} \frac{(\gamma - 1)^2 T_0}{c^2} \Phi_1, \quad \omega_2 = -ck + \frac{i}{2} \frac{(\gamma - 1)^2 T_0}{c^2} \Phi_1, \tag{5.4}$$

where $c = \sqrt{\frac{\gamma R T_0}{m_{mol}}} = \sqrt{\frac{\gamma p_0}{\rho_0}}$ denotes the infinitely small-signal sound speed in the ideal uniform gas. The last term in both the dispersion relations is responsible for amplification of sound in an acoustically active gas (if $\Phi_1 < 0$), which does not depend on the wavenumber k. The amplification increases with the enlargement of $|d\tau/dT|$ and the vibrational nonequilibrium $m(\varepsilon - \varepsilon_{eq})/k_B T$. We will consider weak attenuation and relative dispersion, i.e.

$$|\Phi_1 T_0/c^3 k| \ll 1.$$

This condition provides a weak distortion of the sound wave (caused by both attenuation or amplification) over its period.

The two last roots of the dispersive equation, estimated without limitations, take the forms

$$\omega_3 = i \left(\frac{1}{\tau} + \frac{(\gamma - 1)(\gamma + c^2 k^2 \tau^2) T_0}{c^2 (1 + c^2 k^2 \tau^2)} \Phi_1 \right), \quad \omega_4 = 0. \tag{5.5}$$

The third non-wave mode, which is determined by ω_3, originates from the vibrational relaxation. If $\Phi_1 = 0$, then it corresponds to the relaxation processes with the characteristic time of thermodynamic relaxation τ. Nonzero Φ_1 makes this characteristic time larger or smaller. The fourth root exists in any planar flow of a fluid, not necessarily relaxing or attenuating. It also corresponds to the non-wave motion, but it is stationary. It represents the entropy mode and associates with a weak velocity field in a thermoconducting gas. It has been well established that the nonlinear losses in acoustic energy in a gas with a typical thermoviscous attenuation, lead to the heating of the background, that is, to the isobaric increase in the background temperature and variation in density specifying the entropy mode.

In the linear flow, the overall velocity, pressure, density, and internal energy are also a sum of specific parts: $v(x,t) = \sum_{n=1}^{4} v_n(x,t)$, and so on. According to the dispersion relations (5.4), (5.5), the Fourier transforms of total perturbations also may be represented as a linear combination of four specific Fourier transforms. In turn, all of them are readily expressed in terms of the specific perturbations in density, $\tilde{\rho}_1$, $\tilde{\rho}_2$, $\tilde{\rho}_3$, or $\tilde{\rho}_4$. The Fourier transforms of total perturbations are as follows:

$$\tilde{\rho} = \sum_{n=1}^{4} \tilde{\rho}_n, \quad \tilde{v} = \sum_{n=1}^{4} \tilde{v}_n = \sum_{n=1}^{4} \omega_n \tilde{\rho}_n / k / \rho_0, \quad \tilde{p} = \sum_{n=1}^{4} \tilde{p}_n = \sum_{n=1}^{4} \omega_n^2 \tilde{\rho}_n / k^2,$$

$$\tilde{\varepsilon} = \sum_{n=1}^{4} \tilde{\varepsilon}_n = \frac{T_0 \Phi_1}{\rho_0 c^2} \sum_{n=1}^{2} \tilde{\rho}_n \left(\frac{\gamma \omega_n^2}{k^2} - c^2 \right) / (i\omega_n) + \frac{T_0 \Phi_1}{\rho_0 c^2} \tilde{\rho}_3 \left(\frac{\gamma \omega_3^2}{k^2} - c^2 \right) / (i\omega_3 + 1/\tau)$$
$$+ \frac{\tau T_0 \Phi_1}{\rho_0 c^2} \tilde{\rho}_4 \left(\frac{\gamma \omega_4^2}{k^2} - c^2 \right).$$

The links in the (x,t) space follow from the preceding relations and roots of dispersion relations, Eqs (5.4), (5.5). The operators $\tilde{d}_1^1, \tilde{d}_2^1, \tilde{d}_2^1, \tilde{d}_4^1$ may be readily established. They apply in the Fourier-transform space, decomposing only one constituent of the overall excess density from the vector of total perturbations, for example, corresponding to the rightward progressive sound:

$$\tilde{d}_1^1 \tilde{v} + \tilde{d}_2^1 \tilde{p} + \tilde{d}_3^1 \tilde{\rho} + \tilde{d}_4^1 \tilde{\varepsilon} = \tilde{\rho}_1.$$

The preceding equation in fact contains four algebraic equations determining four unknown quantities uniquely. The matrix of four rows, which projects the vector of perturbations into the vector of specific excess densities, has the leading-order form

$$\begin{pmatrix} \tilde{d}_1^1 & \tilde{d}_2^1 & \tilde{d}_3^1 & \tilde{d}_4^1 \\ \tilde{d}_1^2 & \tilde{d}_2^2 & \tilde{d}_3^2 & \tilde{d}_4^2 \\ \tilde{d}_1^3 & \tilde{d}_2^3 & \tilde{d}_3^3 & \tilde{d}_4^3 \\ \tilde{d}_1^4 & \tilde{d}_2^4 & \tilde{d}_3^4 & \tilde{d}_4^4 \end{pmatrix} \cdot \begin{pmatrix} \tilde{v} \\ \tilde{p} \\ \tilde{\rho} \\ \tilde{\varepsilon} \end{pmatrix} \equiv \tilde{D} \cdot \begin{pmatrix} \tilde{v} \\ \tilde{p} \\ \tilde{\rho} \\ \tilde{\varepsilon} \end{pmatrix} = \begin{pmatrix} \tilde{\rho}_1 \\ \tilde{\rho}_2 \\ \tilde{\rho}_3 \\ \tilde{\rho}_4 \end{pmatrix}, \quad \tilde{D}:$$

$$\begin{pmatrix} \tilde{d}_1^1 \\ \tilde{d}_2^1 \\ \tilde{d}_3^1 \\ \tilde{d}_4^1 \end{pmatrix} = \begin{pmatrix} \frac{\rho_0}{2c} - \frac{i(\gamma-1)^2 \rho_0 T_0}{2c^4 k} \Phi_1 \\ \frac{1}{2c^2} - \frac{i(\gamma-1)(\gamma-3) T_0}{4c^5 k} \Phi_1 \\ -\frac{i(\gamma-1) T_0}{2c^3 k} \Phi_1 \\ -\frac{i(\gamma-1)\rho_0}{2c^3 k \tau} \end{pmatrix}; \quad \begin{pmatrix} \tilde{d}_1^2 \\ \tilde{d}_2^2 \\ \tilde{d}_3^2 \\ \tilde{d}_4^2 \end{pmatrix} = \begin{pmatrix} -\frac{\rho_0}{2c} - \frac{i(\gamma-1)^2 \rho_0 T_0}{2c^4 k} \Phi_1 \\ \frac{1}{2c^2} + \frac{i(\gamma-1)(\gamma-3) T_0}{4c^5 k} \Phi_1 \\ \frac{i(\gamma-1) T_0}{2c^3 k} \Phi_1 \\ \frac{i(\gamma-1)\rho_0}{2c^3 k \tau} \end{pmatrix};$$

$$\begin{pmatrix} \tilde{d}_1^3 \\ \tilde{d}_2^3 \\ \tilde{d}_3^3 \\ \tilde{d}_4^3 \end{pmatrix} = \begin{pmatrix} \frac{i(\gamma-1)^2\rho_0 T_0}{c^4 k}\Phi_1 \\ -\frac{(\gamma-1)T_0\tau}{c^4}\Phi_1 \\ \frac{(\gamma-1)T_0\tau}{c^2}\Phi_1 \\ \frac{(\gamma-1)\rho_0}{c^2} - \frac{(\gamma-1)^2 T_0\rho_0\tau}{c^4}\Phi_1 \end{pmatrix};$$

$$\begin{pmatrix} \tilde{d}_1^4 \\ \tilde{d}_2^4 \\ \tilde{d}_3^4 \\ \tilde{d}_4^4 \end{pmatrix} = \begin{pmatrix} 0 \\ -\frac{1}{c^2} + \frac{(\gamma-1)T_0\tau}{c^4}\Phi_1 \\ 1 - \frac{(\gamma-1)T_0\tau}{c^2}\Phi_1 \\ -\frac{(\gamma-1)\rho_0}{c^2} + \frac{(\gamma-1)^2 T_0\rho_0\tau}{c^4}\Phi_1 \end{pmatrix}.$$

The matrix $\tilde{D}$ is evaluated within an accuracy up to terms proportional to Φ_1^1 and $(1/(ck\tau))^1$, inclusively. The analogous matrix operator D, operating in the (x,t) space, may be readily derived. It includes integrating over x, which corresponds to the factor i/k. The limits of integration depend on the physical context of the problem. Application of any row of matrix D on the total vector of perturbations decomposes the specific excess density corresponding to this row. Analogously, application of these rows individually on the linearized system (5.3) results in a dynamic equation of specific excess density corresponding to this row. With regards to the first acoustic and entropy modes, the dynamic equations for the excess densities corresponding to the first and fourth modes, are

$$\frac{\partial \rho_1}{\partial t} + c\frac{\partial \rho_1}{\partial x} + \frac{(\gamma-1)^2 T_0}{2c^2}\Phi_1\rho_1 = 0, \qquad \frac{\partial \rho_4}{\partial t} = 0.$$

This obviously coincides to the roots of dispersion equation ω_1 and ω_4 established by Eq. (5.4).

In order to account for the effect connected with quadratic nonlinearity, we use leading-order series of excess temperature and vibrational energy:

$$\begin{aligned}
T' =& T_0\left(\frac{p'}{p_0} - \frac{\rho'}{\rho_0} + \frac{\rho'^2}{\rho_0^2} - \frac{p'\rho'}{p_0\rho_0}\right),\\
\frac{d\varepsilon'}{dt} =& -\frac{\varepsilon'}{\tau} + T_0\left(\frac{1}{\tau^2}\frac{d\tau}{dT}\right)_0 \varepsilon'\left(\frac{p'}{p_0} - \frac{\rho'}{\rho_0}\right) + T_0\Phi_1\left(\frac{p'}{p_0} - \frac{\rho'}{\rho_0} + \frac{\rho'^2}{\rho_0^2} - \frac{p'\rho'}{p_0\rho_0}\right)\\
&+ T_0\Phi_2\left(\frac{p'}{p_0} - \frac{\rho'}{\rho_0}\right)^2,\\
\Phi_2 =& T_0\left(-\frac{1}{\tau^2}C_V\frac{d\tau}{dT} - \frac{(\varepsilon_0-\varepsilon_{eq})}{\tau^3}\left(\frac{d\tau}{dT}\right)^2 + \frac{1}{2\tau}\frac{dC_V}{dT} + \frac{(\varepsilon_0-\varepsilon_{eq})}{2\tau^2}\frac{d^2\tau}{dT^2}\right)_0
\end{aligned}$$

The governing dynamic system with an account for quadratic nonlinear terms looks like

$$\frac{\partial v}{\partial t}+\frac{1}{\rho_0}\frac{\partial p'}{\partial x}=-v\frac{\partial v}{\partial x}+\frac{\rho'}{\rho_0^2}\frac{\partial p'}{\partial x},$$

$$\frac{\partial p'}{\partial t}+\gamma p_0\frac{\partial v}{\partial x}-(\gamma-1)\rho_0\frac{\varepsilon'}{\tau}+(\gamma-1)\rho_0 T_0\Phi_1\left(\frac{p'}{p_0}-\frac{\rho'}{\rho_0}\right)$$

$$=-v\frac{\partial p'}{\partial x}-\gamma p'\frac{\partial v}{\partial x}+(\gamma-1)\rho'\left(\frac{\varepsilon'}{\tau}-T_0\Phi_1\left(\frac{p'}{p_0}-\frac{\rho'}{\rho_0}\right)\right)-(\gamma-1)\rho_0$$

$$\times\left(T_0\left(\frac{1}{\tau^2}\frac{d\tau}{dT}\right)_0\varepsilon'\left(\frac{p'}{p_0}-\frac{\rho'}{\rho_0}\right)+T_0\Phi_1\left(\frac{\rho'^2}{\rho_0^2}-\frac{p'\rho'}{p_0\rho_0}\right)+T_0\Phi_2\left(\frac{p'}{p_0}-\frac{\rho'}{\rho_0}\right)^2\right),$$

$$\frac{\partial \rho'}{\partial t}+\rho_0\frac{\partial v}{\partial x}=-v\frac{\partial \rho'}{\partial x}-\rho'\frac{\partial v}{\partial x}, \tag{5.6}$$

$$\frac{\partial \varepsilon'}{\partial t}+\frac{\varepsilon'}{\tau}-T_0\Phi_1\left(\frac{p'}{p_0}-\frac{\rho'}{\rho_0}\right)=T_0\left(\frac{1}{\tau^2}\frac{d\tau}{dT}\right)_0\varepsilon'\left(\frac{p'}{p_0}-\frac{\rho'}{\rho_0}\right)+T_0\Phi_1\left(\frac{\rho'^2}{\rho_0^2}-\frac{p'\rho'}{p_0\rho_0}\right)$$

$$+T_0\Phi_2\left(\frac{p'}{p_0}-\frac{\rho'}{\rho_0}\right)^2-v\frac{\partial \varepsilon'}{\partial x}.$$

Applications of projecting cancel the contributions of all other modes in the linear part of the final dynamic equation, and the quadratic nonlinear terms become distributed between equations in the proper way. As usual, we compare dominative sound with other types of motion.

Applying the row operator $(d_1^1\ \ d_2^1\ \ d_3^1\ \ d_4^1)$ on both sides of equations that form the system (5.6), letting all nonlinear terms be acoustic corresponding to the first (progressive in the positive direction of axis Ox) mode, and expressing all acoustic perturbations in terms of the specific excess density, one can readily derive the dynamic equation governing the dominative sound within an accuracy up to quadratic nonlinear terms. Its nonlinear terms may be considered as certain corrections caused by the interaction of this wave with itself. The row $(\tilde{d}_1^1\ \ \tilde{d}_2^1\ \ \tilde{d}_3^1\ \ \tilde{d}_4^1)$ and the corresponding vector operating in x space may be considerably simplified in view of the inequalities $ck\tau \gg 1$, $|\Phi_1|T_0/(c^3k) \ll 1$. This results in the leading-order formulae

$$\begin{pmatrix}\tilde{d}_1^1\\ \tilde{d}_2^1\\ \tilde{d}_3^1\\ \tilde{d}_4^1\end{pmatrix}=\begin{pmatrix}\frac{\rho_0}{2c}-\frac{i(\gamma-1)^2\rho_0T_0}{2c^4k}\Phi_1\\ \frac{1}{2c^2}-\frac{i(\gamma-1)(\gamma-3)T_0}{4c^5k}\Phi_1\\ -\frac{i(\gamma-1)T_0}{2c^3k}\Phi_1\\ -\frac{i(\gamma-1)\rho_0}{2c^3k\tau}\end{pmatrix},$$

$$\begin{pmatrix} d_1^1 \\ d_2^1 \\ d_3^1 \\ d_4^1 \end{pmatrix} = \begin{pmatrix} \frac{\rho_0}{2c} - \frac{(\gamma-1)^2\rho_0 T_0}{2c^4}\Phi_1 \int dx \\ \frac{1}{2c^2} - \frac{(\gamma-1)(\gamma-3)T_0}{4c^5}\Phi_1 \int dx \\ -\frac{(\gamma-1)T_0}{2c^3}\Phi_1 \int dx \\ -\frac{(\gamma-1)\rho_0}{2c^3\tau} \int dx \end{pmatrix}. \tag{5.7}$$

The following vector ψ_1 represents the linear links for the rightwards progressive sound in the leading order, as it follows from (5.4):

$$\psi_1 = \begin{pmatrix} v_1(x,t) \\ p_1(x,t) \\ \rho_1(x,t) \\ \varepsilon_1(x,t) \end{pmatrix} = \begin{pmatrix} \frac{c}{\rho_0} + \frac{(\gamma-1)^2\Phi_1 T_0}{2c^2\rho_0} \int dx \\ c^2 + \frac{(\gamma-1)^2\Phi_1 T_0}{c} \int dx \\ 1 \\ -\frac{(\gamma-1)\Phi_1 T_0}{c\rho_0} \int dx \end{pmatrix} \rho_1. \tag{5.8}$$

Applying the row $(d_1^1 \quad d_2^1 \quad d_3^1 \quad d_4^1)$ on both sides of the equations from the system (5.6), and accounting for Eq. (5.8), one gets the leading-order dynamic equation for the acoustic excess density:

$$\frac{\partial \rho_1}{\partial t} + c\frac{\partial \rho_1}{\partial x} - cB\rho_1 = -\frac{\gamma+1}{2}\frac{c}{\rho_0}\rho_1\frac{\partial \rho_1}{\partial x}, \tag{5.9}$$

where

$$B = -\frac{(\gamma-1)^2 T_0}{2c^3}\Phi_1.$$

The nonlinear term on the right-hand side of Eq. (5.9) manifests the well-celebrated nonlinearity originating from nonlinearity in the equation of state and hydrodynamic nonlinearity [10,11]; it is of order squared Mach number M^2. Deriving coupling nonlinear equations requires quadratic nonlinear corrections in the linear definition of sound, accounting for terms specific to the Riemann wave [10], which support the adiabaticity of sound in the lossless flow. The corrected links with accuracy up to the quadratic nonlinear terms for the rightwards progressive sound are

$$\Psi_1 = \begin{pmatrix} \frac{c}{\rho_0} - \frac{cB}{\rho_0}\int dx + \frac{(\gamma-3)c}{4\rho_0^2}\rho_1 \\ c^2 - 2c^2 B \int dx + \frac{(\gamma-1)c^2}{2\rho_0}\rho_1 \\ 1 \\ \frac{2Bc^2}{(\gamma-1)\rho_0}\int dx \end{pmatrix} \rho_1.$$

In the context of acoustic heating, sound is dominative, and the corresponding thermal mode caused by it is secondary, with a magnitude of excess density much less than that of sound: $|\rho_{4,A}| \ll |\rho_{1,A}|$. As usual, we consider only the first acoustic mode in the role of acoustic source, assuming that the leftwards one and the third mode remain small compared to the dominative first mode, though they may enlarge their magnitudes in time. The quadratic nonlinear corrections in the sound mode that supports its adiabaticity, should be considered in the course of projecting. We make routine manipulations to decompose the evolution equation for the specific excess density of the entropy mode by means of projecting into the entropy mode, that is, by applying on the system (5.6) by the row $(d_1^4 \quad d_2^4 \quad d_3^4 \quad d_4^4)$ and collecting together terms of the leading order. The equation governing excess density of the entropy mode is

$$\frac{\partial \rho_4}{\partial t} = -\frac{2Bc}{\rho_0}\left(\rho_1^2 + \gamma \frac{\partial \rho_1}{\partial x}\int \rho_1 dx\right). \tag{5.10}$$

An acoustic excess density on the right-hand side must itself satisfy the dynamic equation (5.9). Eq. (5.10) is instantaneous. It applies to both periodic and aperiodic sound. The only limitations that allow us to simplify calculations are: high-frequency sound $\omega_1 \tau \gg 1$, weak attenuation of sound over its period $|B| \ll k$, and weak nonlinearity $M \ll 1$. A simple estimation of the acoustic source in the case of periodic sound may be done by use of a solution of the linear equation, into which the governing equation of sound (5.9) rearranges. Its periodic solution has the form

$$\rho_1 = M\rho_0 \sin(ck(t - x/c))\exp(Bx). \tag{5.11}$$

The acoustic heating is isobaric, $T_4 = -(T_0/\rho_0)\rho_4$, and Eq. (5.10) with excess acoustic density in the form (5.11) results in the leading order in the following equality averaged over the sound period excess temperature:

$$\left\langle \frac{\partial T_4}{\partial t} \right\rangle = -BT_0 c(\gamma - 1)M^2 \exp(2Bx) = -\frac{2BT_0(\gamma - 1)}{\rho_0 c^2} E(x), \tag{5.12}$$

where $\langle \phi(x,t) \rangle = \frac{ck}{2\pi} \int\limits_t^{t+2\pi/(ck)} \phi(x,t')dt'$, $E(x) = \langle p_1 v_1 \rangle$, is the intensity of sound and, T_4 is the excess background temperature, a quantity relating to the thermal mode. Acoustic periodic variations in temperature also contribute to the total perturbation in temperature. On average, they equal zero. The right-hand side of Eq. (5.12) is the acoustic source of heating or cooling depending on the sign of B. Its negative value for positive B guarantees cooling of a gas in the case of acoustically active gas. The temperature production is positive in the equilibrium regime like it takes place in the standard thermoviscous flows. In Newtonian fluids, the approximate periodic acoustic excess density and correspondent heating take the forms

$$\rho_1 = M\rho_0 \sin(ck(t - x/c))\exp(-\alpha x),$$

$$\left\langle \frac{\partial T_4}{\partial t} \right\rangle = \frac{\alpha T_0 c(\gamma - 1)M^2}{2} \exp(-2\alpha x) = \frac{\alpha T_0(\gamma - 1)}{\rho_0 c^2} E(x),$$

where $\alpha = \frac{bk^2}{2c\rho_0}$, and $b = 4\eta/3 + \zeta + \chi(\gamma - 1)^2/(\gamma R m_{mol})$ is the standard attenuation including that caused by the thermal conduction of a fluid. The similarity of the above equation and Eq. (5.12) is obvious if $B < 0$. The nonlinear Eq. (5.9) possesses the exact solution before forming a discontinuity [10]

$$\rho_1 = M\rho_0 \exp(Bx) \sum_{n=1}^{\infty} \frac{2J_n(nK(\exp(Bx)-1))\sin(n\omega(t-x/c))}{nK(\exp(Bx)-1)}, \tag{5.13}$$

where

$$K = \frac{(\gamma+1)M\omega}{2Bc}.$$

The variation in temperature associated with the thermal mode, in accordance to Eq. (5.13), is

$$\left\langle \frac{\partial T_4}{\partial t} \right\rangle = -BT_0 c(\gamma-1)M^2 \exp(2Bx) \sum_{n=1}^{\infty} \left(\frac{2J_n(nK(\exp(Bx)-1))}{nK(\exp(Bx)-1)} \right)^2. \tag{5.14}$$

The distance from the transducer, where the waveform breaks up (it always happens if $B > 0$), equals

$$x_b = \ln(1+1/K)B^{-1}.$$

It is useful to note that Eq. (5.9) simplifies to an equation that may be readily solved by the method of characteristics. Making use of variables

$$\tilde{t} = \exp(cBt) - 1, \quad \tilde{\rho} = \rho_1 \exp(-cBt)$$

and the retarded coordinate $\tilde{x} = x - ct$, one arrives at the leading-order equation that takes the form of the Earnshaw equation

$$\frac{\partial \tilde{\rho}}{\partial \tilde{t}} + \frac{\gamma+1}{2\rho_0}\tilde{\rho}\frac{\partial \tilde{\rho}}{\partial \tilde{x}} = 0. \tag{5.15}$$

The precise dynamics of the initially sawtooth impulse of the total length Λ has been established in the paper by one of the authors [12]. The density perturbation in the initial sawtooth impulse takes the form

$$\rho_1(\tilde{x},0) = \rho_{1,0} \begin{cases} 2\frac{\tilde{x}}{\Lambda}, & 0 \le \frac{\tilde{x}}{\Lambda} < 1/2 \\ 2\frac{\tilde{x}}{\Lambda} - 2, & 1/2 \le \frac{\tilde{x}}{\Lambda} < 1 \end{cases}$$

Its magnitude varies with time as

$$\rho_1(\tilde{x},t) = \rho_{1,0} \begin{cases} \frac{2\exp(cBt)}{1+\frac{\rho_{1,0}(\gamma+1)}{\rho_0 B\Lambda}(\exp(cBt)-1)} \frac{\tilde{x}}{\Lambda}, & 0 \le \frac{\tilde{x}}{\Lambda} < 1/2 \\ \frac{2\exp(cBt)}{1+\frac{\rho_{1,0}(\gamma+1)}{\rho_0 B\Lambda}(\exp(cBt)-1)} \left(\frac{\tilde{x}}{\Lambda} - 1\right), & 1/2 \le \frac{\tilde{x}}{\Lambda} < 1 \end{cases}.$$

In the neutral gases with $B = 0$ it overlaps with the dynamics of an impulse in an ideal gas. An impulse experiences a pure nonlinear attenuation and decays with the enlarging of distance from a transducer. In the case of $B < 0$, it attenuates faster. In the case of $B > 0$, an acoustical activity counteracts attenuation at the shock front, and a magnitude of an impulse tends to some constant (which equals $\frac{B\Lambda\rho_0}{(\gamma+1)}$) at the infinite distances from the transducer. The conclusions concern the finite or infinite series of impulses. The length of an individual impulse remains constant in the course of propagation.

In the case of a simple travelling wave in the role of acoustic source,

$$\rho_1 = M\rho_0\phi(kc(t-x/c)),$$

Equation (5.10) is readily integrated with the leading-order result

$$T_4(\eta) = \frac{2BT_0M^2}{k}\left(\gamma\phi\int_{-\infty}^{\eta}\phi(\eta')d\eta' - (\gamma-1)\int_{-\infty}^{\eta}\phi^2(\eta')d\eta'\right), \tag{5.16}$$

where $\eta = kct - kx$. Perturbations in the travelling waves tend to zero at minus infinity. Figure 5.1a is plotted in accordance to the formula (5.14), which is valid before the formation of discontinuity for $M = 0.2$. In all evaluations, $\gamma = 1.4$. The Figure 5.1b represents the numerical evaluations of an excess temperature caused by two symmetric pulses, $\phi = \exp(-\eta^2)$ (1) and $\phi = -(1+2\eta^2)^{-1}$ (2), which are plotted by the normal lines. The bold lines represent $\frac{kT_4(\eta)}{2BT_0M^2}$.

Taking into account thermal conductivity χ and the total viscosity due to shear, bulk, and thermal attenuation b results in corrections in the roots of the dispersion relation and further analysis. Two acoustic ones, under the condition $\omega\tau \gg 1$, are

$$\omega_1 = ck + \frac{ibk^2}{2\rho_0} + \frac{i}{2}\frac{(\gamma-1)^2T_0}{c^2}(\Phi_1 + Q_T),$$
$$\omega_2 = -ck + \frac{ibk^2}{2\rho_0} + \frac{i}{2}\frac{(\gamma-1)^2T_0}{c^2}(\Phi_1 + Q_T),$$

where $Q_T = (dQ/dT)_0$. The third root remains unchanged, and the last one takes the form

$$\omega_4 = i\frac{\chi k^2(\gamma-1)}{\gamma R m_{mol}\rho_0} + i\frac{(\gamma-1)Q_T T_0}{c^2}.$$

We consider the heat withdrawal, which depends exclusively on temperature T. The linear modes and projectors become also rearranged in accordance to the dispersion relations. Links for the rightwards progressive sound are given by the vector ψ_1 which involves linear terms associated with thermal and mechanical attenuation:

$$\psi_1 = \begin{pmatrix} v_1(x,t) \\ p_1(x,t) \\ \rho_1(x,t) \\ \varepsilon_1(x,t) \end{pmatrix} = \begin{pmatrix} \frac{c}{\rho_0} + \frac{(\gamma-1)^2(\Phi_1+Q_T)T_0}{2c^2\rho_0}\int dx - \frac{b}{2\rho_0^2}\frac{\partial}{\partial x} \\ c^2 + \frac{(\gamma-1)^2(\Phi_1+Q_T)T_0}{c}\int dx + \frac{\chi(\gamma-1)^2 c}{\gamma R m_{mol}\rho_0}\frac{\partial}{\partial x} \\ 1 \\ -\frac{(\gamma-1)\Phi_1 T_0}{c\rho_0}\int dx \end{pmatrix}\rho_1,$$

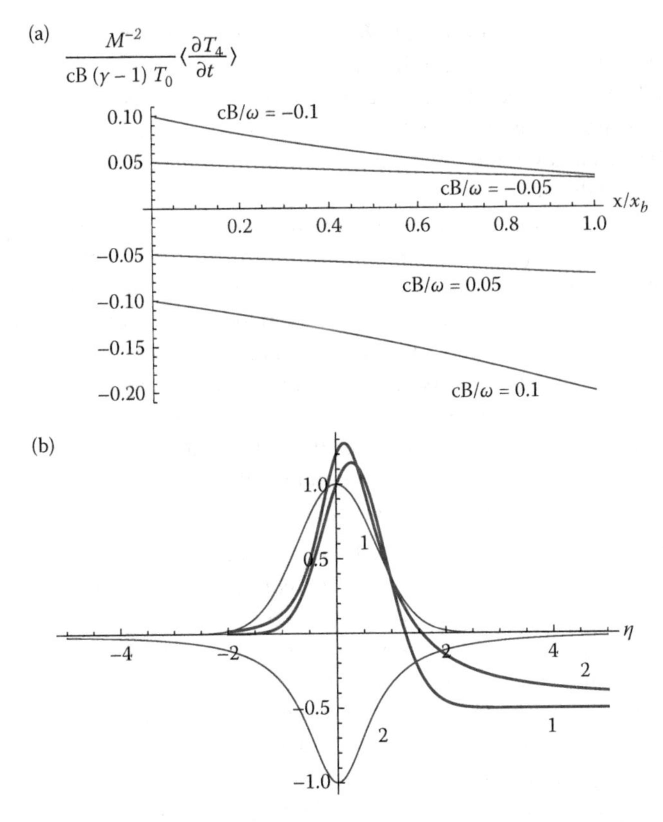

Figure 5.1 (a) Evaluations in accordance with Eq. (5.14) for different B and $M = 0.2$ before the formation of discontinuity in the wave profile. (b) $\frac{kT_4(\eta)}{2BT_0M^2}$ in accordance to Eq. (5.16) for $\phi = \exp(-\eta^2)$ (1) and $\phi = -(1+2\eta^2)^{-1}$ (2)

and the governing equation for sound (5.9) will include the term $-b/(2c^2\rho_0)\partial^2\rho_1/\partial x^2$ in the left-hand side. The equation for an excess density specific to the thermal mode (5.10), will be completed by the term $-\frac{\chi(\gamma-1)}{\gamma R m_{mol}\rho_0}\partial^2\rho_4/\partial x^2 + \frac{(\gamma-1)Q_T T_0}{c^2}\rho_4$ in its linear part. It will include also the quadratic nonlinear terms in the right-hand part, proportional to the thermal, viscous attenuation and Q_T, which are of minor importance.

Equation (5.10) is valid in any time for every type of sound, periodic or aperiodic. We remind the only limitations that have been used in order to simplify calculations. They are $ck\tau \gg 1$, $M \ll 1$, and $|B| \ll k$. The first of them restricts consideration to the high-frequency sound. The nonlinear phenomena produced by the low frequency sound ($ck\tau \ll 1$) were studied by one of the authors in the paper [13]. It was discovered that attenuation or amplification of sound itself is insignificant in this particular case. The nonlinear generation of the thermal mode by sound is also insignificant, at least in the field of periodic sound. As for the third, relaxation mode, it may enlarge efficiently in the field of the low-frequency sound. The acoustic heating (if $B < 0$, or cooling, if $B > 0$) is proportional to B, whereas heating due to the standard attenuation is proportional to the overall attenuation due to first, second viscosity and thermal conduction, b. The standard thermoviscosity always results to sound attenuation and ordinary heating in a nonlinear fluid flow. The larger frequency stipulates the larger thermal and viscous attenuation. One may neglect the standard attenuation comparatively to the nonequilibrium effects if $\omega \ll (\tau\tau_0)^{-1/2}$, where τ_0 is the average time of the molecular free pass. On the other hand, the high-frequency domain imposes inequality $\omega\tau \gg 1$. Both these conditions reduce the domain of sound frequencies where the standard attenuation is insignificant. For O_2 at room temperature, τ equals $10^8\tau_0$, and the condition of validity sounds: $10^4 \gg \omega\tau \gg 1$. In turn, the system of conservation equations in the differential form, which follows from the gaskinetic Boltzmann equation, is valid only under the condition $\omega\tau_0 \ll 1$. Otherwise, the starting point in studies should be the Boltzmann equation. For O_2, this condition gives $\omega\tau \ll 10^8$, which includes the domain above. Acoustic waves may be attenuated even in a nonequilibrium regime if the standard attenuation is large enough; for periodic sound, this is conditioned by the inequality $\frac{b\omega^2}{2\rho_0} + \frac{(\gamma-1)^2 T_0}{2}(\Phi_1 + Q_T) > 0$. With an increase of relaxation time, the amplification coefficient declines; however, a larger magnitude of pumping I is required to maintain the same degree of nonequilibrium, since $\varepsilon - \varepsilon_{eq} \approx I\tau$. This makes the nonequilibrium fluids inhomogeneous with the background temperature being a function of coordinate. The preceding conclusions are no longer valid in the case of intense pumping, because the linearization should be preceded with respect to the background with nonzero spatial gradients. This has impact on the very definition of modes. The perturbations are no longer simple sums of specific harmonics with constant magnitudes. The dispersion relations also could not be established in algebraic form over all the ranges of the wavenumbers. The problem becomes fairly complex mathematically. Some conclusions may be found in Ref. [14], which is devoted to the amplification of sound in a flat layer of nonequilibrium gas. Particularily, the area of instability in the plane pumping intensity and inverse time of relaxation becomes smaller. On the whole, the mathematical difficulties do not allow us to consider the problem in general.

In flows overcoming one dimension, the vortex mode may exist, along with, correspondingly, acoustic streaming, which is induced due to loss in momentum of dominative sound in flow with attenuation. We consider a two-dimensional flow in the plane OXY. The background quantities are constant in the plane OXY, and the transversal background quantities may vary only weakly due to the pumping and removal of energy. This allows us to apply the method of projecting in manner similar to the flows over the uniform background. The governing dynamic system with an account for quadratic nonlinear terms in two dimensions looks like

$$\frac{\partial v_x}{\partial t} + \frac{1}{\rho_0}\frac{\partial p'}{\partial x} = -\left(v_x\frac{\partial}{\partial x} + v_y\frac{\partial}{\partial y}\right)v_x + \frac{\rho'}{\rho_0^2}\frac{\partial p'}{\partial x},$$

$$\frac{\partial v_y}{\partial t} + \frac{1}{\rho_0}\frac{\partial p'}{\partial y} = -\left(v_x\frac{\partial}{\partial x} + v_y\frac{\partial}{\partial y}\right)v_y + \frac{\rho'}{\rho_0^2}\frac{\partial p'}{\partial y},$$

$$\frac{\partial p'}{\partial t}+\gamma p_0\left(\frac{\partial v_x}{\partial x}+\frac{\partial v_y}{\partial y}\right)-(\gamma-1)\rho_0\frac{\varepsilon'}{\tau}+(\gamma-1)\rho_0 T_0\Phi_1\left(\frac{p'}{p_0}-\frac{\rho'}{\rho_0}\right)$$

$$=-\left(v_x\frac{\partial}{\partial x}+v_y\frac{\partial}{\partial y}\right)p'-\gamma p'\left(\frac{\partial v_x}{\partial x}+\frac{\partial v_y}{\partial y}\right)+(\gamma-1)\rho'\left(\frac{\varepsilon'}{\tau}-T_0\Phi_1\left(\frac{p'}{p_0}-\frac{\rho'}{\rho_0}\right)\right) \tag{5.17}$$

$$-(\gamma-1)\rho_0\left(T_0\left(\frac{1}{\tau^2}\frac{d\tau}{dT}\right)_0\varepsilon'\left(\frac{p'}{p_0}-\frac{\rho'}{\rho_0}\right)\right.$$

$$\left.+T_0\Phi_1\left(\frac{\rho'^2}{\rho_0^2}-\frac{p'\rho'}{p_0\rho_0}\right)+T_0\Phi_2\left(\frac{p'}{p_0}-\frac{\rho'}{\rho_0}\right)^2\right),$$

$$\frac{\partial \rho'}{\partial t}+\rho_0\left(\frac{\partial v_x}{\partial x}+\frac{\partial v_y}{\partial y}\right)=-\left(v_x\frac{\partial}{\partial x}+v_y\frac{\partial}{\partial y}\right)\rho'-\rho'\left(\frac{\partial v_x}{\partial x}+\frac{\partial v_y}{\partial y}\right),$$

$$\frac{\partial \varepsilon'}{\partial t}+\frac{\varepsilon'}{\tau}-T_0\Phi_1\left(\frac{p'}{p_0}-\frac{\rho'}{\rho_0}\right)=T_0\left(\frac{1}{\tau^2}\frac{d\tau}{dT}\right)_0\varepsilon'\left(\frac{p'}{p_0}-\frac{\rho'}{\rho_0}\right)+T_0\Phi_1\left(\frac{\rho'^2}{\rho_0^2}-\frac{p'\rho'}{p_0\rho_0}\right)$$

$$+T_0\Phi_2\left(\frac{p'}{p_0}-\frac{\rho'}{\rho_0}\right)^2-\left(v_x\frac{\partial}{\partial x}+v_y\frac{\partial}{\partial y}\right)\varepsilon',$$

where v_x, v_y are components of velocity. The dispersion equation is algebraic of the fifth order:

$$\omega^2\left(\omega^3-i\left(\frac{1}{\tau}+\frac{\gamma(\gamma-1)T_0}{c^2}\Phi_1\right)\omega^2+c^2\widetilde{\Delta}\omega+i\left(\frac{c^2}{\tau}-(\gamma-1)T_0\Phi_1\right)\widetilde{\Delta}\right)=0, \tag{5.18}$$

where

$$\widetilde{\Delta}=-k^2=-k_x^2-k_y^2$$

corresponds to the two-dimensional Laplacian $\Delta=\frac{\partial^2}{\partial x^2}+\frac{\partial^2}{\partial y^2}$. The approximate roots of the dispersion equation for both acoustic branches under conditions of high-frequency sound and weak dispersion are well known [3,7]:

$$\omega_1=c\left(i\sqrt{\widetilde{\Delta}}-iB\right),\quad \omega_2=-c\left(i\sqrt{\widetilde{\Delta}}+iB\right), \tag{5.19}$$

where

$$B=-\frac{(\gamma-1)^2T_0}{2c^3}\Phi_1. \tag{5.20}$$

The non-wave roots of the dispersive equation, estimated without the preceding limitations $\omega\tau\gg 1$ and $|B|\ll k$, sound

$$\omega_3=i\left(\frac{1}{\tau}+\frac{(\gamma-1)(\gamma-c^2\widetilde{\Delta}\tau^2)T_0}{c^2(1-c^2\widetilde{\Delta}\tau^2)}\Phi_1\right),\quad \omega_4=0,\quad \omega_5=0. \tag{5.21}$$

The fifth root, which originates from non-planarity of a flow, associates with the vorticity mode. The relations for both acoustic branches determined by ω_1 and ω_2, are represented by $\tilde{\psi}_1$ and $\tilde{\psi}_2$:

$$\tilde{\psi}_i = \begin{pmatrix} \tilde{v}_{x,i} \\ \tilde{v}_{y,i} \\ \tilde{p}_i \\ \tilde{\varepsilon}_i \\ \tilde{\rho}_i \end{pmatrix} = \begin{pmatrix} -\frac{\omega_i k_x}{\rho_0 \Delta} \\ -\frac{\omega_i k_y}{\rho_0 \Delta} \\ c^2 + \frac{2Bc^3}{i\omega_i + A} \\ -\frac{2Bc^3}{(\gamma-1)\rho_0(i\omega_i + A)} \\ 1 \end{pmatrix} \tilde{\rho}_i, \quad A = \left(\frac{1}{\tau} + \frac{\gamma B c}{(\gamma-1)}\right), \quad i = 1,2. \tag{5.22}$$

The third mode is determined by the vector

$$\tilde{\psi}_3 = \begin{pmatrix} \tilde{v}_{x,3} \\ \tilde{v}_{y,3} \\ \tilde{p}_3 \\ \tilde{\varepsilon}_3 \\ \tilde{\rho}_3 \end{pmatrix} = \begin{pmatrix} -\frac{\omega_3 k_x}{\rho_0 \Delta} \\ -\frac{\omega_3 k_y}{\rho_0 \Delta} \\ -\frac{\omega_3^2}{\Delta} \\ \frac{1}{\rho_0(\gamma-1)}\left(c^2 + \frac{\omega_3^2}{\Delta}\right) \\ 1 \end{pmatrix} \tilde{\rho}_3, \tag{5.23}$$

and the fourth (i.e., the entropy) mode is given by the following vector $\tilde{\psi}_4$

$$\tilde{\psi}_4 = \begin{pmatrix} \tilde{v}_{x,4} \\ \tilde{v}_{y,4} \\ \tilde{p}_4 \\ \tilde{\varepsilon}_4 \\ \tilde{\rho}_4 \end{pmatrix} = \begin{pmatrix} 0 \\ 0 \\ 0 \\ -\frac{\tau T_0}{\rho_0}\Phi_1 \\ 1 \end{pmatrix} \tilde{\rho}_4. \tag{5.24}$$

All four first modes represent the potential velocity field:

$$\vec{\nabla} \times \vec{v}_i = \vec{0}, \; i = 1, \ldots 4. \tag{5.25}$$

The velocity of the fifth vortex mode is solenoidal,

$$\vec{\nabla} \cdot \vec{v}_5 = 0, \; p_5 = 0, \; \varepsilon_5 = 0, \; \rho_5 = 0. \tag{5.26}$$

In order to decompose the vorticity part from the overall vector of velocity, it is sufficient to apply the operator $\tilde{P}_{vort}$ on the vector of the Fourier transforms of velocity components:

$$\tilde{P}_{vort}\begin{pmatrix}\tilde{v}_x\\ \tilde{v}_y\end{pmatrix}=-\frac{1}{\tilde{\Delta}}\begin{pmatrix}k_y^2 & -k_xk_y\\ -k_xk_y & k_x^2\end{pmatrix}\begin{pmatrix}\tilde{v}_x\\ \tilde{v}_y\end{pmatrix}=\begin{pmatrix}\tilde{v}_{x,5}\\ \tilde{v}_{y,5}\end{pmatrix}. \tag{5.27}$$

Operating in the (x,y) space P_{vort} decomposes the vortex velocity

$$P_{vort}\Delta=\begin{pmatrix}\frac{\partial^2}{\partial y^2} & -\frac{\partial^2}{\partial x\partial y}\\ -\frac{\partial^2}{\partial x\partial y} & \frac{\partial^2}{\partial x^2}\end{pmatrix}. \tag{5.28}$$

In the physically interesting (and meaningful in technical applications) case of quasi-planar sound propagating in the direction of axis OX, let us assume that all acoustic perturbations vary much faster in the direction of axis OX than in the transversal direction pointed by OY: $k_x \gg k_y$. This is the standard assumption in the theory of acoustics as regards to acoustic beams [10,15]. This allows us to simplify mathematical content by expanding the relations for sound perturbations in the series of powers of the small parameter $\varepsilon = k_y^2/k_x^2$. For the first acoustic mode, the leading-order relationships (within an accuracy up to terms involving Φ_1^1, ε^1,) take the form

$$\begin{pmatrix}v_{x,1}(x,y,t)\\ v_{y,1}(x,y,t)\\ p_1(x,y,t)\\ \varepsilon_1(x,y,t)\\ \rho_1(x,y,t)\end{pmatrix}=\begin{pmatrix}\frac{c}{\rho_0}+\frac{c}{2\rho_0}\frac{\partial^2}{\partial y^2}\int dx\int dx-\frac{Bc}{\rho_0}\int dx\\ \frac{c}{\rho_0}\frac{\partial}{\partial y}\int dx\\ c^2-2Bc^2\int dx\\ \frac{2Bc^2}{(\gamma-1)\rho_0}\int dx\\ 1\end{pmatrix}\rho_1. \tag{5.29}$$

In view of relations (5.29), the linear equation governing an acoustic excess density in the wave propagating in the positive direction of axis OX in the leading order takes the form

$$\frac{\partial\rho_1}{\partial t}+c\frac{\partial\rho_1}{\partial x}+\frac{c}{2}\frac{\partial^2}{\partial y^2}\int^x\rho_1dx-cB\rho_1=0. \tag{5.30}$$

This coincides with the corresponding dispersion relation, in view of the leading-order equality $\sqrt{\tilde{\Delta}}\approx i\left(k_x+\frac{1}{2}\frac{k_y^2}{k_x}\right)$. The lower limit of integration should correspond to the physical meaning of a problem. In order to describe acoustic streaming, we compare as usual intense sound to all other modes. In particular, this means that the characteristic amplitude of velocity associated with the first branch of sound in the considered domain is much greater than that of other modes:

$$max|v_1|\gg max|v_n|,\ n=2,..,5.$$

We will keep only terms corresponding to the rightwards progressive sound in the nonlinear terms in all formulas below. By use of the links that specify sound and their approximate expressions in the quasi-planer geometry, the leading-order dynamic equations for sound are

General

$$\frac{\partial \rho_1}{\partial t}+c\sqrt{\Delta}\rho_1-cB\rho_1+\frac{1}{2}\left(\gamma\rho_1\vec{\nabla}\cdot\vec{v_1}+(\vec{v_1}\cdot\vec{\nabla})\rho_1\right)=0, \tag{5.31}$$

Quasi-planar

$$\frac{\partial \rho_1}{\partial t}+c\frac{\partial \rho_1}{\partial x}+\frac{c}{2}\frac{\partial^2}{\partial y^2}\int^x \rho_1 dx-cB\rho_1+\frac{(\gamma+1)c}{2\rho_0}\frac{\partial \rho_1}{\partial x}\rho_1=0.$$

In order to decompose the dynamic equations for the velocity of the vorticity mode, it is sufficient to apply the matrix operator P_{vort} (Eq. (5.27)) on the momentum equation (two first equations from the system (5.17)). As a result, all terms, corresponding to the potential velocity vector become reduced in the linear part, and the acoustic quadratic terms form a source of streaming. Application of P_{vort} yields the dynamic equation for the vorticity mode in the field of intense sound, in two equivalent forms,

$$\frac{\partial \vec{v}_v}{\partial t}=-\frac{1}{\rho_0}P_{vort}\left(\rho_a\frac{\partial \vec{v}_a}{\partial t}\right),\quad \frac{\partial \vec{\Omega}}{\partial t}=-\frac{1}{\rho_0}\vec{\nabla}\times\left(\rho_a\frac{\partial \vec{v}_a}{\partial t}\right). \tag{5.32}$$

where $\vec{\Omega}$ is the vorticity of a flow, $\vec{\Omega}=\vec{\nabla}\times\vec{v}_5$. Equations (5.32) are valid for all sound frequencies and for any geometry of the sound wave. Accounting for Eqs (5.22), (5.29), one may rearrange equations (5.32) in the case of high-frequency sound, in the general case, and in the case of the quasi-planar geometry of a flow, respectively:

General

$$\frac{\partial \vec{v}_5}{\partial t}=\frac{2Bc^3}{\rho_0^2}P_{vort}\left(\rho_a\int^t\vec{\nabla}\rho_a dt\right),\quad \frac{\partial \vec{\Omega}}{\partial t}=\frac{2Bc^3}{\rho_0^2}\left(\vec{\nabla}\rho_a\times\int^t\vec{\nabla}\rho_a dt\right), \tag{5.33}$$

Quasi-planar

$$\frac{\partial \vec{v}_v}{\partial t}=-\frac{2Bc^2}{\rho_0^2}P_{vort}\left(\rho_a\int^x\vec{\nabla}\rho_a dx\right),\quad \frac{\partial \vec{\Omega}}{\partial t}=\vec{F}_a=-\frac{2Bc^2}{\rho_0^2}\left(\vec{\nabla}\rho_a\times\int^x\vec{\nabla}\rho_a dx\right).$$

Vector $\vec{F}_a$ represents the acoustic source of the vorticity mode. Equations (5.33), accounting for relations between perturbations in density and velocity in the sound wave, include quadratic acoustic force proportional to B, which reflects that attenuation and nonlinearity are origins of streaming. In order to make progress in the derivation of equations describing effects of quasi-planar sound, we make use of the expansion of P_{vort} in series with respect to the small parameter responsible for a beam's divergence, ε, following from (5.27):

$$P_{vort}\approx\begin{pmatrix}\frac{\partial^2}{\partial y^2}\int dx\int dx & -\frac{\partial}{\partial y}\int dx\\ -\frac{\partial}{\partial y}\int dx & 1\end{pmatrix}. \tag{5.34}$$

Equations (5.33) refer to instantaneous quantities; hence, they are valid in any instant and apply to periodic as well as aperiodic sound. The sign of the right-hand acoustic source depends on sign of B. That differently describes streaming in normal and acoustically active gases. Equations (5.31), and (5.33) are fairly difficult to solve in view of nonlinearity in both equations. A simple estimation of the acoustic source in the case of periodic sound may be undertaken, making use of a solution of the linearized version of Eq. (5.31), which is valid in the case of the sound of an infinitely small magnitude and very weak transversal diffraction of a beam. Under these simplifying conditions, the periodic solution of Eq. (5.31) in an unbounded volume of a gas takes the form

$$\rho_1 = M\rho_0 \sin(\omega(t - x/c)) \exp(Bx - Y^2), \tag{5.35}$$

where $Y = y/L$ denotes the dimensionless coordinate and L designates the characteristic transversal width of the sound beam. Substituting an excess density (5.35) in the last equation in (5.33), in view of $\omega\tau \gg 1$ and making use of the definition of A given by Eq. (5.22), yields the leading-order dynamic equation for vorticity in the field of periodic sound

$$\frac{\partial \Omega_z}{\partial t} = -\frac{4M^2 Bc^2 \omega^2}{(A^2 + \omega^2)L} Y \exp(2(Bx - Y^2)) \approx -\frac{4M^2 Bc^2}{L} Y \exp(2(Bx - Y^2)). \tag{5.36}$$

The acoustic force on the right-hand side of Eq. (5.36) represents the mean field. We did not use averaging over the sound period at any stage during its derivation.

In normal flows, if $B < 0$, the sign of acoustic force on the right-hand side of Eq. (5.36) is positive for $x \geq 0$ and decreases with x. The production of vorticity occurs similarly as in the case of a fluid with standard attenuation due to the nonlinear coupling of sound and vorticity in viscous flows. If $B = 0$, there is no nonlinear generation of vorticity by sound. In acoustically active flows with $B > 0$, the acoustic source is negative and decreases with enlargement in x. This reflects the anomalous character of acoustic streaming compared with the equilibrium regime when $B < 0$. In particular, the direction of streamlines becomes opposite.

The main results of this section are the instantaneous equations governing the vorticity mode, its velocity or vorticity, Eq. (5.33). The only limitations used to simplify evaluations are the domain of high-frequency sound $\omega\tau \gg 1$ and its weak attenuation or amplification over the wavelength, $|B| \ll k$. They are valid in temporally and spatially confined domains, where sound remains dominant with respect to other modes (vorticity, entropy, and vibrational). It is helpful to estimate B for a typical laser mixture $CO_2 : N_2 : He = 1:2:3$ at normal conditions $p_0 = 1atm = 101325Pa$, $T = 300$K. The density of this mixture is $\rho_0 = 0{,}76kg \cdot m^{-3}$, $\tau = 5 \cdot 10^{-5}s$, $\frac{T}{\tau}\frac{d\tau}{dT} = -3{,}4$. The value of B depends on the pumping intensity $I \approx (\varepsilon - \varepsilon_{eq})/\tau$; the threshold quantity is $I_{th} \cdot \rho_0 = 1{,}5 \cdot 10^6 W \cdot m^{-3}$. If $I \cdot \rho_0 = 10^9 W \cdot m^{-3}$, $B = 3{,}3m^{-1}$, if $I \cdot \rho_0 = 10^8 W \cdot m^{-3}$, $B = 0{,}3m^{-1}$. For intensity less than the threshold quantity, B takes negative values. It varies from $B = -0{,}005m^{-1}$ for zero I till $B = 0$ for $I = I_{th}$. Attenuation (if $B < 0$), or amplification (if $B > 0$) and the nonlinear generation of the vorticity mode by it, which were considered in this section, occur exclusively due to relaxation processes. The terms reflecting thermal and viscous (standard, Newtonian) attenuation of a gas should complement the momentum and energy equations in the system (5.2). They originate from the stress tensor and the energy flux associated with thermal conductivity. The larger frequency stipulates the larger thermal and viscous attenuation. Dispersion relations specifying both acoustic branches and vorticity mode depend on viscosity. This last takes the form

$$\omega_5 = i\frac{\eta\tilde{\Delta}}{\rho_0}. \tag{5.37}$$

This corrects the equation governing vorticity in the field of sound,

$$\frac{\partial\vec{\Omega}}{\partial t} - \frac{\eta\Delta\vec{\Omega}}{\rho_0} = \frac{2Bc^3}{\rho_0^2}\left(\vec{\nabla}\rho_a \times \int \vec{\nabla}\rho_a dt\right). \tag{5.38}$$

The dynamics of sound depends on the total diffusivity of sound β due to the first, second viscosities and thermal conductivity:

$$\frac{\partial\rho_1}{\partial t} + c\sqrt{\Delta}\rho_1 - cB\rho_1 - \frac{\beta}{2}\Delta\rho_1 + \frac{1}{2}\left(\gamma\rho_1\vec{\nabla}\cdot\vec{v_1} + (\vec{v_1}\cdot\vec{\nabla})\rho_1\right) = 0. \tag{5.39}$$

Equations (5.38), (5.39) are more difficult to solve than (5.31), (5.33).

5.2 CHEMICALLY REACTING GASES

The close-in-spirit example is the nonllinear phenomena in chemically reacting gases. We follow the discussions and results obtained in Refs [16,17]. In this section, we consider the generation of the entropy and chemical modes in a gas where reversible or irreversible chemical reaction occurs. Two limiting cases of high- or low-frequency sound compared with the inverse time of a chemical reaction, have been considered in Ref. [16] in connection with acoustic heating. In the chemically reacting gases, the nonlinear distortion of low-frequency sound and its nonlinear coupling with other modes are fairly weak. This is very similar to the case of gases with excited vibrational degrees of molecules freedom. In view of that, we will pay attention to the high-frequency sound and nonlinear phenomena in its field.

The momentum, energy, and continuity equations in a gas where a simple exoteric chemical reaction of $A \to B$ type takes place, read in the standard designations of velocity, pressure, density and "frozen" heat capacity at constant volume:

$$\rho\frac{d\vec{v}}{dt} = -\vec{\nabla}p, \tag{5.40}$$

$$\frac{C_{V,\infty}}{R}\frac{dT}{dt} - \frac{T}{\rho}\frac{d\rho}{dt} = Q,$$

$$\frac{d\rho}{dt} + \rho\vec{\nabla}\cdot\vec{v} = 0.$$

In the system of equations (5.40), T is the temperature measured in Joules per molecule, $Q = HmW/\rho$ the heat produced in the medium per one molecule due to a chemical reaction, H denotes the reaction enthalpy per unit mass of reagent A, and m denotes the averaged molecular mass of a gas. The dynamic equation of the mass fraction Y of reagent A and the equation of state complement the system (5.40) is

$$\frac{dY}{dt}=-\frac{W}{\rho},\quad p=\frac{\rho T}{m}. \tag{5.41}$$

We start from studies of one-dimensional flow along axis OX, recalling all steps of standard procedure. Equations (5.40) for velocity and the peturbations of temperature and mass concentration read with an account for Eq. (5.41) within the accuracy of quadratic nonlinear terms:

$$\frac{\partial v}{\partial t}+\frac{T_0}{m\rho_0}\frac{\partial \rho'}{\partial x}+\frac{1}{m}\frac{\partial T'}{\partial x}=-v\frac{\partial v}{\partial x}+\frac{T_0\rho'}{m\rho_0^2}\frac{\partial \rho'}{\partial x}-\frac{T'}{m\rho_0}\frac{\partial \rho'}{\partial x}, \tag{5.42}$$

$$\frac{\partial T'}{\partial t}+(\gamma-1)\left(T_0\frac{\partial v'}{\partial x}-Q_T\frac{Q_0}{T_0}T'-Q_\rho\frac{Q_0}{\rho_0}\rho'-Q_Y\frac{Q_0}{Y_0}Y'\right)=-v'\frac{\partial T'}{\partial x}-(\gamma-1)T'\frac{\partial v'}{\partial x},$$

$$\frac{\partial Y'}{\partial t}+\frac{1}{Hm}\left(Q_T\frac{Q_0}{T_0}T'+Q_\rho\frac{Q_0}{\rho_0}\rho'+Q_Y\frac{Q_0}{Y_0}Y'\right)=-v\frac{\partial Y'}{\partial x},$$

$$\frac{\partial \rho'}{\partial t}+\rho_0\frac{\partial v'}{\partial x}=-v'\frac{\partial \rho'}{\partial x}-\rho'\frac{\partial v'}{\partial x},$$

where γ denotes the frozen adiabatic exponent and the dimensionless quantities Q_T, Q_ρ, Q_Y are proportional to the partial derivatives of the heat produced due to a chemical reaction with respect to its variables:

$$Q_T=\frac{T_0}{Q_0}\left(\frac{\partial Q}{\partial T}\right)_{T_0,\rho_0,Y_0},\quad Q_\rho=\frac{\rho_0}{Q_0}\left(\frac{\partial Q}{\partial \rho}\right)_{T_0,\rho_0,Y_0},\quad Q_Y=\frac{Y_0}{Q_0}\left(\frac{\partial Q}{\partial Y}\right)_{T_0,\rho_0,Y_0}. \tag{5.43}$$

Q_0 denotes the equilibrium value of Q. Four dispersion relations $\omega_i(k)$ with ordering numbers from $i=1$ till 4 specify different types of fluid motion: two wave modes ($i=1,2$), the relaxation mode ($i=3$), and the entropy mode ($i=4$). They take the leading-order forms

$$\omega_1=c(k-iB),\ \omega_2=c(-k-iB),\ \omega_3=i\left(\frac{1}{\tau_c}+\frac{(\gamma-1)Q_0(Q_\rho-Q_T)}{c^2m}\right),\ \omega_4=0, \tag{5.44}$$

where

$$B=\frac{Q_0(\gamma-1)(Q_\rho+(\gamma-1)Q_T)}{2c^3m}, \tag{5.45}$$

$c = \sqrt{\frac{\gamma T_0}{m}}$ denotes the frozen equilibrium sound speed, and

$$\tau_c = \frac{HmY_0}{Q_0 Q_Y} \tag{5.46}$$

is the characteristic duration of a chemical reaction.

The first two roots in (5.44): ω_1, ω_2, are acoustic, and ω_4 corresponds to the thermal (entropy) mode. The third non-acoustic root ω_3 is responsible for the non-wave variation in mass fraction of reagent A. For sound to be a wave process, attenuation (when B is negative, or amplification, if B is positive) should be small compared with the characteristic acoustic wavenumber, that is, $|B| \ll k$. This in fact determines the smallness of $|Q_\rho|$, $|Q_T|$ more precisely depending on the characteristic domain of sound wavenumbers.

The approximate roots of dispersion relations for both acoustic branches in the case of weak attenuation (amplification) of small-scale sound with $kc\tau_c \gg 1$ (or, equivivalently, high-frequency sound with $\omega\tau_c \gg 1$), were first derived in Ref. [18]. We will also consider weak variations of heat release in a chemical reaction with temperature and density, which guarantees also weak attenuation/amplification and a dispersion of sound over its period, that is

$$max(|Q_\rho|, |Q_T|) \ll \frac{\gamma T_0}{Q_0 \tau_c}.$$

All formulae below are derived within an accuracy up to the first powers of small parameters $(kc\tau_c)^{-1}$, $max(|Q_\rho|, |Q_T|)Q_0\tau_c/(\gamma T_0)$. The acoustic wave enhances if

$$Q_\rho + (\gamma - 1)Q_T > 0. \tag{5.47}$$

In this case, a gas is acoustically active as it has been established in Ref. [18]. The corresponding links for the different modes that are established by the dispersion relations take the leading-order forms

$$\psi_1 = \begin{pmatrix} v_1 \\ T_1 \\ Y_1 \\ \rho_1 \end{pmatrix} = \begin{pmatrix} \frac{c}{\rho_0} - \frac{cB}{\rho_0}\int dx \\ \frac{(\gamma-1)T_0}{\rho_0} - \frac{2\gamma T_0 B}{\rho_0}\int dx \\ \frac{2Bc^2}{(\gamma-1)H\rho_0}\int dx \\ 1 \end{pmatrix} \rho_1, \quad \psi_2 = \begin{pmatrix} -\frac{c}{\rho_0} - \frac{cB}{\rho_0}\int dx \\ \frac{(\gamma-1)T_0}{\rho_0} + \frac{2\gamma T_0 B}{\rho_0}\int dx \\ -\frac{2Bc^2}{(\gamma-1)H\rho_0}\int dx \\ 1 \end{pmatrix} \rho_2, \tag{5.48}$$

$$\psi_3 = \begin{pmatrix} \left(\frac{(\gamma-1)Q_0(Q_\rho - Q_T)}{\rho_0 T_0 \gamma_\infty} + \frac{1}{\tau_c \rho_0}\right)\int dx \\ -\frac{T_0}{\rho_0} \\ \frac{c^2}{H(\gamma-1)\rho_0} \\ 1 \end{pmatrix} \rho_3, \quad \psi_4 = \begin{pmatrix} 0 \\ -\frac{T_0}{\rho_0} \\ -\frac{\tau_c Q_0(Q_\rho - Q_T)}{Hm\rho_0} \\ 1 \end{pmatrix} \rho_4.$$

The linear equations for any mode may be decomposed directly from the system (5.42), or they may be derived from the foundation of the dispersion relations.

The dynamic equations that govern the excess densities specifying different modes are as follows:

$$\frac{\partial\rho_1}{\partial t}+c\frac{\partial\rho_1}{\partial x}-cB\rho_1=0, \tag{5.49}$$

$$\frac{\partial\rho_4}{\partial t}=0, \quad \frac{\partial\rho_3}{\partial t}+\left(\frac{Q_0(\gamma-1)(Q_\rho-Q_T)}{T_0\gamma}+\frac{1}{\tau_c}\right)\rho_3=0.$$

Equations (5.49) coincide with the roots of the dispersion equation, Eqs (5.44). The linear links specific for sound should be complemented by the second-order nonlinear terms making it isentropic in the leading order. The corrected links specific for the first acoustic mode are:

$$v_1=\frac{c}{\rho_0}\rho_1-\frac{cB}{\rho_0}\int\rho_1 dx+\frac{(\gamma-3)c}{4\rho_0^2}\rho_1^2, \tag{5.50}$$

$$T_1=\frac{(\gamma-1)T_0}{\rho_0}\rho_1-\frac{2c^2mB}{\rho_0}\int\rho_1 dx+\frac{(\gamma-1)(\gamma-2)T_0}{2\rho_0^2}\rho_1^2.$$

As usual, the nonlinear corrections in Eqs (5.50) establish relationships specific for the planar Riemann wave [10,19] propagating over an ideal gas. One may readily conclude that an excess acoustic density satisfies the equation

$$\frac{\partial\rho_1}{\partial t}+c\frac{\partial\rho_1}{\partial x}+\frac{(\gamma+1)c}{2\rho_0}\rho_1\frac{\partial\rho_1}{\partial x}-cB\rho_1=0, \tag{5.51}$$

which looks similar to the Earnshaw equation [10], but includes the term $-cB\rho_1$ responsible for the attenuation or amplification of sound, depending on the sign of B. It also coincides with Eq. (5.9), which describes the nonlinear distortions of sound in a gas with vibrational relaxation. As usual, we assume that the magnitudes of non-wave modes are small compared with those of sound, $|T_3|,|T_4|\ll max|T_1|$. Applying projecting, that is, multiplying the first equation from the system (5.42) by 0 and the second, third, and fourth equations by corresponding factors:

$$-\frac{\rho_0}{mc^2}+\frac{\rho_0 Q_0(Q_\rho-Q_T)(\gamma-1)\tau_c}{m^2c^4}, \quad \frac{\rho_0 H(\gamma-1)(-mc^2+Q_0(Q_\rho-Q_T)(\gamma-1)\tau_c)}{mc^4}, \tag{5.52}$$

$$1-\frac{1}{\gamma}-\frac{(\gamma-1)^2Q_0(Q_\rho-Q_T)\tau_c}{mc^2\gamma}$$

respectively, and taking the sum of all equations reduces all terms belonging to the first, second, and third modes in the linear part of the final equation. Only acoustic quadratic terms are considered in the quadratic nonlinear part. This yields the equation for an excess temperature attributable to the entropy mode:

$$\frac{\partial T_4}{\partial t}=2c(\gamma-1)B\frac{T_0}{\rho_0^2}\frac{\partial\rho_1}{\partial x}\int\rho_1 dx. \tag{5.53}$$

Equation (5.53) describes heating or cooling generated by an acoustic source, which represents the right-hand side of it. In order to reduce all terms besides those belonging to the third mode in the linear part, one should multiply the first equation of the system (5.42) by the factor

$$\frac{Q_0(\gamma-1)\left(Q_\rho+(\gamma-1)Q_T\right)}{\gamma c^2}\int dx,$$

and the second, third, and fourth equations by the factors

$$\frac{\tau_c Q_0(\gamma-1)(Q_T-Q_\rho)}{\gamma^2 T_0}, \quad -\frac{Hm(\gamma-1)}{\gamma}-\frac{\tau_c Q_0 H(\gamma-1)^2(Q_T-Q_\rho)}{\gamma c^2},$$

$$\frac{\tau_c Q_0(\gamma-1)^2(Q_T-Q_\rho)}{\gamma^2 \rho_0},$$

respectively. Taking the sum of all the equations and keeping only acoustic quadratic terms, one arrives finally to the dynamic equation

$$\frac{\partial T_3}{\partial t}+\left(\frac{1}{\tau_c}+\frac{Q_0(\gamma-1)(Q_\rho-Q_T)}{\gamma T_0}\right)T_3=-\frac{Q_0(\gamma-1)\left[(\gamma-3)Q_\rho-(\gamma-1)^2Q_T\right]}{4\gamma\rho_0^2}\rho_1^2. \tag{5.54}$$

It describes an excess temperature attributable to the non-acoustic third mode, which associates with relaxation during the chemical reaction. An excess acoustic density itself satisfies the nonlinear Eq. (5.51). Its solution, which corresponds to the periodic at the transducer sound, $\rho_1(x=0,t)=M\rho_0\sin(\omega t)$, is represented by Eq. (5.13), which is valid at distances from the transducer that are smaller than the characteristic distance of wave breakup, x_b. Weak attenuation imposes that this be small as compared to the characteristic acoustic wavenumber k, $|B|\ll k$. It is reasonable to evaluate the following integral approximately:

$$\int f(Bx)F(\omega(t-x/c))dx \approx f(Bx)\int F(\omega(t-x/c))dx. \tag{5.55}$$

Eq. (5.14) describes gases with vibrationally excited degrees of a molecule's freedom (with the corresponding coefficient of B). One can readily conclude that the ambient temperature associated with the entropy mode increases in time when coefficient B is negative, and decreases otherwise. The rate of increase (or decrease) in temperature depends on the modulus of B and on the distance from a transducer. In acoustically active flow, when $B>0$, sound enhances in the course of propagation. This is the case of acoustic cooling.

In order to describe the vortex flow, we start to consider two-dimensional flow in the plane OXY. The system, which includes quadratic nonlinear terms, is

$$\frac{\partial v_x}{\partial t}+\frac{T_0}{m\rho_0}\frac{\partial \rho'}{\partial x}+\frac{1}{m}\frac{\partial T'}{\partial x}+(\vec{v}\cdot\vec{\nabla})v_x=\frac{T_0\rho'}{m\rho_0^2}\frac{\partial \rho'}{\partial x}-\frac{T'}{m\rho_0}\frac{\partial \rho'}{\partial x}, \tag{5.56}$$

$$\frac{\partial v_y}{\partial t}+\frac{T_0}{m\rho_0}\frac{\partial \rho'}{\partial y}+\frac{1}{m}\frac{\partial T'}{\partial y}+(\vec{v}\cdot\vec{\nabla})v_y=\frac{T_0\rho'}{m\rho_0^2}\frac{\partial \rho'}{\partial y}-\frac{T'}{m\rho_0}\frac{\partial \rho'}{\partial y},$$

$$\frac{\partial T'}{\partial t}+(\gamma-1)\left[(T_0+T')(\vec{\nabla}\cdot\vec{v})-Q_T\frac{Q_0}{T_0}T'-Q_\rho\frac{Q_0}{\rho_0}\rho'-Q_Y\frac{Q_0}{Y_0}Y'\right]=-(\vec{v}\cdot\vec{\nabla})T',$$

$$\frac{\partial Y'}{\partial t}+\frac{1}{Hm}\left[Q_T\frac{Q_0}{T_0}T'+Q_\rho\frac{Q_0}{\rho_0}\rho'+Q_Y\frac{Q_0}{Y_0}Y'\right]=-(\vec{v}\cdot\vec{\nabla})Y',$$

$$\frac{\partial \rho'}{\partial t}+\rho_0(\vec{\nabla}\cdot\vec{v})+=-(\vec{v}\cdot\vec{\nabla})\rho'-\rho'(\vec{\nabla}\cdot\vec{v}),$$

where $\vec{\nabla}$ in two dimensions denotes $\vec{i}\partial/\partial x+\vec{j}\partial/\partial y$ ($\vec{i}$ and $\vec{j}$ are the corresponding basis vectors of unit length). We consider the high-frequency sound in the role of acoustic source, also in the quasi-planar geometry.

Analysis starts from the establishment of dispersion relations. The general form of dispersion equation is

$$\omega^2\left(\omega^3-\frac{i}{\tau_c}\left(1-\frac{(\gamma-1)Q_0\tau_cQ_T}{T_0}\right)\omega^2+c^2\widetilde{\Delta}\omega\right.$$
$$\left.-\frac{ic^2}{\tau_c}\left(1+\frac{(\gamma-1)Q_0\tau_c(Q_\rho-Q_T)}{\gamma T_0}\right)\widetilde{\Delta}\right)=0. \quad (5.57)$$

The dispersion relations establish modes of a flow. In particular, the relations for both acoustic branches determined by ω_1, ω_2 are represented by $\tilde{\psi}_1$, $\tilde{\psi}_2$ operating in the space of Fourier transforms:

$$\tilde{\psi}_i=\begin{pmatrix}\tilde{v}_{x,i}\\ \tilde{v}_{y,i}\\ \tilde{T}_i\\ \tilde{Y}_i\\ \tilde{\rho}_i\end{pmatrix}=\begin{pmatrix}-\frac{\omega_i k_x}{\widetilde{\Delta}\rho_0}\\ -\frac{\omega_i k_y}{\widetilde{\Delta}\rho_0}\\ \frac{T_0}{\rho_0}\left(\gamma-1+\frac{2\gamma cB\tau_c}{1+i\omega_i\tau_c}\right)\\ -\frac{2Bc^3\tau_c}{(\gamma-1)H\rho_0(1+i\omega_i\tau_c)}\\ 1\end{pmatrix}\tilde{\rho}_i. \quad (5.58)$$

Both approximate high-frequency acoustic roots coincide with Eqs (5.19) with B given by Eq. (5.45). The last non-wave roots of the dispersion equation are zero; they correspond to the entropy ($\omega_4=0$) and vortex ($\omega_5=0$) modes.

The projector that distinguishes the vortex mode takes the form Eq. (5.28), and equations that govern an excess density of sound are Eqs (5.31). As for acoustic streaming, it also is described by Eqs (5.32), (5.33). Analysis of the nonlinear effects related to the low-frequency sound may be found in Ref. [20]. The conclusion is that it is fairly ineffective in generation of non-wave modes. The nonlinear distortions of sound are also insignificant. Note that we have disregarded terms involving second-order derivatives of Q in Eqs (5.42), (5.56), such as $\partial^2Q/\partial T^2$. This restricts the accuracy of final conclusions.

5.2.1 REMARKS ON THE THERMAL SELF-FOCUSING OF SOUND

First, the self-action of the optic waves was discovered. This occurs due to variations in the refraction index of a medium in the course of light-wave propagation. As regards to acoustics, the first theoretical results were reviewed in Ref. [21], and the first experiments confirming the theory were described in Refs [22,23]. The qualitative and quantitative predictions of the theory have been confirmed in many experimental studies [24,25]. If the sound velocity increases with temperature, the normal attenuation leads to the defocusing of a beam becuase losses in sound energy increase the ambient temperature. Considerable attention was paid to the thermal self-action of quasi-harmonic sound. The comprehensive review by Rudenko and Sapozhnikov [26] concentrated on the thermal self-action of beams containing shock fronts in media with quadratic and cubic nonlinearities.

The instantaneous equations (5.10), (5.53) may be used in the evaluations of the thermal self-action of the acoustic beam in the course of its propagation in a relaxing medium, including an acoustically active one. As an example, we reproduce the system that accounts for the thermal self-action in a gas with a chemical reaction. It describes thermal self-action in an axially symmetric flow of chemically reacting thermoconducting gas [16,26]:

$$\frac{\partial}{\partial \tau}\left(\frac{\partial p}{\partial x} - \frac{Td}{c}\frac{\partial p}{\partial \tau} - \frac{\varepsilon}{c^3\rho_0}p\frac{\partial p}{\partial \tau} - Bp\right) = \frac{c}{2}\Delta_\perp p. \tag{5.59}$$

$$\frac{\partial T}{\partial t} - \frac{\chi}{\rho_0 C_p}\Delta_\perp T = \frac{2BT_0(\gamma-1)}{c^3\rho_0^2}\frac{\partial p}{\partial x}\int p'(x,t)\,dx. \tag{5.60}$$

In the usual notation, x and r are cylindrical coordinates, the axis x coincides with the axis of a beam, and $\tau = t - x/c$ is the retarded time. $\Delta_\perp$ denotes the Laplacian with respect to the radial coordinate r, and $d = (\partial c/\partial T)_p/c$ is the thermal coefficient.

$$\varepsilon = \frac{B}{2A} + 1 = \frac{1 - D_1 - D_2}{2} \tag{5.61}$$

is the coefficient of nonlinearity of a medium (A and B are determined by Eqs (3.2)). The system (5.59), (5.60) describes instantaneous thermal self-action, which is valid for any type of quasi-planar acoustic beam. The details of self-action of acoustic beam in a gas with reversible or irreversible chemical reactions are discussed in Ref. [27]. The study considers stationary and non-stationary self-focusing (or defocusing) in the geometrical approach, which eliminates diffraction. The theory describes perturbations in the paraxial region. The Gaussian impulses are considered with the characteristic width at the transducer a_0. The acoustic pressure in a beam is saw-like with the period $2\pi/\omega$. The results of simulations depend on two dimensionless parameters, Π and X_s:

$$\Pi = \frac{6B\chi c^3\rho_0}{(\gamma-1)C_P P_0^2}, \quad X_s = \frac{2\rho_0 c^3 B\pi}{P_0(\gamma+1)\omega}, \tag{5.62}$$

where P_0 is the initial magnitude of acoustic pressure at the axis of a beam. They are both positive or both negative depending on the sign of B. The discontinuity does not form at all if $X_s < -1$. In this case, the assumption of strong nonlinearity is not longer valid, and the sawtooth wave can not be considered. For definiteness, evaluations relate to an ideal gas with $\varepsilon = \frac{\gamma+1}{2}$ and $d = 1/2T_0$. In acoustically active media with $B > 0$, the thermal self-action of a beam occurs anomalously. For times much larger than t_0, where

$$t_0 = \frac{\rho_0 C_p a_0^2}{12\chi}$$

is the characteristic time of temperature establishment, the temporal derivative in the heat transport equation may be put at zero. This corresponds to the stationary thermal self-action of a beam. Figure 5.2 shows the stationary ratio of magnitude of an acoustic beam and its initial value at the axis of a beam in the two cases,

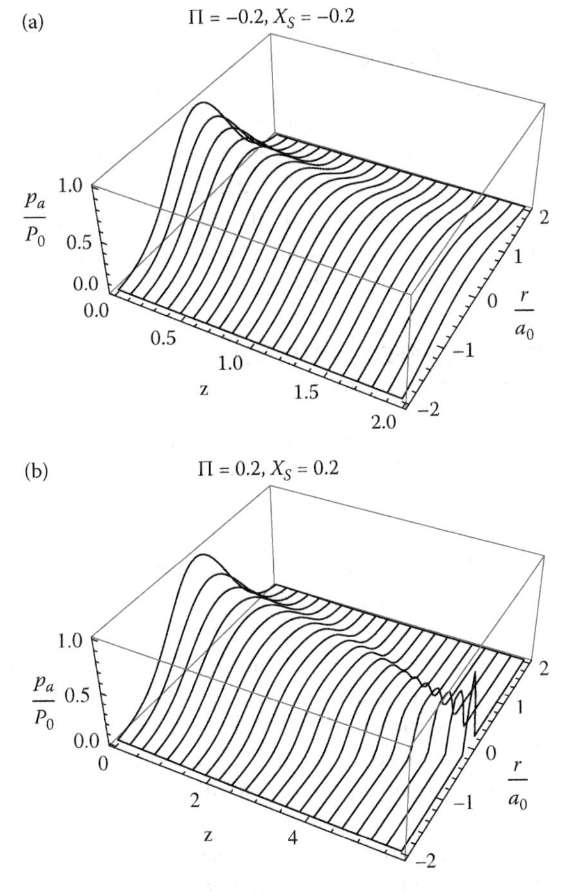

Figure 5.2 Magnitude of acoustic pressure in a beam with initially planar front equilibrium ($\Pi < 0, X_s < 0$) and non-equilibrium ($\Pi > 0, X_s > 0$) stationary regimes

corresponding to the normal attenuation (a) and acoustically active gas (b). A beam is planar at the transducer situated in $z = 0$. The dimensionless variable z is determined to be $z = \frac{x}{x_0}$, where

$$x_0 = \frac{6\chi c_0^3 \rho_0}{(\gamma - 1) C_P P_0^2}$$

is some characteristic length depending on thermodynamical properties of the gas and initial magnitude of acoustic pressure in the sawtooth wave.

Some important features of thermal self-action of the shock sound beams propagating over a gas with relaxation are drawn out in Ref. [27]. The width of a beam always decreases in acoustically active gas (perhaps somewhat increasing in the vicinity of a transducer), but the magnitude of acoustic pressure in a saw-like wave may increase or decrease along the axis of a beam. This reflects two contrary mechanisms: the enlargment in the magnitude of acoustic pressure and the nonlinear attenuation at the fronts of saw-like sound, which becomes stronger with enlargement in sound magnitude. The nonlinear broadening of a beam can be explained by flattening the transverse beam profile due to stronger absorption near the axis. In the nonstationary regime, the thermal lens becomes stronger with time and the focal point moves towards the transducer. Near the nonlinear focus, the width of a beam vanishes and amplitude infinitely grows. In this region, the description becomes inadequate because it does not take into account the divergence that originates from diffraction. These features of sound beams with discontinuities are very similar to those in the majority of Newtonian liquids, which are focusing due to negative d and thermoviscous absorption of the sound energy [24,26].

5.3 THE NONLINEAR EFFECTS OF SOUND IN A LIQUID WITH RELAXATION LOSSES

Two examples that were considered in the previous sections belong to the thermodynamical relaxation of different kinds but fairly similar as concerns to mathematical description of nonlinear distortion of sound and its nonlinear effects. The model of the Maxwell relaxation is also one of the samples with the frequency-dependent linear sound speed and attenuation of sound. The standard Newtonian model may be considered a limit of the Maxwellian one at low frequencies. Various kinds of relaxation follow from the particular type of the stress tensor of a medium and yield different frequency dependence. In turn, the nonlinear phenomena may indicate the kind of relaxation in a fluid making use of sound with different frequences. Relaxation is known to be the most important cause of attenuation in tissue. The following example we consider, is the sound absorption in electrolytes. It attributes an excess sound absorption due to a chemical relaxation. We refer to the classical dynamic equations together with the appropriate thermodynamic and constitutive relations that describe a flow in an electrolyte, making use of Refs [28,29]. The nonlinear phenomena of sound have been considered by Perelomova in Ref. [30]. We do not consider mechanic viscosity and heat conduction, which are well studied in gases. As for liquids and biological tissues, they are small compared to relaxation effects [31].

The starting point is the set of equations for conservation of mass and for conservation of momentum in the usual notations. The Gibbs relation for the rate of variation of entropy s in a fluid describes irreversible thermodynamics:

$$T\frac{Ds}{Dt} = \frac{De}{Dt} - \frac{p}{\rho}\frac{D\rho}{Dt} + \sum_{\nu} A_{\nu}\frac{Dn_{\nu}}{Dt},$$

where A_{ν} are affinities, and n_{ν} represent the number of molecules of species ν per unit mass, as in the equations for the chemical reactions in an electrolyte [29,32]. Following Eigen and Tamm [29], we assume that all relaxation processes are independent or have been expressed in a normal coordinate system, which yields the effective independent mechanisms. The entropy is represented by the sum of an equilibrium part (marked by the upper subscript e), and of an irreversible part:

$$s(p,\rho,\nu_{\nu}) = s^{e}(p,\rho) + \sum_{\nu}\Delta s_{\nu},$$

with

$$\Delta s_{\nu} = \frac{C_p k_{\nu}}{\beta T}\Delta\xi_{\nu}.$$

We make use of quantities $\Delta\xi_{\nu}$, which are defined as

$$\Delta\xi_{\nu} = \frac{n_{\nu} - n_{\nu}^{e}(p,T)}{\partial n_{\nu}^{e}(p,T)/\partial p}.$$

The quantities $\Delta\xi_{\nu}$ satisfy the equation of relaxation:

$$\left(\frac{D}{Dt} + \frac{1}{\tau_{\nu}}\right)\Delta\xi_{\nu} = -\frac{Dp}{Dt}. \quad (5.63)$$

Making use of the equality

$$\rho A_{\nu}\frac{Dn_{\nu}}{Dt} = k_{\nu}\frac{(\Delta\xi_{\nu})^2}{\tau_{\nu}},$$

the entropy-balance equation with an account for the momentum equation takes the leading-order form

$$\frac{Dp}{Dt} - c^2\frac{D\rho}{Dt} - \frac{B}{2}\frac{D\rho^2}{Dt} + \rho c^2\sum_{\nu}\frac{D}{Dt}(k_{\nu}\Delta\xi_{\nu}) = \sum_{\nu}\frac{k_{\nu}\beta c^2}{\tau_{\nu}C_p}(\Delta\xi_{\nu})^2,$$

where c denotes the equilibrium sound speed, b is the thermal expansion determined by Eqs (3.4), and quantities A and B are defined by equalities (3.2). The starting system follows from the conservation equations, which account for only the quadratic nonlinear terms:

$$\frac{\partial\vec{v}}{\partial t} + \frac{\vec{\nabla}p'}{\rho_0} = -(\vec{v}\cdot\vec{\nabla})\vec{v} + \frac{\rho'}{\rho_0^2}\vec{\nabla}p',$$

$$\frac{\partial p'}{\partial t}+\rho_0c^2(\vec{\nabla}\cdot\vec{v})+\rho_0c^2\sum_\nu\frac{\partial}{\partial t}(k_\nu\Delta\xi_\nu)=-(\vec{v}\cdot\vec{\nabla})p'-\frac{A+B}{\rho_0}(\vec{\nabla}\cdot\vec{v})\rho'$$
$$+\sum_\nu\frac{k_\nu\beta c^2}{\tau_\nu C_p}(\Delta\xi_\nu)^2-\rho'c^2\sum_\nu\frac{\partial}{\partial t}(k_\nu\Delta\xi_\nu)-\rho_0c^2\sum_\nu(\vec{v}\cdot\vec{\nabla})(k_\nu\Delta\xi_\nu), \quad (5.64)$$
$$\frac{\partial\rho'}{\partial t}+\rho_0(\vec{\nabla}\cdot\vec{v})=-\rho'\left(\vec{\nabla}\cdot\vec{v}\right)-(\vec{v}\cdot\vec{\nabla})\rho',$$
$$\frac{\partial\Delta\xi_\nu}{\partial t}+\frac{\Delta\xi_\nu}{\tau_\nu}+\frac{\partial p'}{\partial t}=-(\vec{v}\cdot\vec{\nabla})p'-(\vec{v}\cdot\vec{\nabla})\Delta\xi_\nu \quad (\nu=1,\dots N).$$

N is the amount of relaxation modes. The linear version of this system has been derived in Ref. [28]. We repeat the standard procedure of establishing dispertion relations and modes in a flow. The number of dispersion relations equals five plus the number of relaxation processes, N, in three dimensions. For simplicity, only one relaxation process with constant k_ν will be considered, ordered as ν-th with the corresponding time of relaxation. There are two branches of sound ($i=1$ and $i=2$), and four non-wave modes (the entropy mode, $i=3$, the relaxation mode, $i=4$ and two vorticity branches, $i=5$ and $i=6$). The dispersion relations ω_i ($i=1,\dots,6$) are as follows:

$$\omega_{1,2}=\pm c\mathbf{k}+ik_\nu\frac{c^4\mathbf{k}^2\rho_0\tau_\nu}{2(1\pm ic\mathbf{k}\tau_\nu)}, \quad \omega_3=0, \quad \omega_4=\frac{i}{\tau_\nu}\left(1+\frac{c^2\rho_0k_\nu}{1+c^2\mathbf{k}^2\tau_\nu^2}\right), \quad \omega_{5,6}=0, \quad (5.65)$$

where $\mathbf{k}$ denotes the module of the wave vector $\vec{k}$. We consider weakly dispersive flows, which are characterized by the small parameter responsible for relaxation,

$$\alpha=c^2\rho_0k_\nu\ll 1.$$

Both vorticity modes are solenoidal:

$$\vec{\nabla}\cdot\vec{v}_{5,6}=0, \quad p_{5,6}=0, \quad \rho_{5,6}=0.$$

In particular, links between the Fourier transformations of perturbations in density and pressure for both acoustic modes are as follows:

$$\tilde{p}_1=c^2\tilde{\rho}_1+\frac{\alpha c^3\mathbf{k}\tau_\nu}{-i+c\mathbf{k}\tau_\nu}\tilde{\rho}_1; \quad \tilde{p}_2=c^2\tilde{\rho}_2+\frac{\alpha c^3\mathbf{k}\tau_\nu}{i+c\mathbf{k}\tau_\nu}\tilde{\rho}_2.$$

Let us consider the one-dimensional flow along axis OX. In one dimension, there exist four first modal fields; the total small-magnitude perturbations represent their sum, and their Fourier transforms represent the sums of specific Fourier transforms:

$$\tilde{v}=\sum_{i=1}^4\tilde{v}_i=\sum_{i=1}^4\frac{\omega_i\tilde{\rho}_i}{k\rho_0}, \quad \tilde{p}=\sum_{i=1}^4\tilde{p}_i=\sum_{i=1}^4\frac{\omega_i^2\tilde{\rho}_i}{k^2}, \quad \tilde{\rho}=\sum_{i=1}^4\tilde{\rho}_i,$$

$$\widetilde{\Delta\xi_\nu}=\sum_{i=1}^4\widetilde{\Delta\xi_{\nu,i}}=-i\sum_{i=1}^4\frac{\omega_i^4\tilde{\rho}_i}{k^2(i\omega_i+1/\tau_\nu)},$$

where $\tilde{v} = \tilde{v}_x$, $k = k_x$. The projecting rows into the third and fourth modes are readily established by a solution of algebraic equations operating in the space of Fourier transforms:

$$P_3(\tilde{v} \quad \tilde{p} \quad \tilde{\rho} \quad \tilde{\Delta}\xi_v)^T = \tilde{\rho}_3, \quad P_4(\tilde{v} \quad \tilde{p} \quad \tilde{\rho} \quad \tilde{\Delta}\xi_v)^T = \tilde{\rho}_4. \tag{5.66}$$

They take the forms

$$P_3 = \begin{pmatrix} 0 & -\dfrac{1}{c^2} & 1 & -\dfrac{\alpha}{c^2} \end{pmatrix},$$

$$P_4 = \begin{pmatrix} \dfrac{i\alpha c^2 k^3 \rho_0 \tau_v^3}{(1+c^2k^2\tau_v^2)^2} & \dfrac{\alpha \rho_0 c^2 k^4 \tau_v^4}{(1+c^2k^2\tau_v^2)^2} & 0 & \dfrac{\alpha k^2 \tau_v^2}{1+c^2k^2\tau_v^2} \end{pmatrix}.$$

The projecting rows are evaluated with an accuracy up to terms proportional to α^0 and α^1 but without any restrictions concerning the spatial spectrum of perturbations, as well as dispersion relations. The elements of P_4 determine some integro-differential spatial operators in the space (x,t). The equation that governs an excess density in the first planar sound mode may be readily reconstructed by use of Eqs (5.65):

$$\left(1 - c\tau_v \frac{\partial}{\partial x}\right)\left(\frac{\partial \rho_1}{\partial t} + c\frac{\partial \rho_1}{\partial x} + \frac{c\varepsilon}{\rho_0}\rho_1 \frac{\partial \rho_1}{\partial x}\right) = \frac{\alpha c^2 \tau_v}{2}\frac{\partial^2 \rho_1}{\partial x^2}. \tag{5.67}$$

Its approximate and exact solutions in the case of weak or strong relaxation have been discussed in Ref. [15]. The exact solution in the Ref. [15] relates to a stationary waveform that propagates with the equilibrium speed c. Actually, there is an infinite variety of stationary waveforms that take the form of a shock wave and propagate with the speed $\tilde{c}$ greater than c_∞, where c_∞ denotes the frozen sound speed:

$$\tilde{c} > c_\infty, \quad c_\infty = (1 + 0.5\alpha)c.$$

The density jump in a shock wave equals

$$\Delta\rho_{sh} = \frac{2\rho_0(\tilde{c} - c)}{c\varepsilon}.$$

It is always larger than $\alpha\rho_0/\varepsilon$.

As for acoustic streaming, it follows from making use of P_{vort}. Its application at the momentum equation with account of quadratic acoustic terms leads to equations describing vorticity at low-frequency and high-frequency limits,

$$\textbf{low-frequency:} \quad \frac{\partial \vec{\Omega}}{\partial t} = -\frac{\alpha c^3 \tau_v}{\rho_0}\vec{\nabla} \times \left(\rho_1 \cdot \vec{\nabla}\sqrt{\Delta}\rho_1\right),$$

$$\textbf{high-frequency:} \quad \frac{\partial \vec{\Omega}}{\partial t} = \frac{\alpha c}{\rho_0 \tau_v}\vec{\nabla} \times \left(\rho_1 \cdot \vec{\nabla}\Delta^{-1}\rho_1\right).$$

In order to describe acoustic heating, we consider the one-dimensional flow along axis OX. Links for the perturbations in the first acoustic mode take the leading-order forms

$$\tilde{v}_1 = \frac{c}{\rho_0}\left(1+\alpha\frac{ck\tau_v}{2}\right)\tilde{\rho}_1, \quad \Delta\xi_{v,1} = \frac{c^3 k\tau_v}{i-ck\tau_v}\left(1+\frac{\alpha ck\tau_v}{2(-i+ck\tau_v)^2}\right)\tilde{\rho}_1,$$

which allow us to express an acoustic source in terms of the disturbance of density attributable to the acoustic mode. The leading-order low-frequency and high-frequency limits of these relations are, respectively:

$$\textbf{low-frequency:} \quad \tilde{v}_1 = \frac{c}{\rho_0}\tilde{\rho}_1, \quad \widetilde{\Delta\xi}_v = -ic^3\tau_v k\tilde{\rho}_1, \tag{5.68}$$

$$\textbf{high-frequency:} \quad \tilde{v}_1 = \frac{c_\infty}{\rho_0}\tilde{\rho}_1, \quad \widetilde{\Delta\xi}_v = -c_\infty^2\tilde{\rho}_1 - i\frac{c}{k\tau_v}\tilde{\rho}_1.$$

Applying P_3 from (5.66) on the system (5.64) reduces all acoustic, vortex, and relaxation terms in its linear part and yields the leading-order equations that govern acoustic heating:

$$\textbf{low-frequency:} \quad \frac{\partial \rho_3}{\partial t} = -\frac{\alpha\beta c^4\tau_v}{C_p\rho_0}\left(\frac{\partial \rho_1}{\partial x}\right)^2 + \frac{\alpha c}{2\rho_0}\frac{\partial \rho_1^2}{\partial x},$$

$$\textbf{high -frequency:} \quad \frac{\partial \rho_3}{\partial t} = -\frac{\alpha\beta c^2}{\tau_v C_p\rho_0}\rho_1^2.$$

An excess density that specifies the entropy mode decreases with time. This corresponds to the isobaric heating of the background. The first term in the low-frequancy source is of major importance in the case of almost periodic sound. In turn, applying P_4 and making use of relations (5.68), one arrives at the leading-order dynamic equations for the specific relaxation excess density in the low-frequency and high-frequency regimes:

$$\textbf{low-frequency:} \quad \frac{\partial \rho_4}{\partial t} + \frac{1+\alpha}{\tau_v}\rho_4 = 0,$$

$$\textbf{high -frequency:} \quad \frac{\partial \rho_4}{\partial t} + \frac{\rho_4}{\tau_v} = -\frac{2\alpha c}{\tau_v}\left(1+\frac{A+B}{\rho_0 c^2}\right)\rho_1^2.$$

There is insignificant coupling of the relaxation mode with sound in the low-frequency domain.

The considered case of relaxation in liquids differs from the relaxation due to vibrationally excited internal degrees of a molecule's freedom in gases (or relaxation due to exothermic chemical reactions, which parallels that in the vibrationally excited gases). The relaxation equation in liquids, Eq. (5.63), includes on its right-hand side the partial derivative of pressure with respect to time, while the corresponding equation describing a disturbance of apparent vibrational temperatures, T', takes the form

$$\left(\frac{D}{Dt}+\frac{1}{\tau}\right)T' = \frac{T'}{\tau} = \frac{T_0}{\tau}\left(\frac{p'}{p_0}-\frac{\rho'}{\rho_0}\right),$$

whose right-hand side is proportional to excess temperature, T', but not to its temporal derivative (τ denotes the vibrational relaxation time). In view of the algebraic thermodynamic relation linking excess temperature with excess pressure and density, the right-hand side of this equation includes perturbations in pressure and density, but not their derivatives with respect to time. The difference in the types of relaxation in liquids and gases affects both linear and nonlinear characteristics of a flow. This concerns the dispersion relations for sound and the entropy mode, and hence, the linear dynamic equations describing these modes. The modes also look different. As for the high-frequency sound, the dispersion relations of sound in a liquid readily follow from Eqs. (5.65):

$$\omega_{1,2} = \pm(c\mathbf{k} + 0.5k_V c^3 \mathbf{k}\rho_0),$$

but the dispersion relations of sound in a gas with vibrational relaxation determine the frequency-independent attenuation of the high-frequency sound in accordance to Eqs (5.4). We reproduce it in the case of the absence of the inflow of energy:

$$\omega_{1,2} = \pm c\mathbf{k} + 0.5i(\gamma - 1)^2 C_V T_0/(\tau c^2).$$

The dispersion relation that specifies relaxation mode in the gases with vibrational relaxation also looks different when compared with ω_4 from Eqs (5.65):

$$\omega_4 = \frac{i}{\tau} + \frac{i(\gamma-1)(\gamma + c^2\mathbf{k}^2\tau^2)T_0 C_V}{c^2\tau(1 + c^2\mathbf{k}^2\tau^2)}.$$

It determines frequency-dependent behavior differently, especially at intermediate wavenumbers of perturbation. The relaxation mode is isobaric in the case of a gas with excited vibrational degrees of a molecule's freedom, but in the case of relaxation in a liquid, it is not. The Fourier transform of perturbation in pressure, associated with the relaxation mode, takes a leading-order form in terms of the Fourier transform of its excess density:

$$\tilde{p}_4 = -\left(1 + \frac{2c^2\rho_0 k_V}{1 + c^2\mathbf{k}^2\tau_V^2}\right)\frac{\tilde{\rho}_4}{\mathbf{k}^2\tau_V^2}.$$

The nonlinear effects of sound are also different in gases and liquids with relaxation. As for the high-frequency sound in a vibrationally excited gas, the acoustic force of heating is proportional to $-\gamma\frac{\partial\rho_1}{\partial x}\int\rho_1 dx - \rho_1^2$ [33], not to $-\rho_1^2$, as in electrolytes. Hence, heating in the media of these different types of relaxation is fairly similar in the case of periodic sound, but it is different for aperiodic sound, including impulses. The acoustic force of the relaxation mode in gases is proportional to $\rho_1\frac{\partial\rho_1}{\partial x}$ in the high-frequency regime [34], not to ρ_1^2. Hence, the nearly periodic sound in a gas is hardly expected to produce a noticeable perturbation in vibrational energy. As for the low frequencies of sound, the conclusion is that acoustic streaming, acoustic heating, and generation of the relaxation mode due to both types of relaxation are fairly weak effects.

Variations of the thermodynamic quantities associated with nonlinear phenomena of sound, such as excess temperature and bulk velocity of a fluid, may be readily measured. These slow variations may be useful in establishing the details of relaxation processes in a liquid. The results also may be employed to model nonlinear scattering in applications such as image reconstruction in a liquid.

5.4 ON THE NONLINEAR EFFECTS OF MAGNETOACOUSTIC PERTURBATIONS IN A PERFECTLY CONDUCTING VISCOUS AND THERMOCONDUCTING GAS

Nonlinear magnetohydrodynamic phenomena in conducting fluids are of importance in many applications of cosmic physics, geophysics, plasma physics, physics of controlled thermonuclear fusion, and hypersonic aerodynamics. It has been established that sound velocity in an electrically conducting fluid varies in the presence of a magnetic field in the 1950s [35,36]. The first conclusions were that the finite conductivity introduces dispersion and absorption of sound planar waves, which propagate perpendicularly to the magnetic field [37]. Many studies have been devoted to magneto hydrodynamic waves in perfectly conducting gases and also to their nonlinear effects. There are evident successes on the way of establishing exact solutions of the system of PDEs, which describe one-dimensional unsteady planar and cylindrically symmetric motions in magnetohydrodynamics (MHD), involving solutions with discontinuities [38,40]. The structure of MHD shock waves in a viscous compressible plasma has been studied in Ref. [41], and in a viscous non-ideal gas in Ref. [42]. The nonlinear interaction of magnetohydrodynamic waves has been considered by numerous authors [43–45]. Attention to the nonlinear transfer of energy between magnetohydrodynamic waves and other types of waves in a plasma has been attracted by Ponomarev. Three-wave interactions of kinetic Alfvén and magnetosound waves were studied by Brodin in Ref. [46]. Amplification of Alfvén waves due to the nonlinear interaction with a magnetoacoustic wave was discovered recently in Ref. [47]. The review by I.Ballai, which considers also nonlinear resonant waves, summarizes the knowledge on nonlinear waves in solar plasmas [48]. A general description of the dynamics of MHD perturbartions is fairly difficult in view of the strong dependence on the angle between the magnetic strength and particle's velocity. That imposes also a strong dependence on the angle between the magnetic strength and wave vector, which may vary in the space.

The nonlinear theory of MHD is required in many problems of MHD dynamics, in particular, in prediction of nonlinear losses in magnetoacoustic energy and momentum and their transfer to non-wave motions. It has been discovered that the dissipation of sound is difficult to explain by linear damping [49]. In studies of the nonlinear coupling of different modes, we start as usual from weakly nonlinear equations that govern a planar flow of a gas along x-axis. Absorption due to the shear viscosity of plasma η and its thermal conduction σ, is taken into account. We assume the electrical conductivity to be infinite and the magnetic field $\vec{H} = (0,0,H(x,t))$ with the magnetic field strength H, orthogonal to the trajectories of gas particles, that is, to their velocity $\vec{v} = (v(x,t),0,0)$. The starting point represents conservation equations [50]:

$$\frac{\partial \rho}{\partial t} + \frac{\partial (\rho v)}{\partial x} = 0,$$

for mass,

$$\rho\left(\frac{\partial v}{\partial t}+v\frac{\partial v}{\partial x}\right)+\frac{\partial p}{\partial x}+\frac{\partial h}{\partial x}=\frac{4\eta}{3}\frac{\partial^2 v}{\partial x^2},$$

for momentum,

$$\rho T\left(\frac{\partial s}{\partial t}+v\frac{\partial s}{\partial x}\right)=\frac{\partial}{\partial x}\left(\sigma\frac{\partial T}{\partial x}\right)+\frac{4\eta}{3}\left(\frac{\partial v}{\partial x}\right)^2, \tag{5.69}$$

for entropy s, and

$$\frac{\partial h}{\partial t}+v\frac{\partial h}{\partial x}+2h\frac{\partial v}{\partial x}=0,$$

for the magnetic pressure h,

$$h=\mu H^2/2,$$

and μ denotes the magnetic permeability. We suppose that viscosity and thermal conductivity are fixed fractions of their unmagnetized values and obey power-law dependence on temperature:

$$\eta=\kappa\cdot T^{\alpha},\quad \sigma=\chi\cdot T^{\alpha}.$$

In a fully ionized plasma, $\alpha=5/2$ and κ, χ are some constants [51–53]. The last equation from system (5.69) follows from the equations

$$\frac{\partial\vec{H}}{\partial t}-\vec{\nabla}\times(\vec{v}\times\vec{H})=\vec{0},\quad \vec{\nabla}\cdot\vec{H}=0$$

and condition of perpendicularity of magnetic field and velocity of a gas, $\vec{H}\cdot\vec{v}=0$ [39]. The leading-order system of dynamic equations follows from Eqs (5.69); it includes quadratic nonlinear terms:

$$\frac{\partial\rho}{\partial t}+\rho_0\frac{\partial v}{\partial x}=-\rho_0\frac{\partial v}{\partial x}-v\frac{\partial\rho}{\partial x},$$

$$\frac{\partial v}{\partial t}-\delta_1\frac{\partial^2 v}{\partial x^2}+\frac{1}{\rho_0}\frac{\partial p'}{\partial x}+\frac{1}{\rho_0}\frac{\partial h'}{\partial x}=-v\frac{\partial v}{\partial x}+\frac{\rho'}{\rho_0^2}\frac{\partial p'}{\partial x}$$
$$+\frac{\rho'}{\rho_0^2}\frac{\partial h'}{\partial x}-\frac{\delta_1\rho'}{\rho_0}\frac{\partial^2 v}{\partial x^2}+\frac{\alpha\delta_1(\gamma p'-c_0^2\rho')}{\gamma p_0}\frac{\partial^2 v}{\partial x^2},$$

$$\frac{\partial p'}{\partial t}-\frac{\gamma\delta_2}{\gamma-1}\frac{\partial^2 p'}{\partial x^2}+\frac{\delta_2 c_0^2}{\gamma-1}\frac{\partial^2\rho'}{\partial x^2}+c_0^2\rho_0\frac{\partial v}{\partial x}=-v\frac{\partial p'}{\partial x}-\gamma p'\frac{\partial v}{\partial x}+ \tag{5.70}$$
$$\delta_1(\gamma-1)\rho_0\left(\frac{\partial v}{\partial x}\right)^2-\frac{\delta_2}{\rho_0(\gamma-1)}\frac{\partial^2\left(\gamma p'\rho'-c_0^2\rho'^2\right)}{\partial x^2}+\frac{\alpha\delta_2}{(\gamma-1)\gamma p_0}\left(\frac{\partial(\gamma p'-c_0^2\rho')}{\partial x}\right)^2,$$

$$\frac{\partial h'}{\partial t}+2h_0\frac{\partial v}{\partial x}=-v\frac{\partial h'}{\partial x}-2h'\frac{\partial v}{\partial x},$$

where δ_1, δ_2 are coefficients responsible for viscous and thermal effects, and c_0 is the speed of sound of an infinitely small magnitude, which relates to an ideal gas and is evaluated at unperturbed thermodynamic state p_0, ρ_0 in the absence of magnetic field):

$$\delta_1 = \frac{4\kappa T_0^\alpha}{3\rho_0}, \quad \delta_2 = \left(\frac{1}{C_V} - \frac{1}{C_p}\right)\frac{\chi T_0^\alpha}{\rho_0}.$$

The dispersion relations describing all independent modes follow from the linearized version of Eqs (5.69). In a planar flow of a magnetic fluid, the dispersion relations take the leading-order form

$$\omega_{1,2} = \pm c_{m,0}k + \frac{i}{2}\left(\delta_1 + \frac{c_0^2}{c_{m,0}^2}\delta_2\right)k^2, \quad \omega_3 = i\left(\frac{\delta_2}{\gamma-1} + \frac{\delta_2(c_{m,0}^2 - c_0^2)}{c_{m,0}^2}\right)k^2,$$
$$\omega_4 = 0, \tag{5.71}$$

where

$$c_m = \sqrt{c^2 + c_A^2}, \quad c_A = \sqrt{2h/\rho}$$

designate the magnetosonic speed and the Alfvén speed, respectively, and $c_{m,0}$ denotes c_m at the unperturbed state p_0, ρ_0. The first two roots ω_1, ω_2 correspond to the magnetosonic waves of different directions of propagation (fast MHD waves), the third root ω_3 corresponds to the entropy mode, and the last one, ω_4, corresponds to the Alfvén wave in the flow where magnetic field is perpendicular to the particles velocity. The first three roots in Eqs (5.71) are calculated with an accuracy up to terms proportional to the first powers of thermoviscous coefficients δ_1, δ_2. Two last dispersion relations are zero in a fluid without thermal conduction. In this case, there are two degenerate eigenvalues and more than one linearly independent eigenvector corresponding to each of them. Thermal conductivity eliminates this degeneracy. The total perturbation is represented by a sum of perturbations that specifies every mode:

$$v = \sum_{i=1}^{4} v_i = \frac{c_{m,0}}{\rho_0}\rho_1 - \left(\frac{\delta_1}{2\rho_0} + \frac{\delta_2 c_0^2}{2\rho_0 c_{m,0}^2}\right)\frac{\partial \rho_1}{\partial x} - \frac{c_{m,0}}{\rho_0}\rho_2 - \left(\frac{\delta_1}{2\rho_0} + \frac{\delta_2 c_0^2}{2\rho_0 c_{m,0}^2}\right)\frac{\partial \rho_2}{\partial x} -$$
$$\frac{\delta_2(\gamma c_{m,0}^2 - c_0^2(\gamma-1))}{c_{m,0}^2(\gamma-1)\rho_0}\frac{\partial \rho_3}{\partial x},$$
$$p' = \sum_{i=1}^{4} p_i = c_0^2\rho_1 - \delta_2\frac{c_0^2}{c_{m,0}}\frac{\partial \rho_1}{\partial x} + c_0^2\rho_2 + \delta_2\frac{c_0^2}{c_{m,0}}\frac{\partial \rho_2}{\partial x} - (c_{m,0}^2 - c_0^2)\rho_3 + \frac{c_0^2}{\gamma}\rho_4, \tag{5.72}$$
$$h' = \sum_{i=1}^{4} h_i = (c_{m,0}^2 - c_0^2)\rho_1 + (c_{m,0}^2 - c_0^2)\rho_2 + (c_{m,0}^2 - c_0^2)\rho_3 - \frac{c_0^2}{\gamma}\rho_4.$$

The entropy mode is isobaric in a thermoconducting MHD flow, in the sense that summary excess pressure, which is a sum of thermodynamic and magnetic parts, is zero. This applies also to the Alfvén mode. The projecting rows, which distinguish excess densities corresponding to third and fourth roots,

$$d_3 \begin{pmatrix} \rho' \\ v \\ p' \\ h' \end{pmatrix} = \rho_3, \quad d_4 \begin{pmatrix} \rho' \\ v \\ p' \\ h' \end{pmatrix} = \rho_4,$$

take the forms

$$d_3 = \begin{pmatrix} \frac{c_0^2}{\gamma c_{m,0}^2-(\gamma-1)c_0^2} \\ -\frac{\delta_2 \rho_0 c_0^2}{c_{m,0}^2}\frac{\partial}{\partial x} \\ -\frac{1}{c_{m,0}^2} \\ \frac{(\gamma-1)c_0^2}{c_{m,0}^2(\gamma c_{m,0}^2-(\gamma-1)c_0^2)} \end{pmatrix}^T, \quad d_4 = \begin{pmatrix} \frac{\gamma(c_{m,0}^2-c_0^2)}{\gamma c_{m,0}^2-(\gamma-1)c_0^2} \\ 0 \\ 0 \\ -\frac{\gamma}{c_{m,0}^2(\gamma c_{m,0}^2-(\gamma-1)c_0^2)} \end{pmatrix}^T. \tag{5.73}$$

They are evaluated with an accuracy up to terms proportional to the first powers of δ_1, δ_2.

We consider nonlinear effects associated with dominative sound, for definiteness, exclusively with the first acoustic mode. It corresponds to ω_1. As usual, for a correct description of the nonlinear effects of sound, the linear acoustic modes should be completed by the leading-order nonlinear terms, which do not include thermoviscous coefficients. This corrected vector in inviscid and non-heat conducting gas takes the form:

$$\psi_1 + \psi_{1,nl} = \begin{pmatrix} \frac{\rho_0}{c_{m,0}} \\ 1 \\ \frac{\rho_0 c_0^2}{c_{m,0}} \\ \frac{\rho_0(c_{m,0}^2-c_0^2)}{c_{m,0}} \end{pmatrix} v_1 + \begin{pmatrix} A \\ 0 \\ B \\ C \end{pmatrix} v_1^2. \tag{5.74}$$

The unknown constants A, B, and C are readily established. They ensure four equivalent leading-order dynamic nonlinear equations for magnetoacoustic perturbations, when substituted into Eqs (5.70) with zero δ_1, δ_2. Solving algebraic equations, one arrives at

$$A = \frac{c_{m,0}^2 - c_0^2(\gamma-2)}{4c_{m,0}^4}\rho_0, \quad B = \frac{c_0^2(c_{m,0}^2(2\gamma-1) - c_0^2(\gamma-2))}{4c_{m,0}^4}\rho_0,$$

$$C = \frac{(c_{m,0}^2 - c_0^2)(3c_{m,0}^2 + c_0^2(\gamma-2))}{4c_{m,0}^2}\rho_0.$$

These constants coincide with the well-known nonlinear corrections that make the progressive Riemann's wave isentropic [10]. If h_0 tends to zero (as is the case of the Riemann's wave),

$$A = -\frac{(\gamma-3)\rho_0}{4c_0^2}, \quad B = \frac{\gamma+1}{4}\rho_0, \quad C = 0.$$

The nonlinear equation, which includes quadratic nonlinearity and governs velocity in the first magnetoacoustic planar wave, takes the form

$$\frac{\partial v_1}{\partial t} + c_{m,0}\frac{\partial v_1}{\partial x} + \varepsilon_m v_1 \frac{\partial v_1}{\partial x} = 0, \tag{5.75}$$

where

$$\varepsilon_m = \frac{3c_{m,0}^2 + c_0^2(\gamma - 2)}{2c_{m,0}^2}.$$

This coincides with the Earnshaw equation when h_0 tends to zero and ε_m tends to $\varepsilon = \frac{\gamma+1}{2}$ [10]. Waveforms with discontinuity may be established by use of a conservation of total magnetosound momentum. Eq. (5.75) has been derived and used for the description of propagation of a saw-tooth impulse in Ref. [39] (Eq. (12) therein with $m = 0$, which corresponds to the planar geometry of a flow). The equation, which accounts for nonlinear and thermoviscous effects, recalls the Burgers equation [10]:

$$\frac{\partial v_1}{\partial t} + c_{m,0}\frac{\partial v_1}{\partial x} + \varepsilon_m v_1 \frac{\partial v_1}{\partial x} - \frac{1}{2}\left(\delta_1 + \frac{c_0^2}{c_{m,0}^2}\delta_2\right)\frac{\partial^2 v_1}{\partial x^2} = 0. \tag{5.76}$$

Eq. (5.76) coincides with that derived in Ref. [54] for any direction of the magnetic field (this is the case $\theta = \pi/2$ in Eq. (15) therein with zero heating and radiative cooling of the plasma). Chin and co-authors have involved heating and radiative cooling but consider perturbations over stationary and homogeneous background. The validity of this approach should be carefully investigated in view of the fact that the external source violates the temperature uniformity of the background. This follows from the zero-order the energy balance for the thermoconducting fluid. The non-uniformity of the background may essentially affect the definition of modes and the wave dynamics, especially for large-scale perturbations. Thermoviscous attenuation of the magnetoacoustic perturbation in its part, which originates from the thermal conduction, is smaller as compared with that in an unmagnetized Newtonian flow. The damping of magnetoacoustic waves depends in general on magnetic pressure, h_0, by means of $c_{m,0}$. Solutions of Eq. (5.76) may be established by the methods suitable for the Burgers equation. In particular, it transforms into the linear diffusion equation by the Hopf-Cole transformation [10]. It may be readily concluded that the acoustic Reynolds number is always larger in the magnetoacoustic wave as compared to sound for perturbations with equal Mach numbers and frequencies.

The projecting rows d_3 and d_4 being applied at the system (5.70) eliminate foreign terms in the linear part of equations, which describes the dynamics of ρ_3 or ρ_4. Among the variety of quadratic nonlinear terms, only these terms, belonging to the first progressive mode, will be kept. This corresponds to the dominative magnetoacoustic perturbation progressive in the positive direction of axis x. The thermoviscous nonlinear terms form the "magnetoacoustic forces" of the both secondary modes. The result of application of d_3 is the dynamic equation, which governs an excess density in the entropy mode:

$$\frac{\partial \rho_3}{\partial t} - \left(\frac{\delta_2}{\gamma-1} + \frac{\delta_2(c_{m,0}^2 - c_0^2)}{c_{m,0}^2}\right)\frac{\partial^2 \rho_3}{\partial x^2} = -\rho_0(2c_{m,0}^6(\gamma c_{m,0}^2 - c_0^2(\gamma-1)))^{-1}$$

$$\left(c_0^4 c_{m,0}^2(\delta_2(5-8\gamma)+\delta_1(\gamma-1))(\gamma-2)+2c_{m,0}^6\delta_1(\gamma-1)\gamma+4c_0^6\delta_2(\gamma-1)(\gamma-2)\right.$$

$$\left.+c_0^2 c_{m,0}^4(-3\delta_1(\gamma-1)^2+\delta_2\gamma(4\gamma-9))\right)\left(\frac{\partial v_1}{\partial x}\right)^2 \tag{5.77}$$

$$-\rho_0(2c_{m,0}^6(\gamma c_{m,0}^2 - c_0^2(\gamma-1)))^{-1}\left(c_0^2(-c_{m,0}^4(\delta_1(\gamma-1)^2+\delta_2\gamma(7-2\gamma)\right.$$

$$\left.+4c_0^4\delta_2(\gamma-1)(\gamma-2)+c_0^2c_{m,0}^2(\delta_1(\gamma-1)(\gamma-2)+\delta_2(-6\gamma^2+17\gamma-8)))\right)v_1\frac{\partial^2 v_1}{\partial x^2}$$

$$-\frac{\alpha c_0^2\delta_2(\gamma-1)\rho_0}{c_{m,0}^4}\left(\frac{\partial v_1}{\partial x}\right)^2.$$

Equation (5.77) coincides with that which describes the instantaneous acoustic heating in the Newtonian flows of unmagnetized gases [55]. Making use of the leading-order equality for the periodic sound,

$$\overline{\left(\frac{\partial v_1}{\partial x}\right)^2} = -\overline{v_1\frac{\partial^2 v_1}{\partial x^2}},$$

Equation (5.77) simplifies as

$$\overline{\frac{\partial \rho_3}{\partial t}} - \left(\frac{\delta_2}{\gamma-1} + \frac{\delta_2(c_{m,0}^2 - c_0^2)}{c_{m,0}^2}\right)\overline{\frac{\partial^2 \rho_3}{\partial x^2}} = F_{m,h}$$

$$= -\left(\frac{(c_{m,0}^2\delta_1+(1+\alpha)c_0^2\delta_2)(\gamma-1)\rho_0}{c_{m,0}^4}\right)\overline{\left(\frac{\partial v_1}{\partial x}\right)^2}.$$

The equation, which describes the acoustic heating of a Newtonian gas in the absence of a magnetic field (in the limit $h_0 \to 0$, $\alpha \to 0$) is well known in the theory of nonlinear acoustics [11]

$$\overline{\frac{\partial \rho_3}{\partial t}} - \frac{\delta_2}{\gamma-1}\overline{\frac{\partial^2 \rho_3}{\partial x^2}} = F_{N,h} = -\frac{(\delta_1+\delta_2)(\gamma-1)\rho_0}{c_0^2}\overline{\left(\frac{\partial v_1}{\partial x}\right)^2}.$$

It always describes the isobaric increase of temperature of an ideal gas, which associates with variation in its density, ρ_3 [10,56]. The acoustic force of acoustic heating in a Newtonian fluid is proportional to the total attenuation, $\delta_1 + \delta_2$. As compared with the Newtonian case, the equation that governs magnetoacoustic heating, contains a smaller part responsible for viscosity on the acoustic force in its right-hand side. Oppositely, the coefficient by part that associates with the thermal conduction is larger than the unmagnetized value under condition

$$h_0 > 0.5\rho_0 c_0^2(\sqrt{1+\alpha}-1),$$

and equal or smaller otherwise. The conclusion is that the total damping coefficient standing by the magnetoacoustic force of heating may be smaller or larger than that of the absence of magnetic field, depending on unperturbed magnetic pressure, h_0. In contrast to the entropy mode, the specific excess density for the fourth mode equals zero, but perturbations of hydrodynamic and magnetic pressures are fully compensated. The equation for the fourth mode follows from making use of d_4:

$$\frac{\partial \rho_4}{\partial t} =\gamma\rho_0(c_{m,0}^2-c_0^2)(c_{m,0}^2\delta_1+c_0^2\delta_2)(2c_{m,0}^6(\gamma c_{m,0}^2-c_0^2(\gamma-1)))^{-1}$$
$$\left(\left(\frac{\partial v_1}{\partial x}\right)^2+v_1\frac{\partial^2 v_1}{\partial x^2}\right). \tag{5.78}$$

The acoustic force of this mode obviously equals zero in the absence of a magnetic field. The nearly periodic sound results approximately in zero acoustic force. An impulse might weakly contribute in total density and total magnetic and acoustic pressures. Equation (5.78) is readily integrated over time. The approximate result is as follows:

$$\rho_4 = -\gamma\rho_0(c_{m,0}^2-c_0^2)(c_{m,0}^2\delta_1+c_0^2\delta_2)(2c_{m,0}^7(\gamma c_m^2-c_0^2(\gamma-1)))^{-1}v_1\frac{\partial v_1}{\partial x}.$$

Equations (5.77) and (5.78) does not depend on the variance of shear viscosity with temperature.

As for the acoustic streaming, it may exist in a multi-dimentional flow. We consider a two-component vector of velocity $\vec{v}=(v_x(x,y,t),\ v_y(x,y,t),\ 0)$ perpendicular to magnetic field $\vec{H}=(0,\ 0,\ H_z(x,y,t))$. The variability of thermal conductivity and shear viscosity are insignificant in the description of the vortex flow caused by sound in the leading order. These coefficients will be treated as constants in this section. The dynamic equations describing the magnetogysdynamic flow in two dimensions take the leading-order forms:

$$\frac{\partial \rho'}{\partial t}+\rho_0\frac{\partial v_x}{\partial x}+\rho_0\frac{\partial v_y}{\partial y}=-\rho'\frac{\partial v_x}{\partial x}-v_x\frac{\partial \rho'}{\partial x}-\rho'\frac{\partial v_y}{\partial y}-v_y\frac{\partial \rho'}{\partial y}, \tag{5.79}$$

$$\frac{\partial v_x}{\partial t}-\frac{\delta_1}{4}\frac{\partial^2 v_x}{\partial x^2}-\frac{\delta_1}{4}\frac{\partial^2 v_y}{\partial x\partial y}-\frac{3\delta_1}{4}\Delta v_x+\frac{1}{\rho_0}\frac{\partial(p'+h')}{\partial x}$$

$$=-(\vec{v}\cdot\vec{\nabla})v_x+\frac{\rho'}{\rho_0^2}\frac{\partial(p'+h')}{\partial x}-\frac{\delta_1}{4}\frac{\rho'}{\rho_0}\frac{\partial}{\partial x}\vec{\nabla}\cdot\vec{v}-\frac{3\delta_1}{4}\frac{\rho'}{\rho_0}\Delta v_x,$$

$$\frac{\partial v_y}{\partial t}-\frac{\delta_1}{4}\frac{\partial^2 v_y}{\partial y^2}-\frac{\delta_1}{4}\frac{\partial^2 v_x}{\partial x\partial y}-\frac{3\delta_1}{4}\Delta v_y+\frac{1}{\rho_0}\frac{\partial(p'+h')}{\partial y}$$

$$=-(\vec{v}\cdot\vec{\nabla})v_y+\frac{\rho'}{\rho_0^2}\frac{\partial(p'+h')}{\partial y}-\frac{\delta_1}{4}\frac{\rho'}{\rho_0}\frac{\partial}{\partial y}\vec{\nabla}\cdot\vec{v}-\frac{3\delta_1}{4}\frac{\rho'}{\rho_0}\Delta v_y,$$

$$\frac{\partial p'}{\partial t}-\frac{\gamma\delta_2}{\gamma-1}\Delta p'+\frac{\delta_2 c_0^2}{\gamma-1}\Delta\rho'+c_0^2\rho_0\vec{\nabla}\cdot\vec{v}=-\vec{v}\cdot\vec{\nabla}p'-\gamma p'\vec{\nabla}\cdot\vec{v}$$

$$+\frac{3\delta_1}{8}\left(\frac{\partial v_i}{\partial x_k}+\frac{\partial v_k}{\partial x_i}-\frac{2}{3}\delta_{ik}\frac{\partial v_l}{\partial x_l}\right)^2-\frac{\delta_2}{\rho_0(\gamma-1)}\Delta\left(\gamma p'\rho'-c_0^2\rho'^2\right),$$

$$\frac{\partial h'}{\partial t}+2h_0\vec{\nabla}\cdot\vec{v}=-\vec{v}\cdot\vec{\nabla}h'-2h'\vec{\nabla}\cdot\vec{v},$$

where x_1, x_2, x_3 denote correspondingly x, y, and z, Δ and $\vec{\nabla}$ operate in the plane (x,y); $\delta_{i,k}$ is the Cronecker symbol; and i, k, l are integer numbers varying from 1 to 3. As usual, we consider a weakly diffracting magnetoacoustic beam that propagates, for definiteness, in the positive direction of axis x. A small parameter ε accounts for diffraction and measures the ratio of the characteristic scales of perturbations in the longitudinal and transversal directions, $\varepsilon=k_y/k_x$. Considering small attenuation and diffraction, one readily arrives at the leading-order dispersion relations in a magnetogasdynamic flow,

$$\omega_{1,2}=\pm c_{m,0}k_x\left(1+\frac{k_y^2}{2k_x^2}\right)+\frac{i}{2}\left(\delta_1+\frac{c_0^2}{c_{m,0}^2}\delta_2\right)k_x^2,$$

$$\omega_3=i\left(\frac{\delta_2}{\gamma-1}+\frac{\delta_2(c_{m,0}^2-c_0^2)}{c_{m,0}^2}\right)k_x^2,\quad \omega_4=0,\quad \omega_5=i\frac{3}{4}\delta_1\Delta.$$

The new fifth root that appears in two-dimensional flow reflects the existence of the incompressible rotational flow of a gas with velocity whose divergence is zero, $\vec{\nabla}\cdot\vec{v}_5=0$. The dynamic equation which is determined by Eq. (5.28) for solenoidal velocity may be decomposed by applying the operator P_{vort} at the momentum equation from system (5.79):

$$\frac{\partial\vec{v}_5}{\partial t}-\frac{3\delta_1}{4}\Delta\vec{v}_5=-\frac{1}{\rho_0}P_{vort}\left(\rho_a\frac{\partial\vec{v}_a}{\partial t}\right)=\frac{c_{m,0}}{2\rho_0^2}\left(\delta_1+\frac{\delta_2 c_0^2}{c_{m,0}^2}\right)P_{vort}\rho_a\vec{\nabla}\frac{\partial}{\partial x}(\rho_2-\rho_1). \tag{5.80}$$

ρ_a, $\vec{v}_a$ denote summary acoustic perturbation in density, $\rho_a=\sum_{i=1}^{2}\rho_i$, $\vec{v}_a=\sum_{i=1}^{2}\vec{v}_i$. Eq. (5.80) coincides with the instantaneous equation, which governs acoustic streaming in Newtonian flows in the absence of a magnetic field [56,57]. Its average over the sound period leading-order form, in the case of the periodic or nearly periodic magnetoacoustic wave, may be expressed in terms of magnetoacoustic pressure on the right-hand side:

$$\overline{\frac{\partial v_{x,5}}{\partial t}} - \frac{3\delta_1}{4}\overline{\Delta v_{x,5}} = F_{m,s} = \frac{1}{\rho_0^2 c_0^4 c_{m,0}}\left(\delta_1 + \delta_2 \frac{c_0^2}{c_{m,0}^2}\right)\overline{\left(\frac{\partial p_1}{\partial t}\right)^2}.$$

This equations tends to the well-known limit in the case of unmagnetized Newtonian fluid when $c_{m,0}$ tends to c_0. The leading-order equation that describes the longitudinal component of the vortex velocity $v_{x,5}$, in the case of periodic acoustic pressure, takes the form [10,58,59]:

$$\overline{\frac{\partial v_{x,5}}{\partial t}} - \frac{3\delta_1}{4}\overline{\Delta v_{x,5}} = F_{N,s} = \frac{\delta_1 + \delta_2}{\rho_0^2 c_0^5}\overline{\left(\frac{\partial p_1}{\partial t}\right)^2}.$$

The conclusion is that whereas the acoustic Newtonian force of streaming is determined by the summary attenuation due to irreversible mechanical and thermal losses, $\delta_1 + \delta_2$, the acoustic force of magnetoacoustic streaming is always smaller than that in a Newtonian flow, $F_{m,s} < F_{N,s}$. This provides a smaller velocity of streaming as compared with an unmagnetized Newtonian fluid at equal Mach numbers and characteristic frequencies.

Equations (5.77) and (5.80) describe dynamics of magnetoacoustic heating and streaming in the field of sound independently on the spectrum of magnetoacoustic wave. The magnetoacoustic forces of heating and streaming depend on magnetic strength. The slow variations of a temperature and stream's velocity, which associate with the non-wave modes, may be measured remotely. This makes it possible to get information about the processes in plasma analytically, and in particular, to evaluate magnetic pressure in a plasma. Both equations (5.77) and (5.80) are inhomogeneous diffusity equations. They are not averaged over the sound period, and they make use of an instantaneous magnetoacoustic source that relates to periodic and aperiodic sound. In particular, some conclusions concern periodic sound. The acoustic force of magnetoacoustic heating $F_{m,h}$ may be smaller or larger than the Newtonian one $F_{N,h}$. in dependence on unperturbed magnetic pressure h_0. The magnetoacoustic force is not proportional to the summary attenuation, in contrast to the Newtonian acoustic force $F_{N,h}$, but possesses a larger parameter of diffusity. The magnetoacoustic force of streaming is always smaller than the Newtonian one, $F_{m,s} < F_{N,s}$. We get conclusions in the leading order: magnetoacoustic forces of heating and streaming include only quadratic nonlinear terms, and they are proportional to the first powers of damping coefficients. Excitation of the Alfvén mode by sound is described by Eq. (5.78). It is insignificant. This concerns periodic and nearly periodic magnetoacoustic waves and impulses.

There are no restrictions concerning the strength of the magnetic field in this section. Usually, nonlinear interactions are resolved by representing interacting modes as the sum of harmonics and by solving coupling equations for exactly satisfied resonance conditions with some desired accuracy. As a rule, three-wave resonant interactions are considered [46,47]. The authors apply the method of projecting of initial system of conservation equations into the system of coupling equations for interacting modes independently on their initial spectrum.

5.4.1 ON THE NONLINEAR INTERACTIONS IN A PLASMA WITH FINITE ELECTRICAL CONDUCTIVITY

It is remarkable that the inclusion of finite electrical conductivity of a plasma introduces the dispersion and attenuation associated with it. For short discussion, we do not consider any damping caused by mechanical and thermal effects. The last equation in Eqs (5.69) takes the form

$$\frac{\partial h}{\partial t}+v\frac{\partial h}{\partial x}+2h\frac{\partial v}{\partial x}+\Sigma\left(-\frac{\partial^2 h}{\partial x^2}+\frac{1}{2h}\left(\frac{\partial h}{\partial x}\right)^2\right)=0,$$

and other conservation equations remain unchanged; where $\Sigma=(\mu_m\sigma_m)^{-1}$, μ_m is the magnetic permeability, and σ_m is the electrical conductivity of a fluid. The leading-order dispersion relations that specify one-dimensional planar flow, are

$$\omega_{1,2}=\pm c_0 k\pm\frac{(c_0\pm i\Sigma k)h_0 k}{\rho_0(c_0^2+\Sigma^2k^2)},\quad \omega_3=0,\quad \omega_4=i\Sigma k^2-\frac{2i\Sigma h_0 k^2}{\rho_0(c_0^2+\Sigma^2k^2)}. \tag{5.81}$$

The first two roots ω_1, ω_2 correspond to the fast MHD waves the third root ω_3 corresponds to the entropy mode; and the last one, ω_4, corresponds, to the Alfvén wave in the flow where magnetic field is ortogonal to the particles velocity. The only restriction that has been done during evaluation is the smallness of magnetic strength compared with the unperturbed pressure of a gas,

$$h_0\ll\rho_0c_0^2=\gamma p_0.$$

The links of the Fourier transforms of perturbations that correspond to the dispersion relations (5.81) take the form

$$\widetilde{\psi}_1=(\widetilde{\rho}_1\quad \widetilde{v}_1\quad \widetilde{p}_1\quad \widetilde{h}_1)^T=\left(1\quad \frac{c_0}{\rho_0}+\frac{h_0}{\rho_0^2(c_0-i\Sigma k)}\quad c_0^2\quad \frac{2c_0h_0}{\rho_0(c_0-i\Sigma k)}\right)^T\widetilde{\rho}_1,$$

$$\widetilde{\psi}_2=(\widetilde{\rho}_2\quad \widetilde{v}_2\quad \widetilde{p}_2\quad \widetilde{h}_2)^T=\left(1\quad -\frac{c_0}{\rho_0}-\frac{h_0}{\rho_0^2(c_0+i\Sigma k)}\quad c_0^2\quad \frac{2c_0h_0}{\rho_0(c_0+i\Sigma k)}\right)^T\widetilde{\rho}_2,$$

$$\widetilde{\psi}_3=(\widetilde{\rho}_3\quad \widetilde{v}_3\quad \widetilde{p}_3\quad \widetilde{h}_3)^T=(1\quad 0\quad 0\quad 0)^T\widetilde{\rho}_3,$$

$$\widetilde{\psi}_4=(\widetilde{\rho}_4\quad \widetilde{v}_4\quad \widetilde{p}_4\quad \widetilde{h}_4)^T=\left(1\quad \frac{i\Sigma k}{\rho_0}-\frac{2i\Sigma k h_0}{\rho_0^2(c_0^2+\Sigma^2k^2)}\quad c_0^2\quad -c_0^2-\Sigma^2k^2+\frac{2h_0}{\rho_0}\right)^T\widetilde{\rho}_4.$$

These links establish relations of specific perturbations. In particular, the first magnetoacoustic eigenvector takes the form

$$\psi_1=(\rho_1\quad v_1\quad p_1\quad h_1)^T$$

$$= \left(1 \quad \frac{c_0}{\rho_0} + \frac{h_0}{\rho_0^2 \Sigma} \int_{-\infty}^{x} dx' \exp(-\frac{c_0(x-x')}{\Sigma}) \; c_0^2 \; \frac{2c_0 h_0}{\rho_0 \Sigma} \int_{-\infty}^{x} dx' \exp(-\frac{c_0(x-x')}{\Sigma}) \right)^T \rho_1 .$$

With an account for the nonlinear correction to the magnetoacoustic first mode, that governs its specific velocity takes the form

$$\frac{\partial v_1}{\partial t} + c_0 \frac{\partial v_1}{\partial x} + \varepsilon_m v_1 \frac{\partial v_1}{\partial x} + \frac{h_0}{\Sigma \rho_0} v_1(x,t) - \frac{c_0 h_0}{\Sigma^2 \rho_0} \int_{-\infty}^{x} \exp(-(x-x')c_0/\Sigma) v_1(x',t) dx' = 0.$$

It may be readily simplified in the case of low-frequency sound, $\Sigma k \ll c_0$ (this case coincides with the Earnshaw equation):

$$\frac{\partial v_1}{\partial t} + c_m \frac{\partial v_1}{\partial x} + \varepsilon_m v_1 \frac{\partial v_1}{\partial x} = 0,$$

and in the case of high-frequency sound, $\Sigma k \gg c_0$:

$$\frac{\partial v_1}{\partial t} + c_0 \frac{\partial v_1}{\partial x} + \varepsilon_m v_1 \frac{\partial v_1}{\partial x} + \frac{h_0}{\Sigma \rho_0} v_1(x,t) = 0.$$

The low-frequency sound does not experience attenuation in the leading order. The evolution equation which describes sound perturbations may be solved by the method of characteristics. The equation that describes the dynamics of the high-frequency sound, readily rearranges in the new variables

$$\tilde{v}_1 = \exp\left(\frac{h_0 x}{\Sigma \rho_0 c_0}\right) v_1, \quad X = \frac{\Sigma \rho_0 c_0}{h_0}\left(1 - \exp\left(-\frac{h_0 x}{\Sigma \rho_0 c_0}\right)\right), \quad \tau = t - x/c_0$$

into the following leading-order equation:

$$\frac{\partial \tilde{v}_1}{\partial X} - \frac{\varepsilon}{c_0^2} v_1 \frac{\partial \tilde{v}_1}{\partial \tau} = 0,$$

which in turn may be solved by the method of characteristics.

The progressive magnetosonic mode, which is an analogue of the Riemann wave in an ideal inviscid gas, is in the leading order represented by a sum, $\psi_1 + \psi_{1,nl}$. The vector of nonlinear corrections is established by Eq. (5.74). Applying the projecting row d_3 (it is still established by Eq. (5.73)), one arrives at an equation that governs an excess density in the entropy mode:

$$\frac{\partial \rho_3'}{\partial t} = F_{m,h} = \frac{(\gamma-1)h_0}{c_0^3 \Sigma^2} (c_0(\Sigma \partial v_1/\partial x - c_0 v_1) \int_{-\infty}^{x} \exp(-(x-x')c_0/\Sigma) v_1(x',t) dx'$$

$$+ \Sigma v_1 (c_0 v_1 - 2\Sigma \partial v_1/\partial x)).$$

It may be simplified in the case of low-frequency magnetoacoustic perturbations, $\Sigma k \ll c_0$,

$$\frac{\partial \rho_3}{\partial t} = \frac{(\gamma-1)h_0}{c_0^3} v_1(x,t)\frac{\partial v_1(x,t)}{\partial x},$$

and in the case of high-frequency magnetoacoustic perturbations, $\Sigma k \gg c_0$:

$$\frac{\partial \rho_3}{\partial t} = 2\frac{(\gamma-1)h_0}{c_0^3} v_1(x,t)\frac{\partial v_1(x,t)}{\partial x}.$$

The conclusion is that the periodic or nearly periodic sound is not effective in producing heating in both limits. The average over the sound period of the magnetoacoustic force of heating equals approximately zero in the leading order. At least, the harmonic sound also is not effective in producing heating at all magnetoacoustic wavelengths. As for the acoustic streaming, the relative evolution equation in terms of vorticity $\vec{\Omega}$ sounds

$$\begin{aligned}\frac{\partial \vec{\Omega}}{\partial t} &= \frac{1}{\rho_0}\vec{\nabla} \times \left(-\rho_1 \frac{\partial}{\partial t}\vec{v}_1\right) \\ &= \frac{2h_0 c_0}{\rho_0^2 \Sigma^2}\vec{\nabla}\rho_1 \times \left(\int_{-\infty}^{x} \exp\left(-\frac{c_0(x-x')}{\Sigma}\right)\vec{v}_1 dx' - \frac{\Sigma}{c_0}\vec{v}_1\right).\end{aligned}$$

Its leading-order averaged form, in the case of the periodic magnetoacoustic wave, may be expressed in terms of magnetoacoustic pressure on the right-hand side,

$$\frac{\overline{\partial v_{x,vort}}}{\partial t} = F_{m,s} = \frac{2h_0}{\Sigma^2 \rho_0^3 c_0^3} \overline{p_1 \left(\frac{\Sigma}{c_0} p_1 - \int_{-\infty}^{x} \exp\left(-\frac{c_0(x-x')}{\Sigma}\right) p_1 dx'\right)}. \tag{5.82}$$

The limits of large-scale and small-scale magnetoacoustic perturbations may be readily traced. In the case of $\Sigma k \ll c_0$,

$$F_{m,s} = \frac{2h_0}{\rho_0^3 c_0^5} \overline{p_1 \left(\frac{\partial p_1}{\partial x}\right)},$$

and in the case $\Sigma k \gg c_0$,

$$F_{m,s} = \frac{2h_0}{\Sigma \rho_0^3 c_0^4} \overline{p_1^2}.$$

The low-frequency sound is not effective in producing magnetoacoustic streaming. For approximate evaluations of Eq. (5.82) at mediate frequency, we make use of an exemplary periodic waveform with characterestic walelength Σ/c_0, which represents magnetoacoustic pressure:

$$p_1(x,t) = P_0 \sin((x - c_0 t)c_0/\Sigma).$$

It yields the magnetoacoustic force of streaming as follows:

$$F_{m,s} = \frac{h_0 P_0^2}{2\Sigma \rho_0^3 c_0^4}.$$

The longitudinal velocity of streaming (which coincides with the direction of propagation of sound) increases with time. The Alfvén wave is represented by the fourth root of the dispersion equation in Eqs (5.81). The projecting row d_4, which distinguishes perturbation of the specific magnetic pressure h_4,

$$P_4(\rho' \quad v \quad p' \quad h')^T = h_4,$$

takes the form

$$d_4 = \left(0 \quad -\frac{2i\Sigma h_0 k}{c_0^2 + \Sigma^2 k^2} \quad -\frac{2h_0}{(c_0^2 + \Sigma^2 k^2)\rho_0} \quad 1 - \frac{2(c_0^2 - \Sigma^2 k^2) h_0}{(c_0^2 + \Sigma^2 k^2)^2 \rho_0}\right).$$

The dynamic equation for h_4 depends on the spectrum of sound. If $\Sigma k \ll c_0$, it takes the form

$$\frac{\partial h_4}{\partial t} = \frac{\Sigma h_0}{c_0^2}\left((7-\gamma)\left(\frac{\partial v_1(x_1,t)}{\partial x_1}\right)^2 + (5-\gamma) v_1 \frac{\partial^2 v_1(x_1,t)}{\partial x_1^2}\right).$$

This makes the harmonic sound

$$V = V_0 \sin(\omega(t - x/c_m))$$

effective in producing positive perturbation in the magnetic presuure corresponding to the fourth mode:

$$\overline{\frac{\partial h_4}{\partial t}} = \frac{\Sigma \omega^2 h_0 V_0^2}{c_0^4}.$$

The case $\Sigma k \gg c_0$ results in

$$\frac{\partial h_4}{\partial t} = \frac{\Sigma h_0 (5-\gamma)}{c_0^2}\left(\left(\frac{\partial v_1(x_1,t)}{\partial x_1}\right)^2 + v_1 \frac{\partial^2 v_1(x_1,t)}{\partial x_1^2}\right).$$

Hence, the generation of the magnetic perturbations by the periodic sound is insignificant. The same conclusion can be drawn for periodic sound of medium wavelength $k \approx c_0/\Sigma$.

REFERENCES

1. Zeldovich, Y.B., and Y.P. Raizer. 1966. *Physics of Shock Waves and High Temperature Hydrodynamic Phenomena*. Academic Press, New York.
2. Gordiets, B.F., A.I. Osipov, E.V. Stupochenko, and L.A. Shelepin. 1973. Vibrational relaxation in gases and molecular lasers. *Soviet Physics Uspekhi* 15: 759–785.

3. Osipov, A.I., and A.V. Uvarov. 1992. Kinetic and gasdynamic processes in nonequilibrium molecular physics. *Soviet Physics Uspekhi* 35(11): 903–923.
4. Chu, B.T. 1970. Weak nonlinear waves in nonequilibrium flows. In: Wagner, P.P. (Ed.), *Nonequilibrium Flows*, Vol. 1, Part II. Marcel Dekker, New York, p. 33.
5. Parker, D.F. 1972. Propagation of damped pulses through a relaxing gas. *The Physics of Fluids* 15: 256–262.
6. Clarke, J.F., and A. McChesney. 1976. *Dynamics of Relaxing Gases*. Butterworth, London.
7. Molevich, N.E. 2002. Nonstationary self-focusing of sound beams in a vibrationally excited molecular gas. *Acoustical Physics* 48(2): 209–213.
8. Perelomova, A., and P. Wojda. 2012. Generation of the vorticity mode by sound in a vibrationally relaxing gas. *Central European Journal of Physics* 10(5): 1116–1124.
9. Molevich, N.E. 2001. Sound amplification in inhomogeneous flows of nonequilibrium gas. *Acoustical Physics* 47(1): 102–105.
10. Rudenko, O.V., and S.I. Soluyan. 1977. *Theoretical Foundations of Nonlinear Acoustics*. Plenum, New York.
11. Makarov, S., and M. Ochmann. 1996. Nonlinear and thermoviscous phenomena in acoustics, part I. *Acta Acustica United with Acustica* 82: 579–606.
12. Perelomova, A. 2012. Standing acoustic waves and relative nonlinear phenomena in a vibrationally relaxing gas-filled resonator. *Acta Acustica* 98: 713–721.
13. Perelomova, A. 2010. Nonlinear generation of non-acoustic modes by low-frequency sound in a vibrationally relaxing gas. *Canadian Journal of Physics* 88(4): 293–300.
14. Koltsova, E.V., A.I. Osipov, and A.V. Uvarov. 1994. Acoustical disturbances in a nonequilibrium inhomogeneous gas. *Acoustical Physics* 40(6): 859–863.
15. Hamilton, M., and D. Blackstock. 1998 *Nonlinear Acoustics*. Academic Press, New York.
16. Perelomova, A., and W. Pelc-Garska. 2011. Non-wave variations in temperature caused by sound in a chemically reacting gas. *Acta Physica Polonica A* 120(3): 455–461.
17. Perelomova, A., and P. Wojda. 2011. Generation of the vorticity motion by sound in a chemically reacting gas. Inversion of acoustic streaming in the non-equilibrium regime. *Central European Journal of Physics* 9(3): 740–750.
18. Molevich, N.E. 2003. Sound velocity dispersion and second viscosity in media with nonequilibrium chemical reactions. *Acoustical Physics* 49(2): 229–232.
19. Riemann, B. 1953. *The Collected Works of Bernard Riemann*. Dover, New York.
20. Perelomova, A. 2010. Nonlinear generation of non-acoustic modes by low-frequency sound in a vibrationally relaxing gas. *Canadian Journal of Physics* 88(4): 293–300.
21. Bakhvalov, N.S., Y. M. Zhileikin, and E.A. Zabolotskaya. 1987. *Nonlinear Theory of Sound Beams*. American Institute of Physics, New York.
22. Assman, V.A., F.V. Bunkin, A.V. Vernik, G.A. Lyakhov, and K.F. Shipilov. 1985. Observation of a thermal self-effect of a sound beam in a liquid. *JETP Letters* 41(4): 182–184.
23. Andreev V.G., A.A. Karabutov, O.V. Rudenko, and O.A. Sapozhnikov. 1985. Observation of self-focusing of sound. *JETP Letters* 41(9): 466–469.
24. Karabutov, A.A., O.V. Rudenko, and O.A. Sapozhnikov. 1989. Thermal self-focusing of weak shock-waves. *Soviet Physics Acoustics-USSR* 35(1): 40–51.
25. Andreev V.G., A.A. Karabutov, O.V. Rudenko, and Rudenko. O.A. 1985. Observation of the self-induced thermal action of a high-intensity ultrasonic beam in water. *Soviet Physics Acoustics-USSR* 31(5): 393–394.
26. Rudenko, O.V., and O.A. Sapozhnikov. 2004. Self-action effects for wave beams containing shock fronts. *Physics-Uspekhi* 174(9): 973–989.

27. Perelomova, A. 2013. Thermal self-action effects of acoustic beam in a gas with reversible or irreversible chemical reaction. *Acta Acustica United with Acustica* 99: 352–358.
28. Nachman, A., J.F. Smith III, and R.C. Waag. 1990. An equation for acoustic propagation in inhomogeneous media with relaxation losses. *The Journal of the Acoustical Society of America* 88(3): 1584–1595.
29. Eigen, M., and K. Tamm. 1962. Sound absorption in electrolyte solutions as a sequence of chemical reactions. *Zeitschrift fr Elektrochemie* 66(2): 93–121.
30. Perelomova, A. 2015. The nonlinear effects of sound in a liquid with relaxation losses. *Canadian Journal of Physics* 93(11): 1391–1396.
31. Nyborg, W.L. 1978. *Physical Mechanisms for Biological Effects of Ultrasound*. Bureau of Radiological Health, Rockville, MA.
32. Lieberman, L. 1949. Sound propagation in chemically active media. *Physics Review* 76(10): 1520–1524.
33. Perelomova, A. 2013. Interaction of acoustic and thermal modes in the vibrationally relaxing gases. Acoustic cooling. *Acta Physica Polonica A*. 123(4): 681–687.
34. Perelomova, A. 2012. Nonlinear influence of sound on the vibrational energy of molecules in a relaxing gas. *Archives of Acoustics* 37(1): 89–96.
35. Herlofson, N. 1950. Magneto-hydrodynamic waves in a compressible fluid conductor. *Nature* 165: 1020–1021.
36. Truesdell, C. 1950. The effect of the compressibility of the earth on its magnetic field. *Physical Review* 78: 823.
37. Anderson, N.S. 1953. Longitudinal magneto-hydrodynamic waves. *JASA* 23(3): 529–532.
38. Sharma, V.D., L.P. Singh, and R. Ram. 1981. Propagation of discontinuities in magnetogasdynamics. *The Physics of Fluids* 24(7): 1386–1387. (1981)
39. Sharma, V.D., L.P. Singh, and R. Ram. 1987. The progressive wave approach analyzing the decay of a sawtooth profile in magnetogasdynamics. *The Physics of Fluids* 30(5): 1572–1574.
40. Singh, L.P., R. Singh, and S.D. Ram. 2012. Evolution and decay of acceleration waves in perfectly conducting inviscid radiative magnetogasdynamics. *Astrophysics and Space Science* 342: 371–376.
41. Ponomarev, E.A. 1962. On the propagation of low-frequency oscillations along the magnetic field in a viscous compressible plasma. *Soviet Astronomy* 5(5): 673–676.
42. Anand, R.K., and H.C. Yadav. 2014. On the structure of MHD shock waves in a viscous non-ideal gas. *Theoretical and Computational Fluid Dynamics* 28(3): 369–376.
43. Sagdeev, R.Z., and A.A. Galeev. 1969. *Nonlinear Plasma Theory*. Benjamin, New York.
44. Petviashvili, V.I., and O.A. Pokhotelov. 1992. *Solitary Waves in Plasmas and in the Atmosphere*. Gordon and Breach, Berlin.
45. Shukla, P.K., and L. Stenflo. 1999. Nonlinear phenomena involving dispersive Alfven waves. In: Passot, T., and Sulem, P.-L. (Eds), *Part of the Lecture Notes in Physics book series (LNP),* Vol. 536. Springer, Berlin, pp. 1–30.
46. Brodin, G., L. Stenflo, and P.K. Shukla. 2006. Nonlinear interactions between kinetic Alfven and ion-sound waves. arXiv.physics/0604122v1 [physics.plasm-ph].
47. Zavershinsky, D.I., and N.E. Molevich. 2014. Alfven wave amplification as a esult of nonlinear interaction with a magnetoacoustic wave in an acoustically active conducting medium. *Technical Physics Letters* 40(8): 701–703.

48. Ballai, I. 2006. Nonlinear waves in solar plasmas-a review. *Journal of Physics: Conference Series* 44(20): 20–29.
49. Krishna Prasad, S., D. Banerjee, and T. Van Doorsselaere. 2014. Frequency-dependent damping in propagating slow magnetoacoustic waves. arXiv:1406.3565 [astro-ph.SR]
50. Korobeinikov, V.P. 1976. *Problems in the Theory Point Explosion in Gases*. American Mathematical Society, Providence, RI.
51. Landau, L.D., L.P. Pitaevskii, A.M. Kosevich, J.B. Sykes, and R.N. Franklin. 1981. *Physical Kinetics: Volume 10 (Course of Theoretical Physics)*. Heinemann, Butterworth.
52. Fabian, A.C., C.S. Reynolds, G.B. Taylor, and R.J.H. Dunn. 2005. On viscosity, conduction and sound waves in the intracluster medium. *Monthly Notices of the Royal Astronomical Society* 363(3): 891–896.
53. Spitzer, L. 1962. *Physics of Fully Ionized Gases*. Wiley, New York.
54. Chin, R., E. Verwichte, G. Rowlands, and V.M. Nakariakov. 2010. Self-organisation of magnetoacoustic waves in a thermally unstable environment. *Physics of Plasmas* 17(32): 032–107.
55. Perelomova, A. 2004. Heat generation by impulse ultrasound. *Ultrasonics* 43: 92–100.
56. Perelomova, A. 2006. Development of linear projecting in studies of non-linear flow. Acoustic heating induced by non-periodic sound. *Physics Letters A* 357: 42–47. (2006)
57. Perelomova, A., and P. Wojda. 2010. Generation of the vorticity mode by sound in a relaxing Maxwell fluid. *Acta Acustica United with Acustica* 96(5): 807–813.
58. Gusev, V.E., and O.V. Rudenko. 1979. Non-steady quasi-one-dimensional acoustic streaming in unbounded volumes with hydrodynamic nonlinearity. *Soviet Physics-Acoustics* 25: 493–497.
59. Gusev, V.E., and O.V. Rudenko. 1980. Evolution of nonlinear two-dimensional acoustic streaming in the field of highly attenuated sound beam. *Soviet Physics-Acoustics* 27: 481–484.

6 Boundary Layer Problem: Acoustic and Tollmienn-Schlichting Waves

6.1 PRELIMINARY REMARKS

It is a well-established fact now that free-stream and surface disturbances strongly affect the processes in the boundary layer that compose a transition to a turbulent state. This transition in turn determines certain important parameters of a rigid body in fluid mechanics such as skin friction and heat transfer. The classification of the free stream was done first by Kovasznay et al. [1], who showed that within linear approximation a general small-amplitude unsteady disturbance in the free stream can be decomposed into three independent different types: acoustic (A), vortical, and entropy modes. Only the first of these relates to pressure fluctuations propagating with the sound speed; the last two do not cause pressure perturbation. The common idea of many investigations is to pick out length and time scales for each of these disturbances that would make the nonlinear generation of the Tollmienn-Schlichting (T-S) waves possible [2]. This way, the mechanism of T-S wave generation by convecting gusts interacting with sound was justified and numerically investigated [3]. The appropriate scales of boundary roughness were proved to generate T-S waves both theoretically and experimentally [4].

There are abundant efforts (see the introduction and the citations in [3]), devoted to the foundation of an effective control mechanism which supports a cumulative exchange of energy between T-S and acoustic modes. In spite of that, the problem still requires additional efforts. Let us mention three important papers concerning the general, local, and distributed acoustic receptivity [5–7]. The results, however, do not look complete: we revisit the problem in the all-perturbations approach [8–10] starting from the boundary layer (BL) as a background. The perturbations are considered only over the stationary boundary layer; we do not account here for the layer field as a dynamic variable. This scheme gives the possibility to study mutual interactions on the base of a model integrable nonlinear evolution equation. In our approach, the description is local by construction and does not need an averaging procedure [12]. It covers the known results and gives new hope for understanding the related phenomena that appeared in papers [11–13]. The variations in the structure of BL take place. This effect is similar to the heating or streaming generation by acoustic waves [14]; it is a development of the stationary mode and corresponds to an initial stage of the BL reconstruction.

In this chapter, we concentrate our efforts on the mathematical formalism: introducing the complete set of basic modes, we transform the fundamental system of standard conservation laws of fluid mechanics to a set of equivalent equations. In linear approximation, a specific choice of new independent variables splits the system into the set of independent equations for the given modes; the account of nonlinearity naturally introduces the interaction by projecting the fundamental equations set in a vector form. Going to the nonlinear description, we use iterations inside the operator by the small parameter related to amplitude (Mach number for acoustics) and viscosity (Reynolds number). We also analyze the possibilities of the resonant interaction of quasi-plane waves on the level of so-called N-wave systems [15]. Being integrable, such systems admit explicit solutions and plenty of conservation laws. Hence, the detailed investigation of the situation is possible in this approximation.

The mentioned types of waves (T-S and A) are defined by the eigenvectors of the linearized system of dynamic conservation equations for the free flow. Once defined, eigenvectors (or modes) are fixed and independent on time. The process by which the free stream disturbances are internalized to generate boundary-layer instability waves is referred to as receptivity. The basic idea of this chapter is to define the T-S and acoustics modes as eigenvectors of the same system for a viscous flow over a rigid boundary. The eigenvectors of the viscous flow go to the known limit in the free stream over a boundary. Our idea is to fix relations between specific perturbations (velocity components, density, and pressure) for every mode following from the linear equations, to construct projectors and apply them for the nonlinear dynamics investigation. Thus, in the linear dynamics, the overall field may be separated by projectors into the independent modes at any moment. Fixing the relations when going to the nonlinear flow, one goes to a system of coupled evolution equations for the modes. These ideas apply to nonlinear dynamics of exponentially stratified gas and bubbly liquid [10] and other problems [9].

The principal difference between this consideration and the ones from previous chapters is the necessity of the incorporation of variable coefficients that arise from a boundary layer structure [16]. This means that such coordinate dependence impacts the projectors structure: the projector operator should be constructed by nonabelian entries as touched on in Section 2.1.6 for the weak inhomogeneity case. In other words, the matrix elements of the projector matrix will be operator-valued. In fact, we revisit and develop the first attempt in this direction that had been made recently [17,18].

The T-S wave takes the place as a vortical mode in unbounded space; the difference is due to the linearization on the different background. When a stationary boundary-layer flow like the Blausius one appears, a linearization should be preceded by an account of the boundary-layer flow as a background, which would lead to the features of the vortical mode other than those in unbounded space. An important feature of T-S mode is the nonzero disturbance of pressure already in the linear theory. In three-dimensional flow, there exist two T-S modes, two acoustic ones (corrected by background flow), and one entropy mode as well.

The equations of interaction are derived in Section. 6.7 by means of the division of the perturbation field on these subspaces and projecting the system of the basic equations onto the same subspaces. In Chapter 3, the homogeneous background and the particular case of inhomogeneous equilibrium gas were considered. In this section, the background parameters of BL may strongly vary. This transformation, which in fact is nothing but a change of variables, allows to proceed in a choice of adequate approximation. Moreover, we could analyze possibilities of a direct nonlinear resonance interaction account. The results are the following: two-wave and three-wave interactions do not contribute due to the structure of the interaction terms in its minimal possible order. Hence, (in this order) only the four-wave interaction displays a resonance structure. We derive the corresponding four-wave equations and analyze them in Section 6.8.

6.2 BASIC EQUATIONS FOR COMPRESSIBLE FLUID

The mass, momentum, and energy conservation equations read:

$$\frac{\partial \rho}{\partial t}+\vec{\nabla}(\rho\cdot\vec{v})=0, \tag{6.1}$$

$$\rho\left[\frac{\partial \vec{v}}{\partial t}+(\vec{v}\cdot\vec{\nabla})\vec{v}\right]=-\vec{\nabla}p+\eta\Delta\vec{v}+\left(\varsigma+\frac{\eta}{3}\right)\vec{\nabla}(\vec{\nabla}\cdot\vec{v}), \tag{6.2}$$

$$\rho\left[\frac{\partial e}{\partial t}+(\vec{v}\cdot\vec{\nabla})e\right]+p\vec{\nabla}\cdot\vec{v}=\chi\Delta T+\varsigma\left(\vec{\nabla}\cdot\vec{v}\right)^2+\frac{\eta}{2}\left(\frac{\partial v_i}{\partial x_k}+\frac{\partial v_k}{\partial x_i}-\frac{2}{3}\delta_{ik}\frac{\partial v_l}{\partial x_l}\right)^2. \tag{6.3}$$

Here, ρ, p are density and pressure, e, T-internal energy per unit mass and temperature, η, ς, are shear, bulk viscosities, and χ-thermal conductivity coefficient, respectively (all are supposed to be constants), $\vec{v}$ is a velocity vector, x_i-space coordinates. Except for the dynamical equations (6.1–6.3), the two thermodynamic relations are necessary: $e(p,\rho), T(p,\rho)$. To treat a wide variety of substances, let us use the general form of the caloric equation (energy) as an expansion in the Taylor series:

$$\rho_0 e=E_1 p+\frac{E_2 p_0}{\rho_0}\rho+\frac{E_3}{p_0}p^2+\frac{E_4 p_0}{\rho_0^2}\rho^2+\frac{E_5}{\rho_0}p\rho+\frac{E_6}{p_0\rho_0}p^2\rho+\frac{E_7}{\rho_0^2}p\rho^2+\frac{E_8}{p_0^2}p^3+\cdots, \tag{6.4}$$

and the thermic one

$$T=\frac{\Theta_1}{\rho_0 C_v}p+\frac{\Theta_2 p_0}{\rho_0^2 C_v}\rho+\cdots. \tag{6.5}$$

The background values for unperturbed medium are marked by zero; perturbations of pressure and density are denoted by the same characters (no confusion is possible since only perturbations appear below); C_v means the specific heat per unit mass at constant volume; and $E_1,\ldots\Theta_1,\ldots$ are dimensionless coefficients. The two-dimensional flow in the coordinates x (streamwise distance from the plate (model) leading edge), z (wall-normal distance from a model surface) relates to the two-component velocity vector

$$\vec{v} = (u, w) + \vec{u}_0, \tag{6.6}$$

where $\vec{u}_0 = (U_0(z), 0)$ denotes the background streamwise velocity and (u, w) stands for velocity perturbations. The system (6.1–6.3) with account of (6.4–6.6) yields in

$$\begin{gathered}
\rho_0\left(\frac{\partial u}{\partial t} + U_0\frac{\partial u}{\partial x} + w\frac{\partial U_0}{\partial z}\right) + \frac{\partial p}{\partial x} - \eta\Delta u - \\
\left(\varsigma + \frac{\eta}{3}\right)\left(\frac{\partial^2 u}{\partial x^2} + \frac{\partial^2 w}{\partial x \partial z}\right) = -\rho_0 u\frac{\partial u}{\partial x} - \rho_0 w\frac{\partial u}{\partial z} + \frac{\rho}{\rho_0}\frac{\partial p}{\partial x}, \\
\rho_0\left(\frac{\partial w}{\partial t} + U_0\frac{\partial w}{\partial x}\right) + \frac{\partial p}{\partial z} - \eta\Delta w - \left(\varsigma + \frac{\eta}{3}\right)\left(\frac{\partial^2 u}{\partial x \partial z} + \frac{\partial^2 w}{\partial z^2}\right) = \\
-\rho_0 w\frac{\partial w}{\partial z} - \rho_0 u\frac{\partial w}{\partial x} + \frac{\rho'}{\rho_0}\frac{\partial p}{\partial z}, \\
\frac{\partial p}{\partial t} + U_0\frac{\partial p}{\partial x} + c^2\rho_0\left(\frac{\partial u}{\partial x} + \frac{\partial w}{\partial z}\right) = \\
\frac{\chi}{E_1}\left(\frac{\Theta_1}{\rho_0 C_v}\Delta p + \frac{\Theta_2 p_0}{\rho_0^2 C_v}\Delta\rho\right)\left[pZ + \rho c^2 S\right] - u\frac{\partial p}{\partial x} - w\frac{\partial p}{\partial z},
\end{gathered} \tag{6.7}$$

where the constants Z and S are defined by

$$\begin{aligned}
Z &= \left(-1 + 2\frac{1-E_2}{E_1}E_3 + E_5\right)/E_1, \\
S &= \frac{1}{1-E_2}\left(1 + E_2 + 2E_4 + \frac{1-E_2}{E_1}E_5\right).
\end{aligned} \tag{6.8}$$

The constant $c = \sqrt{\frac{p_0(1-E_2)}{\rho_0 E_1}}$ has the sense of linear sound velocity in the medium under consideration when $U_0 = 0$. The right-hand side of the equations involves the quadratic nonlinear and viscous terms as well as the linear ones related to thermal conductivity; no cross viscous-nonlinear terms are accounted for. In fact, the third equation follows from the energy balance and continuity equation. No assumptions on flow compressibility were made yet.

6.3 LINEAR APPROXIMATION

The basic system (6.1–6.3) contains four dynamic equations, and therefore, there are four independent modes of the linear flow: two acoustic ones, vorticity, and heat modes. This is a classification by Kovasznay, who defined the acoustic modes as isentropic and irrotational flow, and who refers to the two last modes as frozen motions, the vorticity one relating to the absence of pressure and density perturbations and the heat one relating to the very density perturbation. The system (6.1–6.3) involves three equations indeed; therefore, only the three modes may be extracted–two acoustic modes and a vorticity one. This is due to the structure of the heat mode where the only density perturbation occurs. Strictly speaking, the term treating a thermal conductivity in the third of the equations (6.3) includes the density perturbations, and the linearized system (6.7) (which defines the modes as possible types of flow) is not closed. If this term did not exist at all, there is a simple linear relation between density and pressure perturbations for both acoustic modes. The presence of thermal conductivity corrects this relation as it was shown in [19]. When the effects of thermal conductivity are small, the corresponding terms may be placed on the right-hand side of equation, together with nonlinear ones, and be accounted further. So the exclusion of the dynamic equation for density serves as just a simplification of a problem suitable in view of its extraordinary complexity when the heat mode is out of the area of interest.

The main idea is to define modes according to the specific relations of the basic perturbation variables following from the linearized system of dynamic equations (6.1–6.3). In general, the procedure is algorithmic and may be expressed as consequent steps: to find the dispersion relation and its roots that determine all possible modes, and to define relations between specific variables for every mode. Later, projectors follow immediately from these relations; they separate every mode from the overall field of the linear flow exactly and serve as a tool for the nonlinear dynamics investigation.

Using the left-hand part of the system (6.7) as a basis for modes definition and introducing the two non-dimensional functions $V_0(z) = U_0(z)/U_\infty$, $\phi(z) = V_{0z}(z)l_0$, we rewrite it in the non-dimensional variables

$$x_* = x/l_0, w_* = w/U_\infty, u_* = u/U_\infty, t_* = tU_\infty/l_0, p_* = p/\rho_0 U_\infty^2. \tag{6.9}$$

The value U_∞ marks the velocity of a flow far from the boundary, and l_0 is the boundary layer width. In the new variables (asterisks will be later omitted), (6.7) with zero on the right side reads:

$$\frac{\partial p}{\partial t} + V_0\frac{\partial p}{\partial x} + \varepsilon^{-2}\left(\frac{\partial u}{\partial x} + \frac{\partial w}{\partial z}\right) = 0, \tag{6.10}$$

$$\frac{\partial u}{\partial t} + V_0\frac{\partial u}{\partial x} + \phi w + \frac{\partial p}{\partial x} - Re^{-1}\Delta u - R^{-1}\left(\frac{\partial^2 u}{\partial x^2} + \frac{\partial^2 w}{\partial x \partial z}\right) = 0, \tag{6.11}$$

$$\frac{\partial w}{\partial t} + V_0\frac{\partial w}{\partial x} + \frac{\partial p}{\partial z} - Re^{-1}\Delta w - R^{-1}\left(\frac{\partial^2 u}{\partial x \partial z} + \frac{\partial^2 w}{\partial z^2}\right) = 0, \tag{6.12}$$

with parameters $\varepsilon = U_\infty/c$ (the Mach number); $Re = U_\infty l_0 \rho_0/\eta$ is the Reynolds number based on the length scale; and $R = U_\infty l_0 \rho_0/(\eta/3 + \varsigma)$.

6.4 THE TOLLMIENN-SCHLICHTING MODE

Formally, the limit of incompressible fluid ($\varepsilon = 0$) corresponds to the vorticity mode. The first relation for velocity components is well known [3]:

$$\partial u/\partial x + \partial w/\partial z = 0. \tag{6.13}$$

From Eqs (6.11 and 6.12), an expression for the pressure perturbation follows:

$$2\phi \partial w/\partial x + \Delta p = 0, \tag{6.14}$$

where $\Delta = \partial^2/\partial z^2 - k^2$ stands for the Laplacian; so far, the $\partial/\partial x$ equivalent operator (multiplier) $-ik$ is used. Both (6.13) and (6.14) define the T-S mode due to the relations of the specific perturbations of pressure and velocity components. Since the geometry of the viscous flow over the boundary supposes strong non-uniformity in the vertical direction, all disturbances may be thought of not as plane waves but as functions like $\Psi(x,z) = \psi(z)\exp(i\omega t - ikx)$. So, one has to leave vertical derivatives that usually are large in comparison with the horizontal ones. The result is rather obvious; hence, we do not introduce a special small parameter. The vector of the T-S mode may be chosen as

$$T = \begin{pmatrix} p_{TS} \\ u_{TS} \\ w_{TS} \end{pmatrix} = \begin{pmatrix} 1 \\ -\frac{1}{2k^2}\frac{\partial}{\partial z}\frac{1}{\phi}\Delta \\ \frac{1}{2ik\phi}\Delta \end{pmatrix} p_{TS}. \tag{6.15}$$

Note also that Eqs (6.10–6.12) yield the well-known equation for the T-S mode [20], when rewritten for the new variable such as the stream function ($u = \partial\Psi/\partial z$, $w = -\partial\Psi/\partial x$), with the restriction to the solenoidal velocity field (6.13):

$$\Delta\partial\Psi/\partial t + V_0\Delta\partial\Psi/\partial x - \partial\Psi/\partial x \cdot \partial\phi/\partial z - Re^{-1}\Delta^2\Psi = 0. \tag{6.16}$$

The well-known Orr-Sommerfeld (OS) equation follows from (6.16):

$$[V_0(z) - c]\left[\partial^2\psi/\partial z^2 - k^2\psi\right] - \psi\partial\phi/\partial z = \frac{i}{Rek}\left[\partial^4\psi/\partial z^4 - 2k^2\partial^2\psi/\partial z^2 + k^4\psi\right]. \tag{6.17}$$

This equation is an initial point of the laminar flow stability theory and for every pair (k, Re) determines an eigenfunction $\psi(z)$ and a complex phase velocity $c = \omega/k = c_r + ic_i$. The sign of c_i is namely a criterion of flow stability: a negative value corresponds to the growth of perturbation and therefore to the instability of the flow.

6.5 ACOUSTIC MODES

The potential flow imposes two acoustic modes with

$$\partial u/\partial z - \partial w/\partial x = 0.$$

In the limit of $Re^{-1} = 0$, $R^{-1} = 0$, $\phi = 0$, (6.10–6.12) naturally goes to the acoustic modes dispersion relation that is directly connected with the wave operator.

$$\varepsilon^2\frac{\partial^2}{\partial t^2} - \Delta. \tag{6.18}$$

We would not consider here the perturbation velocity field changes, forced by the ambient movement of the fluid. It could be accounted for by the perturbation theory to be developed here. Then, two acoustic modes are defined with relations between specific perturbations:

$$A_1 = \begin{pmatrix} p_{A1} \\ u_{A1} \\ w_{A1} \end{pmatrix} = \begin{pmatrix} 1 \\ -i\varepsilon k\Delta^{-1/2} \\ \varepsilon\frac{\partial}{\partial z}\Delta^{-1/2} \end{pmatrix} p_{A1}, A_2 = \begin{pmatrix} p_{A2} \\ u_{A2} \\ w_{A2} \end{pmatrix} = \begin{pmatrix} 1 \\ i\varepsilon k\Delta^{-1/2} \\ -\varepsilon\frac{\partial}{\partial z}\Delta^{-1/2} \end{pmatrix} p_{A2}. \tag{6.19}$$

Here the square root of the operator Δ is defined as an integral operator via the Fourier transform.

6.6 PECULIARITIES OF NON-COMMUTATIVE PROJECTING IN THE INHOMOGENEOUS LINEAR PROBLEM

The T-S and the two acoustic modes are determined by relations of specific perturbations or in the vector form (6.15), (6.19). The superposition of such disturbances appear as all possible types of flow except the heat mode. Every mode is completely defined by one of several specific perturbations—pressure or velocity components since there are strict and local relations between them. Practically, in a linear flow, the overall perturbation may be decoupled into modes, by the corresponding orthogonal projectors. The arbitrary perturbation then is a sum of modes, which, taking into account (6.15), (6.19) looks like

$$\begin{pmatrix} p \\ u \\ w \end{pmatrix} = \begin{pmatrix} p_{A1} + p_{A2} + p_{TS} \\ u_{A1} + u_{A2} + u_{TS} \\ w_{A1} + w_{A1} + w_{TS} \end{pmatrix} = \begin{pmatrix} p_{A1} + p_{A2} + p_{TS} \\ Hp_{A1} - Hp_{A2} + Kp_{TS} \\ Mp_{A1} - Mp_{A2} + Qp_{TS} \end{pmatrix}, \tag{6.20}$$

with operators

$$\begin{aligned} H &= -\varepsilon\Delta^{-1/2} ik, \\ M &= \varepsilon\Delta^{-1/2}\partial/\partial z, \\ K &= -\frac{1}{2k^2}\partial/\partial z\frac{1}{\phi}\Delta, \\ Q &= \frac{1}{2\phi ik}\Delta. \end{aligned} \tag{6.21}$$

The link (6.20) may be considered as a one-to-one map of dynamical variables that in the case of a T-S wave immediately yields projector:

$$P_{TS} = \begin{pmatrix} 0 & -2k^2\Delta^{-1}\phi\partial/\partial z\Delta^{-1} & -2ik^3\Delta^{-1}\phi\Delta^{-1} \\ 0 & \partial^2/\partial z^2\Delta^{-1} & ik\partial/\partial z\Delta^{-1} \\ 0 & ik\partial/\partial z\Delta^{-1} & -k^2\Delta^{-1} \end{pmatrix}. \tag{6.22}$$

For right and left acoustic waves, one has

$$P_{A1} = \frac{\Delta^{1/2}}{2}$$

$$\begin{pmatrix} 1 & 2k^2\Delta^{-1/2}\phi\frac{\partial}{\partial z}\Delta^{-1} - \frac{ik}{\varepsilon}\Delta^{-1/2} & 2ik^3\Delta^{-1/2}\phi\Delta^{-1} + \frac{1}{\varepsilon}\frac{\partial}{\partial z} \\ \frac{-ik}{\varepsilon} & -ik^2\varepsilon\left[2k\Delta^{-1}\phi\frac{\partial}{\partial z}\Delta^{-1} - \frac{i}{\varepsilon}\Delta^{-1/2}\right] & -ik\varepsilon\left[2ik^3\Delta^{-1}\phi\Delta^{-1} + \frac{1}{\varepsilon}\Delta^{-1/2}\frac{\partial}{\partial z}\right] \\ \frac{1}{\varepsilon}\frac{\partial}{\partial z} & k\varepsilon\frac{\partial}{\partial z}\left[2k\Delta^{-1}\phi\frac{\partial}{\partial z}\Delta^{-1} - \frac{ik}{\varepsilon}\Delta^{-1/2}\right] & \varepsilon\frac{\partial}{\partial z}\left[2ik^3\Delta^{-1}\phi\Delta^{-1} + \frac{1}{\varepsilon}\Delta^{-1/2}\frac{\partial}{\partial z}\right] \end{pmatrix}, \tag{6.23}$$

$$P_{A2} = \frac{\Delta^{1/2}}{2}$$

$$\begin{pmatrix} 1 & 2k^2\Delta^{-1/2}\phi\frac{\partial}{\partial z}\Delta^{-1} + \frac{ik}{\varepsilon} & 2ik^3\Delta^{-1/2}\phi\Delta^{-1} - \frac{1}{\varepsilon}\frac{\partial}{\partial z} \\ \frac{ik}{\varepsilon} & 2ik^2\varepsilon\left[k\Delta^{-1}\phi\frac{\partial}{\partial z}z\Delta^{-1} + \frac{i}{\varepsilon}\Delta^{-1/2}\right] & 2ik\varepsilon\left[ik^3\Delta^{-1}\phi\Delta^{-1} - \frac{1}{\varepsilon}\Delta^{-1/2}\frac{\partial}{\partial z}\right] \\ -\frac{1}{\varepsilon}\frac{\partial}{\partial z} & -2k\varepsilon\frac{\partial}{\partial z}\left[\Delta^{-1}\phi\frac{\partial}{\partial z}\Delta^{-1} + \frac{i}{\varepsilon}\Delta^{-1/2}\right] & -2\varepsilon\frac{\partial}{\partial z}\left[ik^3\Delta^{-1}\phi\Delta^{-1} - \frac{1}{\varepsilon}\Delta^{-1/2}\frac{\partial}{\partial z}\right] \end{pmatrix}. \tag{6.24}$$

The projectors possess all properties of orthogonal projectors, and their sum is a unit matrix since all eigenvectors of a linear system are accounted for.

$$P_{TS}^2 = P_{TS}, \cdots, P_{TS} + P_{A1} + P_{A2} = I, P_{TS} \cdot P_{A1} = P_{TS} \cdot P_{A2} = \cdots = 0,$$

where I and 0 are unit and zero matrices, respectively. We would stress that the operators defined by the (6.22–6.24) contain operator-valued matrix elements. Hence, the derivation of their explicit form, as well as the properties mentioned earlier, could be checked by taking the nonabelian nature of the operators into account. In the linear flow, the projectors separate every mode from the overall perturbation, for example:

$$P_{TS}\begin{pmatrix} p \\ u \\ w \end{pmatrix} = \begin{pmatrix} p_{TS} \\ u_{TS} \\ w_{TS} \end{pmatrix} \tag{6.25}$$

and so on. Moreover, acting by a projector on the basic system of dynamic equations (6.10–6.12), yields a linear evolution equation for the mode assigned to this projector. One produces three equations indeed for all specific perturbations that are essentially the same with an account of relations (6.14) and (6.19). For the first rightwards progressive acoustic mode, the evolution equation reads

$$\partial p_{A1}/\partial t + V_0\partial p_{A1}/\partial x + \varepsilon^{-1}\Delta^{1/2}p_{A1} = 0. \tag{6.26}$$

The equation for the second (opposite directed) acoustic mode is produced by acting the projector P_{A2} on the basic system and differs from (6.26) only by the sign before the last term. Combining equations for these directed acoustic modes, one arrives at the wave equation of second order:

$$(\partial/\partial t + V_0\partial/\partial x)^2 p - \varepsilon^{-2}\Delta p = 0.$$

This equation appears as a limit of a more general one: this relates to both acoustic modes and can be found in [13].

6.7 NONLINEAR FLOW: COUPLED DYNAMIC EQUATIONS

In the dimensionless variables introduced by (6.9), the dynamic equations with account of nonlinear terms of the second order look like

$$\begin{aligned} \frac{\partial p}{\partial t} + V_0 \frac{\partial p}{\partial x} + \varepsilon^{-2}\left(\frac{\partial u}{\partial x} + \frac{\partial w}{\partial z}\right) &= \tilde{\varphi}_1, \\ \frac{\partial u}{\partial t} + V_0 \frac{\partial u}{\partial x} + \phi w + \frac{\partial p}{\partial x} - Re^{-1}\Delta u - R^{-1}\left(\frac{\partial^2 u}{\partial x^2} + \frac{\partial^2 w}{\partial x \partial z}\right) &= \tilde{\varphi}_2, \\ \frac{\partial w}{\partial t} + V_0 \frac{\partial w}{\partial x} + \frac{\partial p}{\partial z} - Re^{-1}\Delta w - R^{-1}\left(\frac{\partial^2 u}{\partial x \partial z} + \frac{\partial^2 w}{\partial z^2}\right) &= \tilde{\varphi}_3, \end{aligned} \tag{6.27}$$

with a vector of the second-order nonlinear terms $\tilde{\psi}$ on the right-hand side (a non-dimensional value $\rho_* = \rho/\rho_0$ is used on the right-hand side; the asterisk will be omitted later):

$$\tilde{\varphi} = \begin{pmatrix} \tilde{\varphi}_1 \\ \tilde{\varphi}_2 \\ \tilde{\varphi}_3 \end{pmatrix} = \begin{pmatrix} -u\frac{\partial p}{\partial x} - w\frac{\partial p}{\partial z} + \left(\frac{\partial u}{\partial x} + \frac{\partial w}{\partial z}\right)\left[Z\,p + \varepsilon^{-2}\,S\rho\right] \\ -u\frac{\partial u}{\partial x} - w\frac{\partial u}{\partial z} + \rho\frac{\partial p}{\partial x} \\ -u\frac{\partial w}{\partial x} - w\frac{\partial w}{\partial z} + \rho\frac{\partial p}{\partial z} \end{pmatrix}. \tag{6.28}$$

One could rewrite a system (6.27) in another way:

$$\frac{\partial}{\partial t}\varphi + L\varphi = \tilde{\varphi}, \tag{6.29}$$

where the vector state φ is a specific perturbations column (vector of the fluid state)

$$\varphi = \begin{pmatrix} p \\ u \\ w \end{pmatrix} \tag{6.30}$$

and L is a matrix operator

$$\begin{pmatrix} V_0\partial/\partial x & \varepsilon^{-2}\partial/\partial x & \varepsilon^{-2}\partial/\partial z \\ \partial/\partial x & V_0\partial/\partial x - Re^{-1}\Delta - R^{-1}\partial^2/\partial x^2 & \phi - R^{-1}\partial^2/\partial x\partial z \\ \partial/\partial z & -R^{-1}\partial^2/\partial x\partial z & V_0\partial/\partial x - Re^{-1}\Delta - R^{-1}\partial^2/\partial z^2 \end{pmatrix}. \tag{6.31}$$

All the projectors do commute with operators $\partial/\partial t \cdot I$ and L, so one can act by projectors on the system of equations (6.30) and (6.31) directly, thus obtaining the evolution equation for the corresponding mode. There are three such equations for the specific perturbations p, u, w for every mode (that are equivalent, accounting for the relations between these specific perturbations. So, an independent variable for every mode in a single specific perturbation, such as pressure or one velocity component, may be chosen.

Due to the existing tradition, it is convenient to use the stream function as a basis for the T-S mode and pressure perturbations for the acoustic modes.

Acting by the P_{TS} on the both sides of (6.30), one gets an evolution equation:

$$\Delta\partial\Psi/\partial t + V_0\Delta\partial\Psi/\partial x - \partial\Psi/\partial x\cdot\partial\phi/\partial z - Re^{-1}\Delta^2\Psi = \frac{\partial}{\partial z}[-u\partial u/\partial x \\ -w\partial u/\partial z + \rho\partial p/\partial x] - \frac{\partial}{\partial x}[-u\partial w/\partial x - w\partial w/\partial z + \rho\partial p/\partial z]. \tag{6.32}$$

Next, it should be noted that on the right-hand nonlinear side p, u, w are overall perturbations to be presented as a sum of specific perturbations of all modes:

$$u = \partial\Psi/\partial z + \varepsilon\Delta^{-1/2}\partial p_{A1}/\partial x - \varepsilon\Delta^{-1/2}\partial p_{A2}/\partial x, \\ w = -\partial\Psi/\partial x + \varepsilon\Delta^{-1/2}\partial p_{A1}/\partial z - \varepsilon\Delta^{-1/2}\partial p_{A2}/\partial z, \\ p = 2\partial^2/\partial x^2\Delta^{-1}(\phi\Psi) + p_{A1} + p_{A2}. \tag{6.33}$$

Indeed, there is also a density perturbation in the right-hand nonlinear side that was not involved in the left-hand linear one at all. The continuity equation reads

$$\partial\rho/\partial t + V_0\partial\rho/\partial x + (\partial u/\partial x + \partial w/\partial z) = -\partial(\rho u)/\partial x - \partial(\rho w)/\partial z. \tag{6.34}$$

Comparing the linear left-hand side of Eq. (6.34) with that of the first equation from (6.26), the obvious relations for the both acoustic modes follow:

$$\rho_{A1} = \varepsilon^2 p_{A1},\ \rho_{A2} = \varepsilon^2 p_{A2}.$$

A limit $\varepsilon = 0$ yields in the T-S mode for incompressible flow: $\rho_{TS} = 0$. Therefore, the last relation for the overall density perturbation looks like

$$\rho = \varepsilon^2 p_{A1} + \varepsilon^2 p_{A2} \tag{6.35}$$

Finally, (6.32) goes to

$$\Delta\Psi_t + V_0\Delta\Psi_x - \Psi_x\phi_z - Re^{-1}\Delta^2\Psi = -\Psi_z\Delta\Psi_x + \Psi_x\Delta\Psi_z \\ -\varepsilon(\Delta\Psi\Delta^{1/2}(p_{A1} - p_{A2}) + \Delta\Psi_x\Delta^{-1/2}(p_{A1} - p_{A2})_x \\ +\Delta\Psi_z\Delta^{-1/2}(p_{A1} - p_{A2})_z) + O(\varepsilon^2). \tag{6.36}$$

Derivatives are marked with lower indices. The first two nonlinear terms on the right-hand side of (6.36) expresses the T-S mode self-action, and the last ones—cross acoustic-vorticity terms responsible for the acoustic mode influence on the T-S mode propagation. The structure of the quadratic nonlinear column (6.30) yields the absence of quadratic acoustic terms in (6.36).

In the limit of $V_0 = 0$, $\phi = 0$, the only self-action gives the well-known evolution equation for vorticites transition [20] that follows from (6.36):

$$\Delta\Psi_t - Re^{-1}\Delta^2\Psi + \Psi_z\Delta\Psi_x - \Psi_x\Delta\Psi_z = 0.$$

Let an acoustic field consist of only the first mode. Acting by projector P_{A1} on the system (6.28), we get an evolution equation for this mode:

$$\partial p_{A1}/\partial t + V_0 \partial p_{A1}/\partial x + \Delta^{1/2} p_{A1}/\varepsilon = \frac{1}{2}[-u\partial p/\partial x - w\partial p/\partial z + (\partial u/\partial x + \partial w/\partial z)$$
$$(Zp + Sp\varepsilon^{-2})] + (-\partial^2/\partial x^2 \Delta^{-1}\phi\partial/\partial z \Delta^{-1} + \partial/\partial x(1/2\varepsilon)\Delta^{-1/2})$$
$$[-u\partial u/\partial x - w\partial u/\partial z + \rho\partial p/\partial x] + (\partial^3/\partial x^3 \Delta^{-1}\phi\Delta^{-1} + (1/2\varepsilon)\Delta^{-1/2}\partial/\partial z)$$
$$[-u\partial w/\partial x - w\partial w/\partial z + \rho\partial p/\partial z]. \tag{6.37}$$

Here the constants Z and S were defined earlier by (6.8). The variables p, u, w, are overall perturbations according to (6.33), with $p_{A2} = 0$. So, (6.37) goes to the final version for the directed acoustic mode

$$\varepsilon(\partial p_{A1}/\partial t + V_0 \partial p_{A1}/\partial x) + \Delta^{1/2} p_{A1} = \frac{\varepsilon}{2}[-\Psi_z p_{A1x} + \Psi_x p_{A1z}]$$
$$+ \varepsilon\Delta^{-1/2}[-\Psi_z \Delta^{1/2} p_{A1x} + \Psi_x \Delta^{1/2} p_{A1z} - 2\Psi_{xz}\Delta^{-1/2} p_{A1xx}$$
$$- 2\Psi_{zz}\Delta^{-1/2} p_{A1xz} + 2\Psi_{xx}\Delta^{-1/2} p_{A1xz} + 2\Psi_{xz}\Delta^{-1/2} p_{A1zz}] + \Delta^{-1/2} \tag{6.38}$$
$$\times\left[-(\Psi_{xz})^2 + \Psi_{xx}\Psi_{zz}\right] + \varepsilon\left[-\Psi_z \Delta^{-1}(\phi\Psi)_{xxx} + \Psi_x \Delta^{-1}(\phi\Psi)_{xxz}\right]$$
$$+ \varepsilon\Delta^{-1}\phi\Delta^{-1}\partial^2/\partial x^2 [\Psi_z \Delta\Psi_x - \Psi_x \Delta\Psi_z] + O(\varepsilon^2).$$

Between the nonlinear terms one can recognize interaction (A1–T-S) and generation ones (T-S–T-S). The equation for p_{A2} is obtained by projecting P_{A2}, and it looks very similar. The complete system includes this equation and (6.38) and (6.32). The system covers all possible processes description up to quadratic terms approximation. Here a multimode T-S waves in the (OS) equation solutions basis could be incorporated. The long-wave limit of such a disturbance leads to a coupled KdV system [17].

6.8 RESONANCE INTERACTION OF ACOUSTIC AND T-S MODES

Equations (6.36) and (6.38) form a coupled system of evolution equations for interacting acoustic and T-S modes. In the case of the generation of a T-S mode by an incoming first acoustic mode, the early stage of evolution (for small amplitudes of T-S mode) is defined by a system:

$$\Delta\Psi_t + V_0\Delta\Psi_x - \phi_z \cdot \Psi_x - Re^{-1}\Delta^2\Psi$$
$$= -\varepsilon\left(\Delta\Psi \cdot \Delta^{1/2} p_{A1} + \Delta\Psi_x \cdot \Delta^{-1/2} p_{A1x} + \Delta\Psi_z \cdot \Delta^{-1/2} p_{A1z}\right) \tag{6.39}$$

$$\varepsilon(\partial p_{A1}/\partial t + V_0 \partial p_{A1}/\partial x) + \Delta^{1/2} p_{A1} = \tfrac{\varepsilon}{2}[-\Psi_z p_{A1x} + \Psi_x p_{A1z}]$$
$$+ \varepsilon\Delta^{-1/2}[-\Psi_z \Delta^{1/2} p_{A1x} + \Psi_x \Delta^{1/2} p_{A1z} - 2\Psi_{xz}\Delta^{-1/2} p_{A1xx} \tag{6.40}$$
$$- 2\Psi_{zz}\Delta^{-1/2} p_{A1xz} + 2\Psi_{xx}\Delta^{-1/2} p_{A1xz} + 2\Psi_{xz}\Delta^{-1/2} p_{A1zz}].$$

All quadratic terms relating to T-S–T-S interaction are left out of account. The terms enter only (6.39) and these are relatively small while T-S amplitude just begins to grow. For simplicity, we consider the first incoming acoustic mode exclusively.

As it follows from the discussion in the introduction, let us find a solution in the form:

$$p_{A1}(x,z,t) = A_1(\mu x,\mu t)\pi_1 \exp(i(\omega_1 t - k_1 x)) + A_2(\mu x,\mu t)\pi_2 \exp(i(\omega_2 t - k_2 x)) + c.c. \tag{6.41}$$

$$\begin{aligned}\Psi(x,z,t) =& B_3(\mu x,\mu t)\psi_3(z)\exp(i(\omega_3 t - k_3 x))\\ &+ B_4(\mu x,\mu t)\psi_4(z)\exp(i(\omega_4 t - k_4 x)) + c.c.\end{aligned} \tag{6.42}$$

where $\Pi_1 = \pi_1(k_1,\omega_1,z)\exp(i(\omega_1 t - k_1 x))$, $\Pi_2 = \pi_2(k_2,\omega_2,z)\exp(i(\omega_2 t - k_2 x))$. The amplitude functions A_i, B_i are slow varying envelopes of the wavetrains. The function Π_1 satisfies the linear evolution equation following from (6.26)

$$\partial\Pi_1/\partial t + V_0\partial\Pi_1/\partial x + \varepsilon^{-1}\Delta^{1/2}\Pi_1 = 0, \tag{6.43}$$

this leads to the spectral equation for π_1

$$i\omega_1\pi_1 - ik_1V_0\pi_1 + \varepsilon^{-1}\Delta^{1/2}\pi_1 = 0. \tag{6.44}$$

Suppose the vertical gradients of all wave functions inside the viscous layer are much bigger than horizontal ones. From experiments (e.g. [12]), it is known that the wavelength of T-S mode is much greater than a thickness of the boundary viscous layer for a wide range of values of the Reynolds number. Typical vertical (along z) changes of acoustical field depend on the ambient velocity profile V_0 and its gradient ϕ. If both are not very big (subsonic flow), the vertical scale of the sound field is also much less than the horizontal ones k_i^{-1}. So, taking the mentioned assumptions, the operator $\Delta^{1/2}$ may be evaluated as the generalized operator (Taylor) series with respect to ∂_z/k_1. Hence, the operator radical in the first approximation is evaluated via Gataux derivative as

$$\Delta^{1/2}\Pi_1 = \sqrt[2]{-k_1^2 + \partial_z^2}\Pi_1 = \imath k_1\sqrt[2]{1 - \partial_z^2/k_1^2}\Pi_1 \approx \imath k_1(1 - \partial_z^2/2k_1^2)\Pi_1,$$

and, for (6.44), one arrives at the ordinary differential equation, that we can consider as a spectral problem with the spectral parameter k_1,

$$\left(1 - \frac{k_1}{\omega_1}V_0\right)\pi_1 + \frac{k_1}{\omega_1\varepsilon}\left(1 - \frac{1}{2k_1^2}\partial_z^2\right)\pi_1 = 0. \tag{6.45}$$

The same equation obviously defines π_2; it is enough to change indices $1 \to 2$ in the operator. The functions $\psi_3(z)$, $\psi_4(z)$ are solutions of the OS equation (6.17) suitable for a concrete problem. Generally, the complex functions. $A_1,\ldots,B_4$ are slowly varying functions of x, t; that's why an additional small parameter μ is introduced. Calculating the right-hand nonlinear expressions, we take only first term in series to avoid small terms of the higher order.

Let us discuss the possibility of four-waves resonance. Examining the algebraic relations between parameters yields the appropriate conditions:

$$\omega_1 = \omega_2 - \omega_3, \omega_4 = \omega_3 - \omega_1. \tag{6.46}$$

From which it follows $\omega_4 = \omega_2 - 2\omega_1$. Substituting the formulas (6.41 and 6.42) into (6.39 and 6.43), and picking up the resonant terms only, one goes to the further system of equations (complex conjugate values marked with asterisks, $k_1 - k_2 + k_3 = \Delta k, k_3 - k_1 - k_4 = \Delta k'$):

$$\mu\left(\varepsilon A_{1T}\pi_1 + A_{1X}\left(\varepsilon V_0\pi_1 - ik_1\int_0^z \pi_1 dz\right)\right)$$

$$= \varepsilon A_2 B_3^* \left(1.5\left(ik_3^*\psi_3^*\pi_{2z} + ik_2\psi_{3z}^*\pi_2\right)\right) + \int_0^z \left(ik_3^*\psi_{3z}^*\pi_{2z} + ik_2\psi_{3zz}^*\pi_2\right) dz e^{i(\Delta k)x},$$

$$\mu\left(\varepsilon A_{2T}\pi_2 + A_{2X}\left(\varepsilon V_0\pi_2 - ik_2\int_0^z \pi_2 dz\right)\right)$$

$$= \varepsilon A_1 B_3^* \left(1.5\left(-ik_3\psi_3^*\pi_{1z} + ik_1\psi_{3z}\pi_1\right)\right) + \int_0^z \left(-ik_3\psi_{3z}^*\pi_{1z} + ik_1\psi_{3zz}^*\pi_1\right) dz e^{i(-\Delta k)x},$$

$$\mu\left(B_{3T}\psi_{3zz} + B_{3X}\left[\left(V_0 + 4ik_3Re^{-1}\right)\psi_{3zz} + \left(2k_3\omega_3 - \phi_z\right)\psi_3\right]\right)$$

$$= -\varepsilon A_1 B_4 \left(\psi_{4zz}\pi_{1z} - k_1 k_4 \psi_{4zz}\int_0^z \pi_1 dz + \psi_{4zzz}\pi_1\right) e^{i(\Delta k')x},$$

$$\mu\left(B_{4T}\psi_{4zz} + B_{4X}\left[\left(V_0 + 4ik_4Re^{-1}\right)\psi_{4zz} + \left(2k_4\omega_4 - \phi_z\right)\psi_4\right]\right)$$

$$= -\varepsilon A_1^* B_3^* \left(\psi_{3zz}^*\pi_{1z}^* + k_1^* k_3 \psi_{3zz}^*\int_0^z \pi_1^* dz + \psi_{3zzz}^*\pi_1^*\right) e^{i(-\Delta k')x}.$$

We take into account that acoustic wavenumbers are real, but wavenumbers of both T-S modes may be complex in the general case, namely the points of real values form the neutral curve [20]. The coefficients of the equation depend on z; this it is due to our choice of the only one transverse mode for each "horizontal" one.

We continue the projecting procedure considering the transverse modes as a basis. In such problems, we naturally arrive at two bases. One arises from T-S waves theory, and its origin is from the OS equation (denoted by $\psi_{3,4}$). The other is from the sound problem. Of course, such bases are not orthogonal. In our case, the acoustic wave basis π_i should be incorporated at once. There are lot of publications about the orthogonalization of OS equation solutions. We do not touch here the complete basis construction (hope to do it later) but only formulate a proposal about it. Starting from normalized ψ' and π', one could produce orthogonal vectors $e_{\pm}\psi' \pm \pi'$. We would use such mixed combinations to form the future complete basis. In such a way, we can account for both contributions from the very beginning. Hence, we multiply each equation by its own basic vector and integrate across the boundary layer (such scalar products can be expressed in terms of orthogonal ones $e_{\pm}$). The result is written in "back"-re-scaled variables: we put $\mu = 1$.

$$
\begin{aligned}
A_{1T} + c_{a1}A_{1X} &= n_{a1}A_2B_3^*e^{i(\Delta k)x}, \\
A_{2T} + c_{a2}A_{2X} &= n_{a2}A_1B_3^*e^{i(-\Delta k)x}, \\
B_{3T} + c_{TS1}B_{3X} &= -\varepsilon n_{TS1}A_1B_4e^{i(\Delta k')x}, \\
B_{4T} + c_{TS2}B_{4X} &= -\varepsilon n_{TS1}A_1^*B_3e^{i(-\Delta k')x},
\end{aligned}
\tag{6.47}
$$

where the group velocities and nonlinear constants are expressed via the integrals across the boundary layer with a width δ:

$$
c_{a1} = \frac{\int\limits_0^{\delta}[\pi_1^2 - \varepsilon^{-1}ik_1\pi_1\int_o^{\delta}\pi_1 dz]dz'}{\int_0^{\delta}V_0\pi_1^2 dz},
$$

$$
n_{a1} = \frac{\int_0^{\delta}[1.5(-ik_3^*\psi_3^*\pi_{2z} + ik_2\psi_{3z}^*\pi_{2}t) + \int\limits_0^{z'}\pi_1(-ik_3^*\psi_{3z}^*\pi_{2z} + ik_2\psi_{3zz}^*\pi_2)dz]dz'}{\int_0^{\delta}V_0\pi_1^2 dz},
$$

$$
c_{a2} = \frac{\int\limits_0^{\delta}[\pi_2^2 - \varepsilon^{-1}ik_2\pi_2\int_0^{\delta}\pi_2 dz]dz'}{\int_0^{\delta}V_0\pi_1^2 dz},
$$

$$
n_{a2} = \frac{\int_0^{\delta}\pi_2[(1.5(-ik_3\psi_3\pi_{1z} + ik_1\psi_{3z}\pi_1) + \int\limits_0^{z'}(-ik_3\psi_{3z}\pi_{1z} + ik_1\psi_{3zz}\pi_1)dz)]dz'}{\int_0^{\delta}V_0\pi_1^2 dz}
\tag{6.48}
$$

$$
c_{TS1} = \frac{\int_0^{\delta}\psi_3[(V_0 + i4k_3Re^{-1})\psi_{3zz} + (2k_3\omega_3 + \varphi_z)\psi_3])dz}{\int_0^{\delta}\psi_3\psi_{3zz}dz},
$$

$$
n_{TS1} = \frac{\int_0^{\delta}\psi_3(\psi_{4zz}\pi_{1z} - k_1k_4\psi_{4zz}\int\limits_0^{z'}\pi_1 dz + \psi_{4zzz}\pi_1)dz'}{\int_0^{\delta}\psi_3\psi_{3zz}dz},
$$

$$
c_{TS2} = \frac{\int_0^{\delta}\psi_4[(V_0 + i4k_4Re^{-1})\psi_{4zz} + (2k_4\omega_4 + \varphi_z)\psi_4])dz}{\int_0^{\delta}\psi_4\psi_{4zz}dz},
$$

$$
n_{TS2} = \frac{\int_0^{\delta}\psi_4(\psi_{3zz}\pi_{1z}^* - k_1k_3\psi_{3zz}\int\limits_0^{z'}\pi_1^* dz + \psi_{3zzz}\pi_1^*)dz'}{\int_0^{\delta}\psi_4\psi_{4zz}dz}.
$$

This resulting equation may be considered as four-wave resonance equation but without synchronism condition [15], [21], [22]. A structure of the obtained equations is the particular case of general N-wave system, which may be solved by special technics valid for integrable equations [15]. The four-wave approximation may give rise to such solutions as those that exhibit effective energy exchange between modes. The form of the nonlinearity is typical for a three-wave systems, and even small fluctuations of a T-S field could initiate a rapid growth of both components if the acoustic field is big enough. It is pure nonlinear instability that may be supported by a linear stability curve shift [18]. The solution of the system (6.47) also poses a separate problem. The solution of the problem includes the numerical evaluation of the integrals in constants (6.48) that need rather extended space. Finally, in this section we

1. Fixed the appropriate subspaces of acoustic and T-S waves by relations between basic variables, having thus created a one-to-one mapping to linear evolution of separate modes
2. Constructed projecting operators that combine equations in a way appropriate to the separation of modes. This combination is used in weak nonlinearity account in the first order (in amplitude) approximation, which produces equations of modes interaction.
3. We found conditions in which the interaction effect could be stored; it is four-wave mixing. The corresponding one-dimensional four-wave system is derived and specified in the whole space of solutions.

The resulting system (6.32) and (6.38) and the equation of the opposite directed A-mode could be considered as basic for an all-perturbations description over a BL. Between the particular cases there is also the Lighthill equation describing the generation of acoustic waves by the disturbances of the boundary layer. We also would note that the boundary layer width may depend on x. A slow dependence is usually accepted and does not change the general structure of the expressions. Both resonant and non-resonant processes may be studied with the acoustic waves separated [23] (both modes could be included; sonic or hypersonic flows could be studied if we return to (6.44)). The modes have the different scales; hence, a numerical modeling of the mutual generation and control also could be more effective.

REFERENCES

1. Kovasznay, L.S.G. 1953. Turbulence in supersonic flow. *Journal of the Aeronautical Sciences* 20: 657–682; Chu, B.-T., and L.S.G. Kovasznay. 1958. Non-linear interactions in a viscous heat-conducting compressible gas. *Journal of Fluid Mechanics* 3: 494–514.
2. Tam, C.K.W. 1981. The excitation of Tollmien-Schlichting waves in low subsonic boundary layers by free-stream sound waves. *Journal of Fluid Mechanics* 109: 483–501.
3. Wu, X. 1999. Generation of Tollmien-Schlichting waves by convecting gusts interacting with sound. *Journal of Fluid Mechanics* 397: 285–316.
4. Wu, X. 2001. Receptivity of boundary layers with distributed roughness to vortical and acoustic disturbances: a second-order asymptotic theory and comparison with experiments. *Journal of Fluid Mechanics* 431: 91–133.

5. Ruban, A.I. 1984. On Tollmien-Schlichting waves generation by sound Izv. Akad. Nauk SSSR Mekh Zhid. Gaza (Fluid Dyn. 19, 709–716 (1985)).
6. Goldstein, M.E. 1985. Scattering of acoustic waves into Tollmien-Schlichting waves by small variations in surface geometry. *Journal of Fluid Mechanics* 154: 509–529.
7. Choudhari, M. 1994. Distributed acoustic receptivity in laminar flow control configuration. *Physics of Fluids* 6: 489–506.
8. Leble, S., and A. Perelomova. 2002. Tollmienn-Schlichting and sound waves interacting: nonlinear resonances. In: Rudenko, O.V., and Sapozhnikov, O.A. (Eds), *Nonlinear Acoustics in the Beginning of the 21st Century*, vol. 1. Faculty of Physics, MSU, Moscow, 2002, pp. 203–206.
9. Leble, S.B. 1991. *Nonlinear Waves in Waveguides with Stratification*. Springer-Verlag, Berlin.
10. Perelomova, A.A. 2000. Projectors in nonlinear evolution problem: Acoustic solutions of bubbly liquid. *Applied Mathematics Letters* 13: 93–98; Perelomova, A.A. 1998. Nonlinear dynamics of vertically propagating acoustic waves in a stratified atmosphere. *Acta Acustica* 84(6): 1002–1006.
11. Perelomova, A.A. 2001. Directed acoustic beams interacting with heat mode: coupled nonlinear equations and modified KZK equation. *Acta Acustica* 87: 176–183.
12. Kachanow, J.S., W.W. Kozlov, and W.J. Levchenko: pogranicznom ICO Akad. Nauk SSSR, nr. 13, 1975; Occurrence of Tollmienn-Schlichting waves in the boundary layer under the effect of external perturbations Izv. Akad. Nauk SSSR Mekh Zhid. i Gaza 5: 85–94 (in Russian) Transl. Fluid Dyn. 13, 1979, 704–711.
13. King, R., and K.S. Breuer. 2001. Acoustic receptivity and evolution of two-dimensional and oblique disturbances in a Blasius boundary layer. *Journal of Fluid Mechanics* 432: 69–90.
14. Makarov, S., and M. Ochmann. 1996. Nonlinear and thermoviscous phenomena in acoustics, part I. *Acustica* 82: 579–606.
15. Leble, S. 2000. On binary Darboux transformations and N-wave systems at rings. *Theoretical and Mathematical Physics* 122: 239–250.
16. Perelomova, A. 2005. Modes and projectors in the viscous non-uniform flow. Heating caused by an acoustic beam between two parallel plates. *Archives of Acoustics* 30(2): 217–232.
17. Leble, S. 1988. Nonlinear Waves in Boundary Layer and Turbulence, in Theses of XX Conference Kaliningrad State University, pp. 185–189, Kaliningrad.
18. Leble, S.B., I.Y. Popov, and Y.V. Gugel. 1993. Weak interaction between acoustic and vortex waves in a boundary layer. (Russian) *Gidromekh.* (Kiev) 67: 3–11.
19. Rudenko, O.V., and S.I. Soluyan. 1977. *Theoretical Foundations of Nonlinear Acoustics*. Consultants Bureau, New York.
20. Schlichting, H., and K. Gersten. 2000. *Boundary-Layer Theory*. With contributions by Egon Krause and Herbert Oertel, Jr. Translated from the ninth German edition by Katherine Mayes. Eighth revised and enlarged edition. Springer-Verlag, Berlin. xxiv+799 pp. ISBN: 3-540-66270-7.
21. Lyutikov, M. 2000. Turbulent 4-wave interaction of two type of waves. *Physics Letters A* 265(1–2): 83–90.
22. Qiankang, Z., Y. Dacheng, N. Zhennan, Y. Dachun, S. Weixin, T. Chenkuan, and J. Minjian. 1989. Resonant interactions of Tollmien-Schlichting waves in the boundary layer on a flat plate. *Acta Mechanica Sinica* 21(2): 140–144.
23. Leble, S., and A. Perelomova. 2002. Tollmien-Shlichting and sound waves interaction: general description and nonlinear resonances. Fluid Dynamics. ArxXiv:physics/0207036.

7 1D Electrodynamics

7.1 CAUCHY PROBLEM FOR 1D ELECTRODYNAMICS. POLARIZED HYBRID FIELDS

7.1.1 THE PROBLEM FORMULATION OUTLINE

Our starting point is the set of Maxwell equations for a non-magnetic medium in the Lorentz-Heaviside's unit system

$$\nabla \cdot \mathbf{D} = 0, \tag{7.1}$$

$$\nabla \cdot \mathbf{B} = 0, \tag{7.2}$$

$$\nabla \times \mathbf{E} = -\frac{1}{c}\frac{\partial \mathbf{B}}{\partial t}, \tag{7.3}$$

$$\nabla \times \mathbf{H} = \frac{1}{c}\frac{\partial \mathbf{D}}{\partial t} \tag{7.4}$$

completed by material relations within the choice

$$\mathbf{D} = \mathbf{E} + 4\pi\mathbf{P}, \quad \mathbf{H} = \mathbf{B}. \tag{7.5}$$

The polarization vector $\mathbf{P}$ is the sum of linear and nonlinear terms

$$\mathbf{P} = \mathbf{P}_L + \mathbf{P}_{\mathbf{NL}}. \tag{7.6}$$

In this section, we restrict ourselves to a one-dimensional model such as one of Ref. [1], x-axis, chosen as a direction of a pulse propagation. We assume that $E_x = 0$ and $B_x = 0$, taking into account the only polarization of electromagnetic waves. This allows us to rewrite the stationary Maxwell equations (7.1) as

$$\frac{\partial}{\partial y}D_y + \frac{\partial}{\partial z}D_z = 0, \quad \frac{\partial}{\partial y}B_y + \frac{\partial}{\partial z}B_z = 0, \tag{7.7}$$

which, for one polarization $E_y \neq 0, B_z \neq 0$, simplifies as

$$\frac{\partial}{\partial z}D_z = 0, \quad \frac{\partial}{\partial z}B_z = 0, \tag{7.8}$$

and whose dynamics is determined by the system

$$\begin{aligned} \frac{1}{c}\frac{\partial D}{\partial t} &= -\frac{\partial B}{\partial x}, \quad \text{where } D = E + 4\pi\chi^{(1)}E + 4\pi(P_{NL})_y, \\ \frac{1}{c}\frac{\partial B}{\partial t} &= -\frac{\partial E}{\partial x}. \end{aligned} \tag{7.9}$$

Here and further in this section we use the shorthands $E_y := E$, $D_y = D$, $B_z := B$. In such a model, implying the properties of isotropic optically non-active media [2], the third-order nonlinear part of polarization is supposed to have the form:

$$(P_{NL})_y = \chi^{(3)}_{yyyy} E_y^3 := \chi^{(3)} E^3. \tag{7.10}$$

These assumptions will certainly simplify the further derivation procedure in the part of the nonlinearity account. However, up to this point, we are unable to tell whether the unidirectional approximation would become more or less robust for a description of pulse propagation. To establish a point of reference, we provide steps that lead to an SPE equation through the application of projection operators.

7.1.2 ON DYNAMICAL PROJECTION METHOD APPLICATION: CAUCHY PROBLEM

As a first step to illustrate the method of projecting operators [3,4], we define projection operators for the case of linear isotropic dielectric media where $D = \varepsilon E$ and $\varepsilon = 1 + 4\pi\chi^{(1)}$. In this case, we rewrite Eq. (7.9) as the matrix equation

$$\Psi_t = L\Psi,$$

where the field vector

$$\Psi = \begin{pmatrix} E \\ B \end{pmatrix}, \text{ and matrix operator} L = \begin{pmatrix} 0 & \frac{c}{\varepsilon}\partial_x \\ c\partial_x & 0 \end{pmatrix}, \tag{7.11}$$

enter the matrix operator equation as

$$\begin{pmatrix} \frac{\partial E}{\partial t} \\ \frac{\partial B}{\partial t} \end{pmatrix} = \begin{pmatrix} 0 & \frac{c}{\varepsilon}\partial_x \\ c\partial_x & 0 \end{pmatrix} \begin{pmatrix} E \\ B \end{pmatrix} = \begin{pmatrix} \frac{c}{\varepsilon}\frac{\partial}{\partial x}B \\ c\frac{\partial}{\partial x}E \end{pmatrix}. \tag{7.12}$$

Consider now a Cauchy problem for the system Eq. (7.12). Besides the system (7.12), the problem formulation includes the initial conditions, which we write also in vector form:

$$\Psi(x,0) = \begin{pmatrix} E(x,0) \\ B(x,0) \end{pmatrix} = \begin{pmatrix} \phi(x) \\ \psi(x) \end{pmatrix}. \tag{7.13}$$

Applying to the fields E, B, a Fourier transformation on x, ($E = \frac{1}{\sqrt{2\pi}} \int exp(ikx)\tilde{E}dx$) one goes to the system of ordinary differential equations, parametrized by k

$$\begin{pmatrix} \frac{d\tilde{E}}{dt} \\ \frac{d\tilde{B}}{dt} \end{pmatrix} = \begin{pmatrix} \frac{ick}{\varepsilon}\tilde{B} \\ ick\tilde{E} \end{pmatrix}. \tag{7.14}$$

This is a system of differential equations with constant coefficients that has an exponential $\exp(i\omega t)$ type solution, where $\omega = \pm\frac{ck}{\sqrt{\varepsilon}}$. We hence arrive at a 2×2 matrix eigenvalue problem that can be solved by the simplest projection operators application (see, e.g. Chapter 1).

Namely, let us search for matrices P_i such that $P_1\Psi = \Psi_1$ and $P_2\Psi = \Psi_2$ are eigenvectors of the evolution matrix in Eq. (7.14). The standard properties of orthogonal projecting operators

$$P_iP_j = 0, \quad P_i^2 = P_i, \quad \sum_i P_i = 1 \tag{7.15}$$

are implied. For Eq. (7.14), the matrices P_i get the form

$$P_1 = \frac{1}{2}\begin{pmatrix} 1 & \frac{1}{\sqrt{\varepsilon}} \\ \sqrt{\varepsilon} & 1 \end{pmatrix}, \quad P_2 = \frac{1}{2}\begin{pmatrix} 1 & -\frac{1}{\sqrt{\varepsilon}} \\ -\sqrt{\varepsilon} & 1 \end{pmatrix}. \tag{7.16}$$

Technically we solve the problem in the k-space and, next, perform the inverse Fourier transform, which yields the x-representation of the operators; see Chapter 1, or [3]. In this happy case, the matrix elements of the projecting operators (7.16) in fact do not depend on k (this is not the case if dispersion is accounted for; see the previous chapters and the next section); hence, its x-representation coincides with the k-representation Eq. (7.16).

Applying the projection operators to the vector Ψ (7.11), we can introduce new variables $\Lambda = \frac{1}{2}E + \frac{1}{2\sqrt{\varepsilon}}B$ and $\Pi = \frac{1}{2}E - \frac{1}{2\sqrt{\varepsilon}}B$, e.g. as

$$P_1\Psi = \begin{pmatrix} \frac{1}{2}E + \frac{1}{2\sqrt{\varepsilon}}B \\ \frac{1}{2}\sqrt{\varepsilon}E + \frac{1}{2}B \end{pmatrix} = \begin{pmatrix} \Pi \\ \sqrt{\varepsilon}\Pi \end{pmatrix}, \tag{7.17}$$

which correspond to the left and right directions of wave propagation [3]. Comparing new variables Λ and Π to the ones presented by Kinsler et al. [5], ours have similar form. Kinsler [6] coined the name hybrid amplitude for such combinations of electric and magnetic fields. Our form of Eq. (7.17) is exactly determined by the dispersion relation $\omega = \pm\frac{ck}{\sqrt{\varepsilon}}$ from Eq. (7.14) and hence allows us to account for its arbitrary form. Moreover, it allows us to present in an algorithmic way both electric and magnetic field [7], in an the simplest example we trace now,

$$E = \Lambda + \Pi \quad B = \sqrt{\varepsilon}(\Pi - \Lambda). \tag{7.18}$$

This correspondence Eq. (7.17 and 7.18) is a one-to-one local map and hence allows us to determine the initial conditions in the Cauchy problem for both left and right wave variables (Λ, Π). It also gives us the possibility to follow waves, extracting data in a each time t by the corresponding projecting in a very general situation.

If we account for the nonlinearity Eq. (7.10), then Eq. (7.12) is generalized as

$$\frac{\partial}{\partial t}\Psi - L\Psi = N(\Psi). \tag{7.19}$$

Plugging the nonlinear term into Eq. (7.9) as $4\pi\frac{\partial}{\partial t}\frac{1}{\varepsilon}\chi^{(3)}E^3$ [1,10] yields

$$\frac{\partial}{\partial t}\begin{pmatrix} E \\ B \end{pmatrix} - \begin{pmatrix} 0 & \frac{c}{\varepsilon}\partial_x \\ c\partial_x & 0 \end{pmatrix}\begin{pmatrix} E \\ B \end{pmatrix} = 4\pi\frac{\partial}{\partial t}\begin{pmatrix} \frac{1}{\varepsilon}\chi^{(3)}E^3 \\ 0 \end{pmatrix}. \tag{7.20}$$

Applying the operators P_i Eq. (7.16) to Eq. (7.20) and take into account the projector operators property $[P_i, \left(\frac{\partial}{\partial t} - L\right)] = 0$, yields

$$\left(\frac{\partial}{\partial t} - L\right) P_i \Psi = P_i N(\Psi). \tag{7.21}$$

At this point, we have to admit that Λ is strictly connected to the dispersion relation $\omega = \frac{-ck}{\sqrt{\varepsilon}}$, and, analogously, Π is related to $\omega = \frac{ck}{\sqrt{\varepsilon}}$. Hence, this determines a sign, which stands at the parameter c in our equations. Finally, plugging in E from (7.18), the result

$$\begin{pmatrix} \frac{\partial}{\partial t}\Pi \\ \frac{\partial}{\partial t}\sqrt{\varepsilon}\Pi \end{pmatrix} - \begin{pmatrix} \frac{c}{\varepsilon}\frac{\partial}{\partial x}\sqrt{\varepsilon}\Pi \\ c\frac{\partial}{\partial x}\Pi \end{pmatrix} = 4\pi\frac{\partial}{\partial t}\begin{pmatrix} \frac{1}{\varepsilon}\chi^{(3)}(\Lambda+\Pi)^3 \\ \frac{1}{\sqrt{\varepsilon}}\chi^{(3)}(\Lambda+\Pi)^3 \end{pmatrix} \tag{7.22}$$

presents (twice!) the evolution equation for the right wave interacting with the left one.

Repeating our calculations from Eqs (7.19) to (7.22) with use of the second projecting operator Eq. (7.32), we obtain a closed system of equations that describes an interaction between two waves propagating in opposite directions. The system of equations has the form

$$\begin{cases} \frac{\partial}{\partial t}\Pi - \frac{c}{\sqrt{\varepsilon}}\frac{\partial}{\partial x}\Pi = \frac{2\pi}{\varepsilon}\frac{\partial}{\partial t}\chi^{(3)}(\Lambda+\Pi)^3 \\ \frac{\partial}{\partial t}\Lambda + \frac{c}{\sqrt{\varepsilon}}\frac{\partial}{\partial x}\Lambda = \frac{2\pi}{\varepsilon}\frac{\partial}{\partial t}\chi^{(3)}(\Lambda+\Pi)^3 \end{cases} \tag{7.23}$$

In the conditions of the derivation, this system is equivalent to the Maxwell one and hence describes all linear polarized waves with arbitrary frequency. Let us discuss the result.

7.1.3 THE EFFECT OF A CUMULATIVE PART OF INTERACTION

A solution of such coupled equations (7.23) contains two contributions [3]. One follows with an initial perturbation (extra self-action via reverse wave generation) and the second (rest) relates to a reflection from the exited matter as a region with changed refraction index. We also should have in mind that a mode definition relies upon the linear projecting; an alternative could account for nonlinearity [9]. It, however, would change a mode definition on a level of measurement procedure.

This may be understood via formal integration by characteristics. The result generally depends on initial (boundary) conditions. Consider a case with a right- dominative stationary solitary wave $\Pi(t - \frac{x}{c_0})$, $c_0 = \frac{c}{\sqrt{\varepsilon}}$. As a first iteration, take the second equation from Eq. (7.23) neglecting the term Λ, still without accounting for dispersion ($\varepsilon = const$)

$$\frac{\partial}{\partial t}\Lambda + c_0\frac{\partial}{\partial x}\Lambda = \frac{2\pi}{\varepsilon}\chi^{(3)}\frac{\partial}{\partial t}(\Pi^3) \tag{7.24}$$

and go to the variables

$$\xi = t - \frac{x}{c_0}, \quad \eta = t + \frac{x}{c_0}. \tag{7.25}$$

The equation rewrites as

$$\frac{\partial}{\partial \xi}\Lambda = \frac{\pi}{\varepsilon}\chi^{(3)}\left[\frac{\partial}{\partial \xi}(\Pi^3) + \frac{\partial}{\partial \eta}(\Pi^3)\right]. \tag{7.26}$$

Accounting for only the stationary contribution, $\Pi_\eta = 0$ yields an ordinary differential equation

$$\frac{\partial}{\partial \xi}\Lambda = \frac{\pi}{\varepsilon}\chi^{(3)}\frac{\partial \Pi^3}{\partial \xi}, \tag{7.27}$$

integrated by ξ from $-\infty$ to ξ with an account for the zero boundary condition on Π at minus infinity, gives

$$\Lambda = \frac{\pi}{\varepsilon}\chi^{(3)}\Pi^3. \tag{7.28}$$

In the second iteration, we should insert the result into the equation for Π to obtain the closed equation for such a kind of initial problem:

$$\frac{\partial}{\partial t}\Pi - c_0\frac{\partial}{\partial x}\Pi = \frac{2\pi}{\varepsilon}\chi^{(3)}\frac{\partial}{\partial t}\left[(\Pi + \frac{\pi}{\varepsilon}\chi^{(3)}\Pi^3)^3\right]. \tag{7.29}$$

This equation accounts for the cumulation of interactions with opposite waves. It essentially contributes when the higher nonlinearities should be accounted for. Let us note that this equation is of Hopf type and demonstrates the phenomenon of wave breaking. In more realistic models (waveguide or matter dispersion account), its steepness growth could be stopped by dispersion or dissipation.

Note also that if the velocity of the solitary wave Π does not coincide with c_0, we would solve the general system (7.23); the result of integration is more complicated.

The example we consider is simple, but contains all the principle ingredients of this book method. A more complicated example in the next section shows what changes when the projection operator matrix elements in a k-representation depend on k.

7.1.4 DISPERSION ACCOUNT, AN EXAMPLE

A dispersion account drastically changes the form of projection operators. The relation between the basic vectors of electrodynamics $\vec{D}$ and $\vec{E}$ in a matter is generally an integral tensor operator. The integral operator for an isotropic medium is investigated in the forthcoming section. If, for the considered case of a 1D isotropic medium and the previous section notations, one neglects space dispersion, the k-representations of the vectors are expressed by a rather simple scalar relation:

$$\tilde{D} = \varepsilon(k)\tilde{E}.$$

Following [1] we focus on the propagation of light in the infrared range with $\lambda = 1600 - 3000$nm. In this range, we can approximate

$$\chi^{(1)}(k) \approx \chi_0^{(1)} + \chi_2^{(1)}\left(\frac{c}{k}\right)^2. \tag{7.30}$$

If we plug Eq. (7.30) into $\varepsilon(k)$, and into Eq. (7.16), then the right projection operator in the k space gets the form

$$P_1 = \frac{1}{2}\begin{pmatrix} 1 & \frac{1}{\sqrt{1+4\pi\chi_0^{(1)}+4\pi\chi_2^{(1)}\frac{c^2}{k^2}}} \\ \sqrt{1+4\pi\chi_0^{(1)}+4\pi\chi_2^{(1)}\frac{c^2}{k^2}} & 1 \end{pmatrix}, \tag{7.31}$$

with the dependence on k that arises from dispersion account. A similar form has the second operator

$$P_2 = \frac{1}{2}\begin{pmatrix} 1 & -\frac{1}{\sqrt{1+4\pi\chi_0^{(1)}+4\pi\chi_2^{(1)}\frac{c^2}{k^2}}} \\ -\sqrt{1+4\pi\chi_0^{(1)}+4\pi\chi_2^{(1)}\frac{c^2}{k^2}} & 1 \end{pmatrix}. \tag{7.32}$$

To proceed with the basic system, it is necessary to present these operators in x-representation. If we take (7.32) literally, the matrix elements of the operators $P_{1,2}(x)$ are integral ones. The Taylor series representation of $\sqrt{\varepsilon}$ over $\frac{1}{k}$ allows us to consider the first two terms, assuming that $\chi_2^{(1)}\frac{c^2}{k^2}$ is small. Then, the x-representation for the electric permittivity looks like an operator:

$$\hat{\varepsilon} = 1+4\pi\chi_0^{(1)} - 4\pi\chi_2^{(1)}c^2\partial_x^{-2}. \tag{7.33}$$

With new projection operators, we introduce new variables:o

$$\begin{aligned} \Pi &= \frac{1}{2}\left(\sqrt{1+4\pi\chi_0^{(1)}}E + \frac{2\pi\chi_2^{(1)}c^2}{\sqrt{1+4\pi\chi_0^{(1)}}}(i\partial_x)^{-2}E + B\right), \\ \Lambda &= \frac{1}{2}\left(-\sqrt{1+4\pi\chi_0^{(1)}}E - \frac{2\pi\chi_2^{(1)}c^2}{\sqrt{1+4\pi\chi_0^{(1)}}}(i\partial_x)^{-2}E + B\right), \end{aligned} \tag{7.34}$$

where ∂_x^{-2} is an integral operator reciprocal to ∂_x^2. With an analogy to Eq. (7.17), taking the first terms of Taylor series for inverse operators, we obtain a relation between new variables and electric and magnetic fields:

$$\begin{aligned} \Pi + \Lambda &= B \\ \left(\frac{1}{\sqrt{1+4\pi\chi_0^{(1)}}} - \frac{2\pi\chi_2^{(1)}c^2}{(1+4\pi\chi_0^{(1)})^{3/2}}(i\partial_x)^{-2}\right)(\Pi - \Lambda) &= E. \end{aligned} \tag{7.35}$$

We get the approximate pseudodifferential relations between an electric/magnetic field and the hybrid amplitudes of left/right modes for directed electromagnetic waves.

7.2 GENERAL DYNAMICS EQUATIONS, SPE SYSTEM

The nonlinear term Eq. (7.10) of polarization may be left unchanged, but the linear one has an extra term from Eq. (7.33) [1]. Then, Eq. (7.12) may be rewritten by means of the new operator $\hat{\varepsilon}$; see (7.33):

$$\frac{\partial}{\partial t}\Psi - \mathbb{L}\Psi = \mathbb{N}(\Psi), \tag{7.36}$$

or, in details,

$$\frac{\partial}{\partial t}\begin{pmatrix} E \\ B \end{pmatrix} - \begin{pmatrix} 0 & c\hat{\varepsilon}^{-1}\partial_x \\ c\partial_x & 0 \end{pmatrix}\begin{pmatrix} E \\ B \end{pmatrix} = 4\pi\frac{\partial}{\partial t}\begin{pmatrix} \hat{\varepsilon}^{-1}\chi^{(3)}E^3 \\ 0 \end{pmatrix}. \tag{7.37}$$

Applying P_i Eq. (7.31) to the both sides of Eq. (7.37) and respecting the projectors property $[P_i, \left(\frac{\partial}{\partial t} - \mathbb{L}\right)] = 0$, yields

$$\left(\frac{\partial}{\partial t} - \mathbb{L}\right)P_i\Psi = P_i\mathbb{N}(\Psi). \tag{7.38}$$

With an analogy to the previous example, we strictly connected new variable Λ with dispersion relation $\omega = -ck\sqrt{\varepsilon^{-1}}$, and analogously Π is related to $\omega = ck\sqrt{\varepsilon^{-1}}$. Hence, as before, this determines a sign, which stands at the parameter c in our equations. Finally, the P_1 action presents the result:

$$\begin{pmatrix} \frac{\partial}{\partial t}\Pi \\ \frac{\partial}{\partial t}\sqrt{\hat{\varepsilon}}\Pi \end{pmatrix} - \begin{pmatrix} c\hat{\varepsilon}^{-1/2}\frac{\partial}{\partial x}\Pi \\ c\frac{\partial}{\partial x}\Pi \end{pmatrix} = 4\pi\frac{\partial}{\partial t}\begin{pmatrix} \hat{\varepsilon}^{-1}\chi^{(3)}(\Lambda+\Pi)^3 \\ \sqrt{\hat{\varepsilon}^{-1}}\chi^{(3)}(\Lambda+\Pi)^3 \end{pmatrix}, \tag{7.39}$$

Note that $\hat{\varepsilon}$ and ∂_x commute. Repeating our calculations from Eqs (7.36) to (7.39) with the use of second projector operator Eq. (7.32), we obtain a *general system of equations that describes the interaction between two waves propagating in opposite directions*. This system of equations has the form

$$\begin{cases} \frac{\partial}{\partial t}\Pi - c\sqrt{\hat{\varepsilon}^{-1}}\frac{\partial}{\partial x}\Pi = 2\pi\frac{\partial}{\partial t}\hat{\varepsilon}^{-1}\chi^{(3)}(\Lambda+\Pi)^3 \\ \frac{\partial}{\partial t}\Lambda + c\sqrt{\hat{\varepsilon}^{-1}}\frac{\partial}{\partial x}\Lambda = 2\pi\frac{\partial}{\partial t}\hat{\varepsilon}^{-1}\chi^{(3)}(\Lambda+\Pi)^3 \end{cases}, \tag{7.40}$$

where $\hat{\varepsilon} = 1 + 4\pi\chi_0^{(1)} + 4\pi\chi_2^{(1)}c^2(i\partial_x)^{-2}$ is the approximate operator in the x-domain, whose inverse and other functions we understand via corresponding Taylor series expansions. Multiplying both equations by the operator $\hat{\varepsilon}$ and using the expansions, we arrive at

$$\begin{gathered} \left(1 + 4\pi\chi_0^{(1)} - 4\pi\chi_2^{(1)}c^2\partial_x^{-2}\right)\frac{\partial}{\partial t}\Pi \\ -c\left(\sqrt{1+4\pi\chi_0^{(1)}} - \frac{2\pi\chi_2^{(1)}c^2}{\sqrt{1+4\pi\chi_0^{(1)}}}\partial_x^{-2}\right)\frac{\partial}{\partial x}\Pi = 2\pi\chi^{(3)}\frac{\partial}{\partial t}(\Lambda+\Pi)^3, \\ \left(1 + 4\pi\chi_0^{(1)} + 4\pi\chi_2^{(1)}c^2(\partial_x)^{-2}\right)\frac{\partial}{\partial t}\Lambda \\ +c\left(\sqrt{1+4\pi\chi_0^{(1)}} - \frac{2\pi\chi_2^{(1)}c^2}{\sqrt{1+4\pi\chi_0^{(1)}}}\partial_x^{-2}\right)\frac{\partial}{\partial x}\Lambda = 2\pi\chi^{(3)}\frac{\partial}{\partial t}(\Lambda+\Pi)^3. \end{gathered} \tag{7.41}$$

This system **describes the interaction between two waves propagating in opposite directions for the dispersion specified in Ref. [1].**

7.2.1 THE SHAFER-WAYNE (SPE) AND GENERALIZATIONS

Let us choose the second equation from the system (7.41). If we mean a pulse launched from the right end of a fiber, we can consider the only direction of propagation. Hence, we have chosen Λ as a dominating left wave. Assume therefore that $\Pi = 0$; then, differentiating twice with respect to x produces the generalized SP equation

$$4\pi\chi_2^{(1)}c^2\frac{\partial\Lambda}{\partial t} - \left(1+4\pi\chi_0^{(1)}\right)\frac{\partial^3\Lambda}{\partial x^2\partial t} - c\sqrt{1+4\pi\chi_0^{(1)}}\frac{\partial^3\Lambda}{\partial x^3} - \frac{\pi\chi_2^{(1)}c^3}{\sqrt{1+4\pi\chi_0^{(1)}}}\frac{\partial\Lambda}{\partial x} = -2\pi\frac{\partial^3}{\partial x^2\partial t}\chi^{(3)}\Lambda^3. \tag{7.42}$$

Following [1] we apply a multiple scales ansatz expansion:

$$\Pi(x,t) = \left(\kappa A_0(\phi,x_1,x_2) + \kappa^2 A_1(\phi,x_1,x_2) + \cdots\right) \tag{7.43}$$

where $\phi = \frac{t-x}{\kappa}$ and $x_n = \kappa^n x$. If we consider all terms $O(\kappa^0)$, then we reach the generalized SP equation

$$\left(2(1+4\pi\chi_0^{(1)}) + 3c\sqrt{1+4\pi\chi_0^{(1)}}\right)\frac{\partial^2}{\partial\phi\partial x_1}A_0 - \left(4\pi\chi_2^{(1)}c^2 - \frac{2\pi\chi_2^{(1)}c^3}{\sqrt{1+4\pi\chi_0^{(1)}}}\right)A_0 = 2\pi\chi^{(3)}\frac{\partial^2}{\partial\phi^2}A_0^2A_0^* \tag{7.44}$$

At this point we obtain a second-order differential equation that describes an ultra-short pulse launched in one direction. To fulfil this description, we have to present initial-boundary conditions; that is an immediate corollary from the projection fixed by Eq. (7.34) within the choice $\Pi = 0$. It reads as the correlated action of magnetic and electric fields.

7.2.2 DISCUSSION AND CONCLUSIONS

As we derive the system of equations Eq. (7.23) describing the interaction between two waves propagating in two directions, we would like to remark that such an interaction is weak in the case of a long optical fiber exited from both ends. The significance of this phenomena is obvious: it gives us a new opportunity to investigate a nonlinearity and measure nonlinear constants. The fundamental importance accounting for the opposite waves interaction phenomena is quite clear in the case of optical resonators, when the effects of interaction are accumulated. Going down to the unidirectional case, we acquire a SPE equation with modified coefficients within the demonstration of the general method [3]. One of the purposes of this publication is to draw the attention of researchers in optics and other fields to the projecting operators method. Some interesting features of the projection method applications to optical problems can be also seen in the paper of Kolesik et al. [8]. In acoustics there are many interesting applications, shown in preceding chapters of this book and an important development with nonlinearity account in projecting operators published in Ref. [9] with references therein.

We have also demonstrated how to obtain SPE with a projection operators method for unidirectional pulse propagation in isotropic, dispersionless and dispersive media containing nonlinearity of the third order. The achieved results give generalized SPE and corrections to coefficients in it. Its generalizations, which account for polarization and other important effects, are also performed by the method.

7.3 BOUNDARY REGIME PROPAGATION IN 1D ELECTRODYNAMICS

7.3.1 STATEMENT OF PROBLEM

Problems of excitation of a wave in a medium with dispersion by a transition through a plane surface $x = 0$ that divides a space into two half-spaces are modeled by the boundary regime problem with the following formulation. Our starting point is the Maxwell equations for a homogeneous linear isotropic but dispersive dielectric medium without sources in the SI unit system (for the convenience of the reader we use this alternative option as well as the Gauss one in Section 7.1):

$$\mathrm{div}\vec{D}(\vec{r},t) = 0, \tag{7.45}$$

$$\mathrm{div}\vec{B}(\vec{r},t) = 0, \tag{7.46}$$

$$\frac{\partial \vec{B}(\vec{r},t)}{\partial t} = -\mathrm{rot}\vec{E}(\vec{r},t), \tag{7.47}$$

$$\mathrm{rot}\vec{H}(\vec{r},t) = \frac{\partial \vec{D}(\vec{r},t)}{\partial t}. \tag{7.48}$$

In this section, we restrict ourselves to an instructive one-dimensional model, similar to those of Chung et al. [28] and Kuszner and Leble [4], in which the x-axis is chosen as the direction of a wave, propagation. For a plane wave; $D_{x,z} = 0$ and $B_{x,y} = 0$, taking into account only one polarization of electromagnetic waves. The Maxwell system simplifies as:

$$\begin{aligned} \frac{\partial H_z}{\partial x} &= -\frac{\partial D_y}{\partial t}, \\ \frac{\partial E_y}{\partial x} &= -\frac{\partial B_z}{\partial t}, \end{aligned} \tag{7.49}$$

having in mind x-evolution description. The boundary regime is settled by

$$\begin{aligned} E_y(0,t) &= \mu(t), \\ B_z(0,t) &= \nu(t). \end{aligned} \tag{7.50}$$

Omitting the index write the Fourier transformations for D_y as

$$D(x,t) = \frac{1}{\sqrt{2\pi}} \int_{-\infty}^{\infty} \mathscr{D}(x,\omega) \exp(i\omega t) d\omega, \tag{7.51}$$

where $\mathscr{D}$ is the Fourier transform of the field D_y named as ω-representation. The reality of the field D implies

$$D(x,t)^* = \frac{1}{\sqrt{2\pi}} \int_{-\infty}^{\infty} \mathscr{D}^*(x,-\omega) \exp(i\omega t) d\omega, \tag{7.52}$$

which yields

$$\mathscr{D}^*(x,-\omega) = \mathscr{D}(x,\omega) \tag{7.53}$$

and defines the transform for negative ω. Similar conditions take place for all the remaining three fields E, B, H.

If we continue the fields anti-symmetrically to the negative t-plane ($t < 0$),

$$E(x,-t) = \frac{1}{2\sqrt{\pi}} \int_{-\infty}^{\infty} \mathscr{E}(x,\omega) e^{-i\omega t} d\omega = \frac{1}{2\sqrt{\pi}} \int_{-\infty}^{\infty} \mathscr{E}(x,-\omega') e^{i\omega' t} d\omega'$$

$$= -\frac{1}{2\sqrt{\pi}} \int_{-\infty}^{\infty} \mathscr{E}(x,\omega) e^{i\omega t} d\omega, \tag{7.54}$$

while $\omega' = -\omega$, then the transforms are also odd (now in ω)

$$\mathscr{E}(x,\omega) = \mathscr{E}(x,-\omega). \tag{7.55}$$

It immediately gives (look to (7.53))

$$\mathscr{E}(x,\omega) = -\mathscr{E}(x,-\omega) = \mathscr{E}^*(x,-\omega), \tag{7.56}$$

or all transforms are purely imaginary. This is not very convenient from the point of material relation in a frequency domain. It is better to choose a symmetrical (even) continuation that leads to the real transforms $\mathscr{E}$ and $\mathscr{D}$ if the dielectric permittivity is real (absence of absorption).

The concept we rely upon in the subsequent sections is as follows. The empiric or quantum consideration yields the material equations in a frequency domain:

$$\mathscr{D} = \varepsilon_0 \varepsilon(\omega) \mathscr{E}, \tag{7.57}$$

$$\mathscr{B} = \mu_0 \mu(\omega) \mathscr{H}. \tag{7.58}$$

Here:
$\varepsilon(\omega)$—dielectric permittivity of medium, ε_0—dielectric permittivity of vacuum.

$\mu(\omega)$—magnetic permeability of medium, and μ_0—magnetic permeability of vacuum. $\mathscr{B}$—analogue of function B in ω-representation. Then:

$$D(x,t) = \frac{\varepsilon_0}{\sqrt{2\pi}} \int_{-\infty}^{\infty} \varepsilon(\omega)\mathscr{E}(x,\omega)\exp(i\omega t)d\omega, \tag{7.59}$$

$$B(x,t) = \frac{\mu_0}{\sqrt{2\pi}} \int_{-\infty}^{\infty} \mu(\omega)\mathscr{B}(x,\omega)\exp(i\omega t)d\omega. \tag{7.60}$$

The transforms define the fields in the time domain, using the conventional continuation of the fields to the half space $t < 0$ [10] and the condition of causality [2].

7.3.2 OPERATORS OF DIELECTRIC PERMITTIVITY AND MAGNETIC PERMEABILITY

The Maxwell equations are written in t-representation. The material equations (8.11) in its original form are in ω-representation. In the t-representation, the functions ε and μ become operators. Let's find them. We write inverse Fourier transformations for $\varepsilon(\omega)$, $\mathscr{E}(x,\omega)$, $\mu(\omega)$, and $\mathscr{H}(x,\omega)$ as

$$\varepsilon(\omega) = \frac{1}{\sqrt{2\pi}} \int_{-\infty}^{\infty} \tilde{\varepsilon}(t)\exp(-i\omega t)dt, \tag{7.61}$$

$$\mathscr{E}(x,\omega) = \frac{1}{\sqrt{2\pi}} \int_{-\infty}^{\infty} E(x,t)\exp(-i\omega t)dt, \tag{7.62}$$

$$\mu(\omega) = \frac{1}{\sqrt{2\pi}} \int_{-\infty}^{\infty} \tilde{\mu}(t)\exp(-i\omega t)dt, \tag{7.63}$$

$$\mathscr{H}(x,\omega) = \frac{1}{\sqrt{2\pi}} \int_{-\infty}^{\infty} H(x,t)\exp(-i\omega t)dt. \tag{7.64}$$

By definition, an arbitrary Fredholm integral operator $\widehat{k}$ introduces as

$$\widehat{k}\varphi(x) = \int_{a}^{b} k(x,s)\varphi(s)ds, \tag{7.65}$$

with the kernel $k(x,s)$. Plugging (7.61) into (7.59) and (7.60) gives

$$D(x,t) = \frac{\varepsilon_0}{2\pi} \int_{-\infty}^{\infty}\int_{-\infty}^{\infty} \mathscr{E}(x,\omega)\exp[i\omega(t-\tau)]\tilde{\varepsilon}(\tau)d\tau d\omega, \tag{7.66}$$

next, plugging (7.62) yields the integral operator link:

$$D(x,t) = \frac{\varepsilon_0}{(2\pi)^{3/2}} \int\limits_{-\infty}^{\infty}\int\limits_{-\infty}^{\infty}\int\limits_{-\infty}^{\infty} E(x,s)\exp[i\omega(t-\tau-s)]\tilde{\varepsilon}(\tau)d\tau d\omega ds, \tag{7.67}$$

which represents the material relation:

$$D(x,t) = \hat{\varepsilon}E(x,t) = \int\limits_{-\infty}^{\infty} E(x,s)\bar{\varepsilon}(t-s)ds. \tag{7.68}$$

Separately, we introduce operators $\widehat{\varepsilon}$ and $\widehat{\mu}$ as integral ones:

$$\widehat{\varepsilon}\psi(x,t) = \int\limits_{-\infty}^{\infty} \bar{\varepsilon}(t-s)\psi(x,s)ds, \tag{7.69}$$

where the kernel

$$\begin{aligned}\bar{\varepsilon}(t-s) &= \frac{\varepsilon_0}{(2\pi)^{3/2}} \int\limits_{-\infty}^{\infty}\int\limits_{-\infty}^{\infty} \exp[i\omega(t-\tau-s)]\tilde{\varepsilon}(\tau)d\tau d\omega \\ &= \frac{\varepsilon_0}{(2\pi)^{1/2}} \int\limits_{-\infty}^{\infty} \delta(t-\tau-s)\tilde{\varepsilon}(\tau)d\tau = \frac{\varepsilon_0}{(2\pi)^{1/2}}\tilde{\varepsilon}(t-s),\end{aligned} \tag{7.70}$$

defines the integral operator of convolution type. Similarly, for magnetic fields we obtain

$$B(x,t) = \frac{\mu_0}{\sqrt{2\pi}} \int\limits_{-\infty}^{\infty} \tilde{\mu}(t,s)H(x,s)ds, \tag{7.71}$$

or, shortly

$$B(x,t) = \hat{\mu}H, \tag{7.72}$$

where

$$\widehat{\mu}\psi(x,t) = \frac{\mu_0}{\sqrt{2\pi}} \int\limits_{-\infty}^{\infty} \tilde{\mu}(t,\beta)\psi(x,\beta)d\beta. \tag{7.73}$$

Here

$$\tilde{\mu}(t,\beta) = \tilde{\mu}(t-\beta),$$

where the function $\tilde{\mu}(t)$ is defined by the inverse to (7.63). Let us discuss some terminology details. For concision, we'll call operators of dielectric permittivity and magnetic permeability *dielectric* and magnetic operators correspondingly. Generally, the dielectric and magnetic operators do not commute.

7.3.3 INVERSE DIELECTRIC AND MAGNETIC OPERATORS

From (7.68) the inverse link $E(x,t)$ with $D(x,t)$ can also be written as integral operators:

$$E(x,t) = \hat{\varepsilon}^{-1} D(x,t), \tag{7.74}$$

and, similarly,

$$H(x,t) = \hat{\mu}^{-1} B(x,t). \tag{7.75}$$

This means that we introduce integral operators $\widehat{\varepsilon}^{-1}$ and $\widehat{\mu}^{-1}$ as inverse operators to $\widehat{\varepsilon}$ and $\widehat{\mu}$:

$$\widehat{\varepsilon}^{-1} \psi(x,t) = \int_{-\infty}^{\infty} \mathfrak{e}(t,\beta)\psi(x,\beta)d\beta, \tag{7.76}$$

$$\widehat{\mu}^{-1} \psi(x,t) = \int_{-\infty}^{\infty} \mathfrak{m}(t,\beta)\psi(x,\beta)d\beta, \tag{7.77}$$

$\mathfrak{e}$ and $\mathfrak{m}$-integral kernels of them. Next, (7.4) reads

$$E(x,t) = \int_{-\infty}^{\infty} \mathfrak{e}(t,s)D(x,s)ds = \int_{-\infty}^{\infty} \mathfrak{e}(t,s) \int_{-\infty}^{\infty} E(x,p)\bar{\varepsilon}(s-p)dpds. \tag{7.78}$$

The identity (7.78) implies

$$\int_{-\infty}^{\infty} \mathfrak{e}(t,s)\bar{\varepsilon}(s-p)ds = \delta(p-t), \tag{7.79}$$

which gives

$$\mathfrak{e}(t,s)\bar{\varepsilon}(s-p) = \frac{1}{2\pi}\exp[is(p-t)], \tag{7.80}$$

or

$$\mathfrak{e}(t,s) = \frac{1}{2\pi\bar{\varepsilon}(s-p)}\exp[is(p-t)], \tag{7.81}$$

On the other hand, the material relation (7.57) may be read as

$$\mathscr{E} = \frac{1}{\varepsilon_0 \varepsilon(\omega)}\mathscr{D}. \tag{7.82}$$

Jumping to x-domain, one has

$$\frac{1}{\sqrt{2\pi}}\int_{-\infty}^{\infty} \mathscr{E}(x,\omega)\exp(i\omega t)d\omega = E(x,t) = \frac{1}{\sqrt{2\pi}}\int_{-\infty}^{\infty} \frac{1}{\varepsilon_0\varepsilon(\omega)}\mathscr{D}(x,\omega)\exp(i\omega t)d\omega. \tag{7.83}$$

Plugging the inverse Fourier transform in for $\mathscr{D}(x,\omega)$ gives

$$E(x,t) = \frac{1}{2\pi}\int_{-\infty}^{\infty}\int_{-\infty}^{\infty}\frac{1}{\varepsilon_0\varepsilon(\omega)}D(x,s)\exp[i\omega(t-s)]d\omega ds. \tag{7.84}$$

Comparing to (7.76) results in the kernel

$$\mathfrak{e}(t,s) = \frac{1}{2\pi}\int_{-\infty}^{\infty}\frac{1}{\varepsilon_0\varepsilon(\omega)}\exp[i\omega(t-s)]d\omega, \tag{7.85}$$

which is alternative to (7.80).

Next, for magnetic variables, the corresponding link

$$H(x,t) = \frac{1}{2\pi}\int_{-\infty}^{\infty}\int_{-\infty}^{\infty}\frac{1}{\mu_0\mu(\omega)}B(x,s)\exp[i\omega(t-s)]d\omega ds, \tag{7.86}$$

defines the inverse to $\hat{\mu}$ integral operator (7.77) with the kernel:

$$\mathfrak{m}(t,s) = \frac{1}{2\pi}\int_{-\infty}^{\infty}\frac{1}{\mu_0\mu(\omega)}\exp[i\omega(t-s)]d\omega. \tag{7.87}$$

7.3.4 PROJECTING OPERATORS IN 1D ELECTRODYNAMICS WITH UNIQUE POLARIZATION: BOUNDARY REGIME PROPAGATION

Let's use transformations (7.59) and (7.60). We substitute them in initial system of Eq. (7.49). Accounting for material equations (7.57) and (7.58), we have

$$\frac{\partial}{\partial t}\left(\frac{\varepsilon_0}{\sqrt{2\pi}}\int_{-\infty}^{\infty}\varepsilon(\omega)\mathscr{E}(x,\omega)\exp(i\omega t)d\omega\right) = -\frac{\partial H}{\partial x}. \tag{7.88}$$

The magnetic field H is expressed similarly via (7.75):

$$H(x,t) = \frac{1}{\mu_0\sqrt{2\pi}}\int_{-\infty}^{\infty}\frac{\mathscr{B}(x,\omega)}{\mu(\omega)}\exp(i\omega t)d\omega.$$

In ω-representation, the first equation (7.88) reads as

$$\frac{\partial\mathscr{B}}{\partial x} = -i\omega\mu_0\varepsilon_0\mu(\omega)\varepsilon(\omega)\mathscr{E}. \tag{7.89}$$

The second equation of the system (7.49) goes to

$$\frac{\partial\mathscr{E}}{\partial x} = -i\omega\mathscr{B}. \tag{7.90}$$

The matrix representation of the system may be written in a form similar to (7.14), but with x-derivatives on the L.H.S.:

$$\frac{\partial \tilde{\Psi}}{\partial x} = \widehat{L}\tilde{\Psi}. \tag{7.91}$$

The definition of $\widehat{L}$ and $\tilde{\Psi}$ is

$$\tilde{\Psi} = \begin{pmatrix} \mathscr{B} \\ \mathscr{E} \end{pmatrix} \tag{7.92}$$

$$\widehat{L} = \begin{pmatrix} 0 & -i\omega\mu_0\varepsilon_0\varepsilon(\omega)\mu(\omega) \\ -i\omega & 0 \end{pmatrix}. \tag{7.93}$$

Let us introduce the designation:

$$\mu_0\varepsilon_0\varepsilon(\omega)\mu(\omega) \equiv a^2(\omega). \tag{7.94}$$

Hence:

$$\frac{\partial \mathscr{B}}{\partial x} = -i\omega a^2(\omega)\mathscr{E}, \tag{7.95}$$

$$\frac{\partial \mathscr{E}}{\partial x} = -i\omega\mathscr{B}. \tag{7.96}$$

This is a system of differential equations that have exponential-type solutions. Hence, as in the previous section, we arrive at a 2×2 eigenvalue problem. Namely, let us search for matrices $P^{(i)}, i = \overline{1,2}$ such that $P^{(i)}\Psi_i = \Psi_i$ are eigenvectors of the evolution matrix (8.151). The standard properties of orthogonal projecting operators

$$\begin{aligned} P^{(1)}P^{(2)} &= 0, i \neq j, \\ P^{(i)} \cdot P^{(i)} &= P^{(i)}, \\ P^{(1)} + P^{(2)} &= I \end{aligned} \tag{7.97}$$

are implied. The dynamical projecting operators $\widehat{\mathbf{P}}^{(i)}$ in $\omega-$representation take the form

$$\widehat{\mathbf{P}}^{(1)}(\omega) = \frac{1}{2}\begin{pmatrix} 1 & -a(\omega) \\ -\frac{1}{a(\omega)} & 1 \end{pmatrix}, \tag{7.98}$$

$$\widehat{\mathbf{P}}^{(2)}(\omega) = \frac{1}{2}\begin{pmatrix} 1 & a(\omega) \\ \frac{1}{a(\omega)} & 1 \end{pmatrix}. \tag{7.99}$$

Using the inverse Fourier transformation, we return to t-representation. There, the elements of projectors are proportional to the operators, which action on arbitrary functions f, g looks as

$$\widehat{P}_{11}^{(1)} f(x,t) = \int_{-\infty}^{\infty} \tilde{P}_{11}^{(1)}(t,t')f(x,t')dt', \tag{7.100}$$

$$\widehat{P}_{12}^{(1)} g(x,t') = \int_{-\infty}^{\infty} \tilde{P}_{12}^{(1)}(t,t')g(x,t')dt'. \tag{7.101}$$

The functions $\tilde{P}_{11}^{(1)}(t,t')$ and $\tilde{P}_{12}^{(1)}(t,t')$ are kernels of corresponding integral operators:

$$\tilde{P}_{11}^{(1)}(t,t') = \frac{1}{2\pi}\int_{-\infty}^{\infty} \exp(i\omega t - i\omega t')d\omega, \tag{7.102}$$

$$\tilde{P}_{12}^{(1)}(t,t') = -\frac{1}{2\pi}\int_{-\infty}^{\infty} a(\omega)\exp(i\omega t - i\omega t')d\omega. \tag{7.103}$$

For the rest of the elements

$$\widehat{P}_{21}^{(1)} f(x,t) = \int_{-\infty}^{\infty} \tilde{P}_{21}^{(1)}(t,t')f(x,t')dt', \tag{7.104}$$

$$\widehat{P}_{22}^{(1)} g(x,t) = \int_{-\infty}^{\infty} \tilde{P}_{22}^{(1)}(t,t')g(x,t')dt'. \tag{7.105}$$

The kernels $\tilde{P}_{21}^{(1)}(t,t')$ and $\tilde{P}_{22}^{(1)}(t,t')$ are

$$\tilde{P}_{21}^{(1)}(t,t') = -\frac{1}{2\pi}\int_{-\infty}^{\infty} \frac{1}{a(\omega)}\exp(i\omega t - i\omega t')d\omega, \tag{7.106}$$

$$\tilde{P}_{22}^{(1)}(t,t') = \frac{1}{2\pi}\int_{-\infty}^{\infty} \exp[i\omega(t-t')]d\omega. \tag{7.107}$$

Operators $\widehat{P}_{11}^{(2)}$, $\widehat{P}_{12}^{(2)}$, $\widehat{P}_{21}^{(2)}$, $\widehat{P}_{22}^{(2)}$ are defined similarly:

$$\widehat{P}_{11}^{(2)} f(x,t) = \int_{-\infty}^{\infty} \tilde{P}_{11}^{(2)}(t,t')f(x,t')dt', \tag{7.108}$$

$$\widehat{P}_{12}^{(2)} g(x,t) = \int_{-\infty}^{\infty} \tilde{P}_{12}^{(2)}(t,t')g(x,t')dt', \tag{7.109}$$

$$\widehat{P}_{21}^{(2)} f(x,t) = \int_{-\infty}^{\infty} \tilde{P}_{21}^{(2)}(t,t')f(x,t')dt', \tag{7.110}$$

$$\widehat{P}_{22}^{(2)} g(x,t) = \int_{-\infty}^{\infty} \tilde{P}_{22}^{(2)}(t,t')g(x,t')dt'. \tag{7.111}$$

Integral kernels for them are

$$\tilde{P}_{11}^{(2)}(t,t') = \frac{1}{2\pi}\int_{-\infty}^{\infty} \exp[i\omega(t-t')]d\omega, \tag{7.112}$$

$$\tilde{P}_{12}^{(2)}(t,t') = \frac{1}{2\pi}\int_{-\infty}^{\infty} a(\omega)\exp[i\omega(t-t')]d\omega, \tag{7.113}$$

$$\tilde{P}_{21}^{(2)}(t,t') = \frac{1}{2\pi}\int_{-\infty}^{\infty} \frac{1}{a(\omega)}\exp[i\omega(t-t')]d\omega, \tag{7.114}$$

$$\tilde{P}_{22}^{(2)}(t,t') = \frac{1}{2\pi}\int_{-\infty}^{\infty} \exp[i\omega(t-t')]d\omega. \tag{7.115}$$

Hence, we have constructed the dynamical projectors:

$$\widehat{\mathbf{P}}^{(1)}(t) = \frac{1}{2}\begin{pmatrix} 1 & \widehat{P}_{12}^{(1)} \\ \widehat{P}_{21}^{(1)} & 1 \end{pmatrix}, \tag{7.116}$$

$$\widehat{\mathbf{P}}^{(2)}(t) = \frac{1}{2}\begin{pmatrix} 1 & \widehat{P}_{12}^{(2)} \\ \widehat{P}_{21}^{(2)} & 1 \end{pmatrix}; \tag{7.117}$$

its matrix elements are operators $\widehat{P}_{12}^{(1)}$, $\widehat{P}_{21}^{(1)}$, $\widehat{P}_{12}^{(2)}$, $\widehat{P}_{21}^{(2)}$, defined by relations (7.108, 7.109, 7.110, 7.111).

7.3.5 ON INTEGRAL KERNELS DETAILS

The kernel (7.103) is equivalent to delta-function $\delta(t-t')$. After plugging it into the operator action (8.40), finally we find:

$$\widehat{P}_{11}^{(1)} f(x,t) = f(x,t). \tag{7.118}$$

From the structure of the operators $\widehat{\mathbf{P}}_1$ and $\widehat{\mathbf{P}}_2$ we see

$$\widehat{P}_{11}^{(2)} f(x,t) = \widehat{P}_{11}^{(1)} f(x,t), \tag{7.119}$$

and, similarly,

$$\widehat{P}_{22}^{(1,2)} f(x,t) = f(x,t). \tag{7.120}$$

Here, again, $f(x,t)$ is arbitrary function. It's easy to check that operators $\widehat{\mathbf{P}}_1$ and $\widehat{\mathbf{P}}_2$, commute with $\widehat{L}$. After substituting some explicit expressions in (7.57) and (7.58) we expand the kernel of operator $\widehat{P}_{12}^{(1)}$ into a sum of rational functions as, e.g. in (7.30). For simplicity, let's look at a term in the integrand before $\exp(i\omega t - i\omega t')d\omega$:

$$a(\omega) = c^{-2}\varepsilon(\omega)\mu(\omega). \tag{7.121}$$

Note: We have marked the product $\mu_0\varepsilon_0$ as c^{-2}. On the other hand, the projector $\widehat{P}_{12}^{(2)}$ is equivalent to the operator $\widehat{\varepsilon}\widehat{\mu}$ (by definition (8.120)). Let's compare the operator (7.101):

$$\widehat{P}_{12}^{(2)} g(x,t) = \frac{c^{-2}}{2\pi} \int_{-\infty}^{\infty} g(x,t') \int_{-\infty}^{\infty} \varepsilon(\omega)\mu(\omega)\exp(i\omega(t-t')dt'd\omega \tag{7.122}$$

with the integral operators' product action to the same function:

$$\widehat{\varepsilon}\widehat{\mu} g(x,t) = \frac{c^{-2}}{2\pi} \int_{-\infty}^{\infty} \tilde{\varepsilon}(t-\alpha) \int_{-\infty}^{\infty} \tilde{\mu}(\alpha-\beta)g(x,\beta)d\beta d\alpha \tag{7.123}$$

From the inverse to (7.61), (7.63) we find

$$\begin{aligned} \tilde{\varepsilon}(t-\alpha) &= \frac{1}{\sqrt{2\pi}} \int_{-\infty}^{\infty} \varepsilon(\omega)\exp[i\omega(t-\alpha)]d\omega, \\ \tilde{\mu}(\alpha-\beta) &= \frac{1}{\sqrt{2\pi}} \int_{-\infty}^{\infty} \mu(\omega')\exp[i\omega'(\alpha-\beta)]d\omega', \end{aligned} \tag{7.124}$$

Plugging it into (7.123) yields

$$\begin{aligned} \widehat{\varepsilon}\widehat{\mu} g(x,t) = &\frac{1}{(c2\pi)^2} \int_{-\infty}^{\infty}\int_{-\infty}^{\infty} \varepsilon(\omega) \\ &\times \exp[i\omega(t-\alpha)]d\omega \int_{-\infty}^{\infty}\int_{-\infty}^{\infty} \mu(\omega')\exp[i\omega'(\alpha-\beta)]d\omega' g(x,\beta)d\beta d\alpha \end{aligned} \tag{7.125}$$

and integrating by α and, after that, by ω', absorbing the δ-function,

$$\begin{aligned} \widehat{\varepsilon}\widehat{\mu} g(x,t) = &\frac{1}{2\pi c^2} \int_{-\infty}^{\infty}\int_{-\infty}^{\infty} \varepsilon(\omega)\mu(\omega) \\ &\times \exp[i\omega(t-\beta)]g(x,\beta)d\beta d\omega \end{aligned} \tag{7.126}$$

which is equivalent to (7.122).

Finally, the operator $\widehat{P}_{12}^{(1)}$ action we can rewrite as

$$\widehat{P}_{12}^{(2)} g(x,t) = \frac{c^{-2}}{2\pi} \iint_{-\infty}^{\infty} \varepsilon(t-\alpha)\mu(\alpha-t')\exp(-i\omega(t-t'))g(x,t')d\alpha dt'. \tag{7.127}$$

The integral operator $\widehat{P}_{21}^{(1)}$ is inverse to the operator $\widehat{P}_{12}^{(1)}$. It could be compared with the operators product $\widehat{\mu}^{-1}\widehat{\varepsilon}^{-1}$:

$$\widehat{P}_{21}^{(1)} g(x,t) = \frac{c^2}{2\pi} \int_{-\infty}^{\infty} g(x,t') \int_{-\infty}^{\infty} \frac{1}{\varepsilon(\omega)\mu(\omega)} \exp(i\omega t - i\omega t')dt'd\omega, \tag{7.128}$$

$$\widehat{\mu}^{-1}\widehat{\varepsilon}^{-1}g(x,t)=\frac{c^2}{2\pi}\int_{-\infty}^{\infty}\mathfrak{m}(t,\beta)\int_{-\infty}^{\infty}\mathfrak{e}(t,\alpha)g(x,\alpha)d\alpha d\beta, \tag{7.129}$$

see the definitions of the inverse integral operators (7.76,7.77) with the kernel $\mathfrak{e},\mathfrak{m}$.

7.3.6 POLARIZED HYBRID FIELDS. EQUATIONS FOR LEFT AND RIGHT WAVES

In $t-$representation, acting by the operator $\widehat{\mathbf{P}}^{(1)}(t)$ (7.116) on Eq. (7.91) with operator $\widehat{L}$ (8.151) written in t-representation:

$$\widehat{\mathbf{P}}^{(1)}\frac{\partial\Psi}{\partial x}=\widehat{\mathbf{P}}^{(1)}L(t)\tilde{\Psi} \tag{7.130}$$

We can commute $\widehat{\mathbf{P}}^{(1)}$ and $\frac{\partial}{\partial x}$, because the dynamic projectors for boundary regime problem don't depend on x.

$$\frac{\partial}{\partial x}\widehat{\mathbf{P}}^{(1)}(t)\Psi=\widehat{\mathbf{P}}^{(1)}(t)L\Psi=L\widehat{\mathbf{P}}^{(1)}(t)\Psi. \tag{7.131}$$

Introduce the designations:

$$\partial_t\equiv\frac{\partial}{\partial t},\ \partial_x\equiv\frac{\partial}{\partial x}, \tag{7.132}$$

and after the dynamic projectors action, we find

$$\partial_x\begin{pmatrix}\frac{1}{2}B+\frac{1}{2}\widehat{P}^{(1)}_{12}E\\ \frac{1}{2}\widehat{P}^{(1)}_{21}B+\frac{1}{2}E\end{pmatrix}=L\begin{pmatrix}\frac{1}{2}B+\frac{1}{2}\widehat{P}^{(1)}_{12}E\\ \frac{1}{2}\widehat{P}^{(1)}_{21}B+\frac{1}{2}E\end{pmatrix}. \tag{7.133}$$

As in previous cases, applying the projection operators to the vector Ψ in Eq. (8.55), we can introduce new variables Λ and Π as

$$\widehat{\mathbf{P}}^{(1)}(t)\Psi=\begin{pmatrix}\frac{1}{2}B+\frac{1}{2}\widehat{P}^{(1)}_{12}E\\ \frac{1}{2}\widehat{P}^{(1)}_{21}B+\frac{1}{2}E\end{pmatrix}=\begin{pmatrix}\Lambda\\ \widehat{P}^{(1)}_{21}\Lambda\end{pmatrix}, \tag{7.134}$$

where the hybrid field Λ for the left propagating wave is defined by

$$\Lambda\equiv\frac{1}{2}(B+\widehat{P}^{(1)}_{12}E), \tag{7.135}$$

via the integral operator (7.103). And, the second field arises from

$$\widehat{\mathbf{P}}^{(2)}(t)\Psi=\begin{pmatrix}\frac{1}{2}B+\frac{1}{2}\widehat{P}^{(2)}_{12}E\\ \frac{1}{2}\widehat{P}^{(2)}_{21}B+\frac{1}{2}E\end{pmatrix}=\begin{pmatrix}\Pi\\ \widehat{P}^{(2)}_{21}\Pi\end{pmatrix}, \tag{7.136}$$

whereas a similar hybrid Π-field is obtained as

$$\Pi \equiv \frac{1}{2}(B + \widehat{P}_{12}^{(2)} E) \tag{7.137}$$

From (8.56 and 8.58) it follows that

$$\begin{pmatrix} \partial_x \Pi \\ \partial_x \widehat{P}_{21}^{(1)} \Pi \end{pmatrix} = L \begin{pmatrix} \Pi \\ \widehat{P}_{21}^{(1)} \Pi \end{pmatrix}. \tag{7.138}$$

Formally, we extract four equations:

$$\begin{aligned} \partial_x \Lambda &= L_{11}\Lambda + L_{12}\widehat{P}_{21}^{(1)}\Lambda, \\ \partial_x \widehat{P}_{21}^{(1)}\Lambda &= L_{21}\Lambda + L_{22}\widehat{P}_{21}^{(1)},\Lambda \\ \partial_x \Pi &= L_{11}\Pi + L_{12}\widehat{P}_{21}^{(2)}\Pi, \\ \partial_x \widehat{P}_{21}^{(2)}\Pi &= L_{21}\Pi + L_{22}\widehat{P}_{21}^{(2)}\Pi, \end{aligned} \tag{7.139}$$

but only two of them can be chosen as independent.

We rewrite the operator $\widehat{L}$ (7.93) in t-representation:

$$L = \begin{pmatrix} 0 & -\partial_t \mu_0 \varepsilon_0 \widehat{a} \\ -\partial_t & 0 \end{pmatrix}. \tag{7.140}$$

As we see, $L_{11} = L_{22} = 0$. Also, $\mu_0 \varepsilon_0 = c^{-2}$. Hence, the simplest of Eq. (7.139) are

$$\begin{aligned} \partial_x \widehat{P}_{21}^{(2)}\Pi &= -\partial_t \Pi, \\ \partial_x \widehat{P}_{21}^{(1)}\Lambda &= -\partial_t \Lambda. \end{aligned} \tag{7.141}$$

Using relations (7.128), we find

$$\begin{aligned} \partial_x \widehat{\mu}^{-1}\widehat{\varepsilon}^{-1}\Pi(x,t) &= -\partial_t \Pi, \\ \partial_x \widehat{\mu}^{-1}\widehat{\varepsilon}^{-1}\Lambda(x,t) &= -\partial_t \Lambda, \end{aligned} \tag{7.142}$$

or, the equivalent

$$\begin{aligned} \partial_x \Pi(x,t) &= -\widehat{\mu}\widehat{\varepsilon}\partial_t \Pi, \\ \partial_x \Lambda(x,t) &= \widehat{\mu}\widehat{\varepsilon}\partial_t \Lambda. \end{aligned} \tag{7.143}$$

Otherwise, in terms of $\varepsilon(\omega)$ and $\mu(\omega)$:

$$\begin{aligned} \partial_x \Pi(x,t) &= -\frac{c^{-2}}{2\pi}\int_{-\infty}^{\infty}\int_{-\infty}^{\infty} \varepsilon(\omega)\mu(\omega)\exp(i\omega(t-t'))[\partial_t \Pi(x,t)]_{t=t'}dt'd\omega \\ \partial_x \Lambda(x,t) &= \frac{c^{-2}}{2\pi}\int_{-\infty}^{\infty}\int_{-\infty}^{\infty} \varepsilon(\omega)\mu(\omega)\exp(i\omega(t-t'))[\partial_t \Lambda(x,t)]_{t=t'}dt'd\omega. \end{aligned} \tag{7.144}$$

Kinsler's wave equation (written in our notations, and in t-representation for linear case) [6] looks like

$$\partial_x \Pi(x,t) = -\partial_t \widehat{\mu}\widehat{\varepsilon}\Pi - \frac{1}{2}\partial_t \varepsilon_c \mu(\Pi - \Lambda) \tag{7.145}$$

As we see, Kinsler failed to decouple left and right waves. As we see, the left and right hybrid waves could be decoupled more effectively by the regular procedure of dynamical projecting.

7.4 POLARIZATION ACCOUNT

7.4.1 GENERAL REMARKS

Expanding our thesis to another degree of freedom, we get the unidirectional propagation of light pulses in a medium with cubic nonlinearity. To describe such a model we use nonlinear Schrödinger equation, which was derived by Zakharov [11]. In our interest remain ultra-short pulses whose length is shorter than a few cycles of the center frequency. Such pulses were investigated by other authors [1,6,12,13]. In the derivation of equations describing the propagation of ultra-short pulses, there are two basic approximations for an electromagnetic field, which are described in the review by Caputo and Maimistov [14], namely the slowly varying envelope approximation and the unidirectional wave approximation. For ultra-short pulses whose length does not exceed a few picoseconds, the first approximation leads to inaccurate results. Therefore, the validity of the application of NLSE to describe the propagation of ultra-short pulses of light is questionable by different groups [15–17]. Other approaches to the problem use slowly evolving wave approximation SEWA [17] or an application of Bloch equations [18]. Another approach is to introduce new variables of the form

$$\psi^{\pm} = \varepsilon\frac{1}{2}E_i \pm \mu\frac{1}{2}H_j,$$

as did Fleck [12] and Kinsler [5,6,19] in their works. This is a method in a sense similar to the projection operator method, which was described in the book [3], in which many examples in different fields of science. The dynamic projecting operator method is a powerful tool, which was initially set for the Cauchy problem and which allows for detailed analysis of the field and its modes. Applying the projection operators to the one-dimensional model of the light pulse propagation has led to the generalized SP equation [4], in which the physical form of cofactors has been shown and the interaction between waves in two directions has been taken into consideration. The case of the interaction of two counterpropagating waves was the topic of experimental studies [20], in which authors presented the evidence of the interaction of a counterpropagating pulses with the same polarizations. In the other case, the authors of Ref. [21] proved the existence of the wave propagating backwards in optical resonators. The most fitting case is a paper about backward pulse generation in a trasmission line [22], which lacks of theoretical description of the phenomenon. In the next section, we mainly follow the results of Ref. [23].

7.4.2 THEORY OF INITIAL DISTURBANCE PROPAGATION, CAUCHY PROBLEM FORMULATION

To create a description of a one-dimensional pulse propagation in both directions [6,19,24] and both polarizations, developing results of Ref. [4], we start with the standard linearized Maxwell equations setup:

$$\begin{aligned} \frac{1}{c}\frac{\partial D_y}{\partial t} &= -\frac{\partial H_z}{\partial x}, \\ \frac{1}{c}\frac{\partial D_z}{\partial t} &= \frac{\partial H_y}{\partial x}, \\ \frac{1}{c}\frac{\partial B_z}{\partial t} &= -\frac{\partial E_y}{\partial x}, \\ \frac{1}{c}\frac{\partial B_y}{\partial t} &= \frac{\partial E_z}{\partial x}, \end{aligned} \tag{7.146}$$

while material relations for an isotropic dispersive medium in operator form, which repeat the results of the previous section $D_y = \hat{\varepsilon} E_y$ and $D_z = \hat{\varepsilon} E_z$, $\hat{\mu} H_y = B_y$, $\hat{\mu} H_z = B_z$ mathematically, close the description.

From the point of view of the initial problem the operators may by introduced on a base of Fourier transformation in x-variable to be provided for the variable $k \in (-\infty, \infty)$ as

$$D_y(x,t) = \int_{-\infty}^{\infty}\int_{-\infty}^{\infty} \varepsilon(k) e^{-ik(x-x')} dk E_y(x',t) dx' \quad = \quad \hat{\varepsilon} E_y, \tag{7.147}$$

$$B_y(x,t) = \int_{-\infty}^{\infty}\int_{-\infty}^{\infty} \mu(k) e^{-ik(x-x')} dk H_y(x',t) dx' \quad = \quad \hat{\mu} H_y, \tag{7.148}$$

where, omitting t-dependence,

$$D_y(x) = \frac{1}{\sqrt{2\pi}} \int_{-\infty}^{\infty} \tilde{D}_y(k) e^{ikx} dk, \tag{7.149}$$

and

$$E_y(x) = \frac{1}{\sqrt{2\pi}} \int_{-\infty}^{\infty} \tilde{E}_y(k) e^{ikx} dk, \tag{7.150}$$

which implies a simple link between transforms

$$\tilde{D}_y(k) = \varepsilon(k) \tilde{E}_y(k), \tag{7.151}$$

which characterizes material properties of a medium of propagation used for example in the seminal work [1]. The magnetic fields are connected by a similar integral operator; we traditionally choose the magnetic induction as a basic one, excluding $H_y = \hat{\mu}^{-1} B_y$ and $H_z = \hat{\mu}^{-1} B_z$.

The Cauchy problem needs four initial conditions

$$\begin{aligned} E_y(x,0) &= \eta_1(x), \quad E_z(x,0) = \eta_2(x), \\ B_y(x,0) &= \xi_1(x), \quad B_z(x,0) = \xi_2(x) \end{aligned} \tag{7.152}$$

to make the solution of (7.146) unique.

7.4.3 THE PROJECTION METHOD FOR THE CAUCHY PROBLEM

From the point of view of the initial problem to be considered, we transform the system (7.146) to a more convenient splitted form (mode representation [3]) and specify the corresponding initial conditions for such mode variables. We introduce projection operators in x-space that have to fulfill the standard properties of orthogonality and completeness

$$P_iP_j = 0, \quad P_i^2 = P_i, \quad \sum_i P_i = 1. \tag{7.153}$$

To construct projection operators for a specific problem such as (7.146), it is convenient to find eigenvectors for the linear evolution operator L [7] that enter an evolution equation of general form

$$\psi_t = L\psi. \tag{7.154}$$

In the case we consider, Eq. (7.146) can be rewritten in such a manner that the vector of state and evolution operators are written as

$$\psi = \begin{pmatrix} E_z \\ B_y \\ E_y \\ B_z \end{pmatrix}, L = \begin{pmatrix} 0 & c\hat{\varepsilon}^{-1}\hat{\mu}^{-1}\partial_x & 0 & 0 \\ c\partial_x & 0 & 0 & 0 \\ 0 & 0 & 0 & -c\hat{\varepsilon}^{-1}\hat{\mu}^{-1}\partial_x \\ 0 & 0 & -c\partial_x & 0 \end{pmatrix}. \tag{7.155}$$

Entering elements from Eq. (7.155) to the matrix operator equation general form (7.154) gives

$$\begin{pmatrix} (E_z)_t \\ (B_y)_t \\ (E_y)_t \\ (B_z)_t \end{pmatrix} = \begin{pmatrix} 0 & c\hat{\varepsilon}^{-1}\hat{\mu}^{-1}\partial_x & 0 & 0 \\ c\partial_x & 0 & 0 & 0 \\ 0 & 0 & 0 & -c\hat{\varepsilon}^{-1}\hat{\mu}^{-1}\partial_x \\ 0 & 0 & -c\partial_x & 0 \end{pmatrix} \begin{pmatrix} E_z \\ B_y \\ E_y \\ B_z \end{pmatrix}. \tag{7.156}$$

Let us apply the Fourier transformation by x in a form of (7.150) to Eq. (7.156), which yields the system of ordinary differential equations with material constants ε, μ as functions of wave-number k:

$$\begin{pmatrix} (\tilde{E}_z(k,t))_t \\ (\tilde{B}_y(k,t))_t \\ (\tilde{E}_y(k,t))_t \\ (\tilde{B}_z(k,t))_t \end{pmatrix} = i \begin{pmatrix} ck\varepsilon^{-1}\mu^{-1}\tilde{B}_y(k,t) \\ ck\tilde{E}_z(k,t) \\ -ck\varepsilon^{-1}\mu^{-1}\tilde{B}_z(k,t) \\ -ck\tilde{E}_y(k,t) \end{pmatrix}, \tag{7.157}$$

whose particular solutions have the form

$$\tilde{E}_{y,z} = \breve{E}_{y,z}\exp(i\omega t), \quad \tilde{B}_{y,z} = \breve{B}_{y,z}\exp(i\omega t). \tag{7.158}$$

Plugging the solutions into the system (7.157) yields a system of homogeneous linear algebraic equations:

$$\begin{pmatrix} \omega\breve{E}_z(k,\omega) \\ \omega\breve{B}_y(k,\omega) \\ \omega\breve{E}_y(k,\omega) \\ \omega\breve{B}_z(k,\omega) \end{pmatrix} = \begin{pmatrix} -ck\varepsilon^{-1}\mu^{-1}\breve{B}_y(k,\omega) \\ -ck\breve{E}_z(k,\omega) \\ ck\varepsilon^{-1}\mu^{-1}\breve{B}_z(k,\omega) \\ ck\breve{E}_y(k,\omega) \end{pmatrix}, \tag{7.159}$$

that have a sense of an eigenvalue problem under the solvability condition

$$\omega^2 = k^2c^2\mu^{-1}\varepsilon^{-1}. \tag{7.160}$$

There is an important dependency hidden in the preceding equation. The dispersion relation is described by the dielectric parameter ε, which, in this case, is a function of k. The function may be recalculated to the ω−dependent by means of (7.160). Strictly from those equations one can construct projection operators by a standard scheme of Section 1. Hence, we can present a direct relation between $\breve{E}_z$ and $\breve{H}_y$ as well as between $\breve{E}_y$ and $\breve{H}_z$,

$$P_{11} = \frac{1}{2}\begin{pmatrix} 1 & -\mu^{-\frac{1}{2}}\varepsilon^{-\frac{1}{2}} & 0 & 0 \\ -\varepsilon^{\frac{1}{2}}\mu^{\frac{1}{2}} & 1 & 0 & 0 \\ 0 & 0 & 0 & 0 \\ 0 & 0 & 0 & 0 \end{pmatrix}, \quad P_{12} = \frac{1}{2}\begin{pmatrix} 0 & 0 & 0 & 0 \\ 0 & 0 & 0 & 0 \\ 0 & 0 & 1 & \mu^{-\frac{1}{2}}\varepsilon^{-\frac{1}{2}} \\ 0 & 0 & \varepsilon^{\frac{1}{2}}\mu^{\frac{1}{2}} & 1 \end{pmatrix}, \tag{7.161}$$

$$P_{21} = \frac{1}{2}\begin{pmatrix} 1 & \mu^{-\frac{1}{2}}\varepsilon^{-\frac{1}{2}} & 0 & 0 \\ \varepsilon^{\frac{1}{2}}\mu^{\frac{1}{2}} & 1 & 0 & 0 \\ 0 & 0 & 0 & 0 \\ 0 & 0 & 0 & 0 \end{pmatrix}, \quad P_{22} = \frac{1}{2}\begin{pmatrix} 0 & 0 & 0 & 0 \\ 0 & 0 & 0 & 0 \\ 0 & 0 & 1 & -\mu^{-\frac{1}{2}}\varepsilon^{-\frac{1}{2}} \\ 0 & 0 & -\varepsilon^{\frac{1}{2}}\mu^{\frac{1}{2}} & 1 \end{pmatrix}, \tag{7.162}$$

which, acting on ψ, gives us four eigenstates

$$\breve{\psi}_{11} = \begin{pmatrix} 1 \\ -\varepsilon^{\frac{1}{2}}\mu^{\frac{1}{2}} \\ 0 \\ 0 \end{pmatrix}\breve{E}_z, \quad \breve{\psi}_{12} = \begin{pmatrix} 0 \\ 0 \\ 1 \\ \varepsilon^{\frac{1}{2}}\mu^{\frac{1}{2}} \end{pmatrix}\breve{E}_y, \tag{7.163}$$

and

$$\breve{\psi}_{21} = \begin{pmatrix} 1 \\ \varepsilon^{\frac{1}{2}}\mu^{\frac{1}{2}} \\ 0 \\ 0 \end{pmatrix}\breve{E}_z, \quad \breve{\psi}_{22} = \begin{pmatrix} 0 \\ 0 \\ 1 \\ -\varepsilon^{\frac{1}{2}}\mu^{\frac{1}{2}} \end{pmatrix}\breve{E}_y \tag{7.164}$$

of the evolution matrix.

After the inverse Fourier transformation of [7.161,7.162], the projection operators obtain the form of matrix-integral ones quite similar to Section 7.3.3:

$$\hat{P}_{11} = \frac{1}{2}\begin{pmatrix} 1 & -\hat{\mu}^{-\frac{1}{2}}\hat{\varepsilon}^{-\frac{1}{2}} & 0 & 0 \\ -\hat{\varepsilon}^{\frac{1}{2}}\hat{\mu}^{\frac{1}{2}} & 1 & 0 & 0 \\ 0 & 0 & 0 & 0 \\ 0 & 0 & 0 & 0 \end{pmatrix}, \quad \hat{P}_{12} = \frac{1}{2}\begin{pmatrix} 0 & 0 & 0 & 0 \\ 0 & 0 & 0 & 0 \\ 0 & 0 & 1 & \hat{\mu}^{-\frac{1}{2}}\hat{\varepsilon}^{-\frac{1}{2}} \\ 0 & 0 & \hat{\varepsilon}^{\frac{1}{2}}\hat{\mu}^{\frac{1}{2}} & 1 \end{pmatrix}, \tag{7.165}$$

$$\hat{P}_{21} = \frac{1}{2}\begin{pmatrix} 1 & -\hat{\mu}^{-\frac{1}{2}}\hat{\varepsilon}^{-\frac{1}{2}} & 0 & 0 \\ \hat{\varepsilon}^{\frac{1}{2}}\hat{\mu}^{\frac{1}{2}} & 1 & 0 & 0 \\ 0 & 0 & 0 & 0 \\ 0 & 0 & 0 & 0 \end{pmatrix}, \quad \hat{P}_{22} = \frac{1}{2}\begin{pmatrix} 0 & 0 & 0 & 0 \\ 0 & 0 & 0 & 0 \\ 0 & 0 & 1 & -\hat{\mu}^{-\frac{1}{2}}\hat{\varepsilon}^{-\frac{1}{2}} \\ 0 & 0 & -\hat{\varepsilon}^{\frac{1}{2}}\hat{\mu}^{\frac{1}{2}} & 1 \end{pmatrix}, \tag{7.166}$$

where the integral operator $\hat{\varepsilon}$ is given by Eq. (7.147). The operators acting on ψ gives us four subspaces, generated by the vectors

$$\psi_{11} = \begin{pmatrix} 1 \\ -\hat{\varepsilon}^{\frac{1}{2}}\hat{\mu}^{\frac{1}{2}} \\ 0 \\ 0 \end{pmatrix} E_z, \quad \psi_{12} = \begin{pmatrix} 0 \\ 0 \\ 1 \\ \hat{\varepsilon}^{\frac{1}{2}}\hat{\mu}^{\frac{1}{2}} \end{pmatrix} E_y,$$
$$\psi_{21} = \begin{pmatrix} 1 \\ \hat{\varepsilon}^{\frac{1}{2}}\hat{\mu}^{\frac{1}{2}} \\ 0 \\ 0 \end{pmatrix} E_z, \quad \psi_{22} = \begin{pmatrix} 0 \\ 0 \\ 1 \\ -\hat{\varepsilon}^{\frac{1}{2}}\hat{\mu}^{\frac{1}{2}} \end{pmatrix} E_y. \tag{7.167}$$

The result of identity operator expansion $I = \hat{P}_{11} + \hat{P}_{12} + \hat{P}_{21} + \hat{P}_{22}$ by action on the vector ψ defines the following transition to new variables:

$$\Pi_1 = \frac{1}{2}E_z - \frac{1}{2}\hat{\mu}^{-\frac{1}{2}}\hat{\varepsilon}^{-\frac{1}{2}}B_y, \qquad \Lambda_1 = \frac{1}{2}E_z + \frac{1}{2}\hat{\mu}^{-\frac{1}{2}}\hat{\varepsilon}^{-\frac{1}{2}}B_y, \tag{7.168}$$

$$\Pi_2 = \frac{1}{2}E_y - \frac{1}{2}\hat{\mu}^{-\frac{1}{2}}\hat{\varepsilon}^{-\frac{1}{2}}B_z, \qquad \Lambda_2 = \frac{1}{2}E_y + \frac{1}{2}\hat{\mu}^{-\frac{1}{2}}\hat{\varepsilon}^{-\frac{1}{2}}B_z, \tag{7.169}$$

$$\Lambda_1 + \Pi_1 = E_z, \qquad \Lambda_2 + \Pi_2 = E_y, \tag{7.170}$$

$$\hat{\varepsilon}^{\frac{1}{2}}\hat{\mu}^{\frac{1}{2}}(\Lambda_1 - \Pi_1) = B_y, \qquad \hat{\varepsilon}^{\frac{1}{2}}\hat{\mu}^{\frac{1}{2}}(\Lambda_2 - \Pi_2) = B_z. \tag{7.171}$$

From the definition of the projectors it follows that $[L, P_i] = 0$. The result of the operator $\hat{P}_{11}$ application to Eq. (7.159) is as follows:

$$\hat{P}_{11}\begin{pmatrix} (E_z)_t \\ (B_y)_t \\ (E_y)_t \\ (B_z)_t \end{pmatrix} - L\hat{P}_{11}\begin{pmatrix} E_z \\ B_y \\ E_y \\ B_z \end{pmatrix} = 0. \tag{7.172}$$

Further evaluation yields

$$\begin{pmatrix} (\Pi_1)_t \\ (-\hat{\varepsilon}^{\frac{1}{2}}\hat{\mu}^{\frac{1}{2}}\Pi_1)_t \\ 0 \\ 0 \end{pmatrix} - \begin{pmatrix} 0 & c\hat{\varepsilon}^{-1}\hat{\mu}^{-1}\partial_x & 0 & 0 \\ c\partial_x & 0 & 0 & 0 \\ 0 & 0 & 0 & -c\hat{\varepsilon}^{-1}\hat{\mu}^{-1}\partial_x \\ 0 & 0 & -c\partial_x & 0 \end{pmatrix}$$
$$\times \begin{pmatrix} \Pi_1 \\ -\hat{\varepsilon}^{\frac{1}{2}}\hat{\mu}^{\frac{1}{2}}\Pi_1 \\ 0 \\ 0 \end{pmatrix} = 0, \tag{7.173}$$

or, after the matrix operator action,

$$\begin{pmatrix} (\Pi_1)_t \\ (-\hat{\varepsilon}^{\frac{1}{2}}\hat{\mu}^{\frac{1}{2}}\Pi_1)_t \\ 0 \\ 0 \end{pmatrix} - \begin{pmatrix} -c\hat{\varepsilon}^{-\frac{1}{2}}\hat{\mu}^{-\frac{1}{2}}(\Pi_1)_x \\ c(\Pi_1)_x \\ 0 \\ 0 \end{pmatrix} = 0, \tag{7.174}$$

which reads as one of main integro-differential equation

$$(\hat{\varepsilon}^{\frac{1}{2}}\hat{\mu}^{\frac{1}{2}}\Pi_1)_t + c(\Pi_1)_x = 0, \tag{7.175}$$

with the initial condition specified by (7.168):

$$\Pi_1(x,0) = \frac{1}{2}\eta_2(x) - \frac{1}{2}\hat{\mu}^{-\frac{1}{2}}\hat{\varepsilon}^{-\frac{1}{2}}\xi_1(x). \tag{7.176}$$

Similar results one obtains acting to Eq. (7.159) by the rest of operators. Resulting equations define independent linear modes propagation that correspond to left and right waves, each with two fixed orthogonal polarizations and initial conditions as Eq. 7.176.

7.4.4 NONLINEARITY ACCOUNT, INTERACTION OF POLARIZED WAVES: GENERAL RELATIONS

Following the general scheme of this book, originating from Ref. [3], we have shown in the work [4] that new variables, namely Eqs (7.168 and 7.169), may be introduced into Eq. (7.156) by acting with projection operators of Eq. (7.162) such as

$$\frac{\partial}{\partial t}\hat{P}_{11}\Psi - L\hat{P}_{11}\Psi = \hat{P}_{11}\mathbb{N}(E). \tag{7.177}$$

As this action will be repeated with the rest of projection operators, a system of four equations will be derived:

$$\frac{\partial}{\partial t}\hat{P}_{11}\Psi - L\hat{P}_{11}\Psi = \hat{P}_{11}\mathbb{N}(E), \tag{7.178}$$

$$\frac{\partial}{\partial t}\hat{P}_{12}\Psi - L\hat{P}_{12}\Psi = \hat{P}_{12}\mathbb{N}(E), \tag{7.179}$$

$$\frac{\partial}{\partial t}\hat{P}_{21}\Psi - L\hat{P}_{21}\Psi = \hat{P}_{21}\mathbb{N}(E), \tag{7.180}$$

$$\frac{\partial}{\partial t}\hat{P}_{22}\Psi - L\hat{P}_{22}\Psi = \hat{P}_{22}\mathbb{N}(E). \tag{7.181}$$

Stepping to a more precise model, we look closer to a description of propagation of electromagnetic pulses in silica fiber. According to our main target, we restrict ourselves by $\mu = 1$ because we would ignore contributions to dispersion and nonlinear effects from the magnetic field. As it was obvious how to present $(P_{NL})_i$ in the case of unique polarization (7.10) [4], we analyze now the more general form of third-order optical susceptibility $\chi^{(3)}$ as, e.g., in Refs [25,26]. We have based our approach on the interaction of two polarization modes, described by the terms $(P_{NL})_y$ and $(P_{NL})_z$, taking into account the directions of propagation.

Both citations [25,26] present the nonlinearity of a Kerr type as follows:

$$\mathbb{N}(E_y,E_z) = -4\pi\frac{\partial}{\partial t}\hat{\varepsilon}^{-1}\begin{pmatrix} 3\chi_{yyyy}(E_y)^3 + (\chi_{yyzz}+\chi_{yzzy}+\chi_{yzyz})E_y(E_z)^2 \\ 0 \\ 3\chi_{zzzz}(E_z)^3 + (\chi_{zzyy}+\chi_{zyyz}+\chi_{zyzy})E_z(E_y)^2 \\ 0 \end{pmatrix}. \tag{7.182}$$

Using links between the nonlinear constants for isotropic material as in Refs [25,26] we arrive at

$$\mathbb{N}(E_y,E_z) = -4\pi\frac{\partial}{\partial t}\begin{pmatrix} \hat{\varepsilon}^{-1}\chi_{yyyy}(3(E_y)^3 + E_y(E_z)^2) \\ 0 \\ \hat{\varepsilon}^{-1}\chi_{zzzz}(3(E_z)^3 + E_z(E_y)^2) \\ 0 \end{pmatrix}. \tag{7.183}$$

In the frame of weak nonlinearity theory, one applies the (linear!) projection operators to third-order nonlinearities, which are so common in many materials, for example in the silica used to make optical fibers. Acting with Eq. (7.161) on Eq. (7.182), after the transition to new variables via (7.170) yields

$$P_{11}\mathbb{N}(E_y,E_z) = \mathbb{N}(\Lambda_1,\Lambda_2,\Pi_1,\Pi_2) =$$

$$-2\pi\frac{\partial}{\partial t}\begin{pmatrix} \hat{\varepsilon}^{-1}\chi_{yyyy}(3(\Pi_1-\Lambda_1)^3 + (\Pi_1+\Lambda_1)(\Pi_2-\Lambda_2)^2) \\ -\hat{\varepsilon}^{-\frac{1}{2}}\chi_{yyyy}(3(\Pi_1-\Lambda_1)^3 + (\Pi_1+\Lambda_1)(\Pi_2+\Lambda_2)^2) \\ 0 \\ 0 \end{pmatrix}, \tag{7.184}$$

the third-order nonlinearity has been projected on the directed polarization mode. More than that, we obtain an equation for bidirectional pulse propagation with Kerr effect.

Equation (7.184) is just one part of our result. The second part we can obtain by acting with projection operator P_{12} from Eq. (7.162) on Eq. (7.182):

$$P_{12}\mathbb{N}(E_y, E_z) = -2\pi\frac{\partial}{\partial t}\begin{pmatrix} 0 \\ 0 \\ \hat{\varepsilon}^{-1}\chi_{yyyy}\left(3(\Pi_2-\Lambda_2)^3+(\Pi_2+\Lambda_2)(\Pi_1-\Lambda_1)^2\right) \\ \hat{\varepsilon}^{-\frac{1}{2}}\chi_{yyyy}\left(3(\Pi_2-\Lambda_2)^3+(\Pi_2+\Lambda_2)(\Pi_1+\Lambda_1)^2\right) \end{pmatrix}. \tag{7.185}$$

With the above result, the full formula of Eq. (7.177) can be presented as

$$\hat{\varepsilon}(\Pi_1)_t + c\hat{\varepsilon}^{\frac{1}{2}}(\Pi_1)_x = 2\pi\frac{\partial}{\partial t}\chi_{yyyy}\left(3(\Pi_1-\Lambda_1)^3+(\Pi_1+\Lambda_1)(\Pi_2-\Lambda_2)^2\right). \tag{7.186}$$

If we consider a unidirectional approach to our result, neglecting the opposite wave (which is supposed to be absent in initial disturbance), we assume that $\Lambda_1 = 0$, $\Lambda_2 = 0$, which implies that

$$\begin{aligned} \hat{\varepsilon}(\Pi_1)_t + c\hat{\varepsilon}^{\frac{1}{2}}(\Pi_1)_x &= -2\pi\tfrac{\partial}{\partial t}\left(\chi_{yyyy}(\Pi_1(3\Pi_1^2+\Pi_2^2))\right), \\ \hat{\varepsilon}(\Pi_2)_t + c\hat{\varepsilon}^{\frac{1}{2}}(\Pi_2)_x &= -2\pi\tfrac{\partial}{\partial t}\left(\chi_{yyyy}(\Pi_2(3\Pi_2^2+\Pi_1^2))\right). \end{aligned} \tag{7.187}$$

The resulting equations describe interaction between unidirected electromagnetic pulses with mutually orthogonal polarizations.

7.5 COMPARISON OF RESULTS OBTAINED WITH THE MULTIPLE SCALE METHOD

To compare our general system with ones of Refs [1,24] we would develop a method of nonsingular perturbation theory (see also [3,7]). The sense of the method is an expansion of the evolution operator (e.g. in (7.177)) with respect to a small dimensionless parameter. In our case, such a parameter may be chosen as a unified dispersion-nonlinear one.

The resulting equation type depends on the dispersion form. The dielectric susceptibility coefficient $\varepsilon(k)$ originated from Ref. [1], where authors focused on the propagation of light in the infrared range with $\lambda = 1600-3000$nm and durations correspond to femtoseconds. In this range, it can be approximated as

$$\varepsilon = 1 + 4\pi\chi^{(1)}(\lambda) \approx 1 + 4\pi(\chi_0^{(1)} + \chi_2^{(1)}\lambda^2), \tag{7.188}$$

where λ is equal to $\frac{2\pi}{k}$. In further calculations, it has been treated as a parameter, as we consider propagation in dielectric media. An alternative assumption will open further investigation for those materials that qualify as metamaterials (see the next chapter of this book). At this point, we consider permittivity a function of k

$$\varepsilon = 1 + 4\pi\chi^{(1)}(\lambda) \approx 1 + 4\pi(\chi_0^{(1)} + \chi_2^{(1)}4\pi^2\frac{1}{k^2}), \tag{7.189}$$

which, in fact is the Taylor series at the vicinity of $k=\infty$. Having in mind the calculations of integral operators such as (7.147) and their functions such as (7.155) will be continued in the x-representation, it is crucial to present the expansion of ε^{α}, $\alpha=\frac{1}{2},-\frac{1}{2},-1$ in the same vicinity:

$$\varepsilon(k)^{\frac{1}{2}} \approx \sqrt{1+4\pi\chi_0}+\frac{8\pi^3\chi_2}{\sqrt{1+4\pi\chi_0}k^2}, \tag{7.190}$$

$$\varepsilon(k)^{-1} \approx \frac{1}{1+4\pi\chi_0}-\frac{16\left(\pi^3\chi_2\right)}{(1+4\pi\chi_0)^2k^2}, \tag{7.191}$$

$$\varepsilon(k)^{-\frac{1}{2}} \approx \frac{1}{\sqrt{1+4\pi\chi_0}}-\frac{8\left(\pi^3\chi_2\right)}{(1+4\pi\chi_0)^{3/2}k^2}. \tag{7.192}$$

After an inverse Fourier transformation, the corresponding operators are built as

$$\hat{\varepsilon}(x)^{\frac{1}{2}} \approx \sqrt{1+4\pi\chi_0}+\frac{8\pi^3\chi_2}{\sqrt{1+4\pi\chi_0}}(i\partial_x)^{-2}, \tag{7.193}$$

because the factor k^{-1} is represented by primitive $(i\partial_x)^{-1}=-i\int dx$

$$\hat{\varepsilon}(x)^{-1} \approx \frac{1}{1+4\pi\chi_0}-\frac{16\left(\pi^3\chi_2\right)}{(1+4\pi\chi_0)^2}(i\partial_x)^{-2}, \tag{7.194}$$

similarly

$$\hat{\varepsilon}(x)^{-\frac{1}{2}} \approx \frac{1}{\sqrt{1+4\pi\chi_0}}-\frac{8\left(\pi^3\chi_2\right)}{(1+4\pi\chi_0)^{3/2}}(i\partial_x)^{-2}. \tag{7.195}$$

By means of the projection operators in corresponding approximation we rewrite relations Eq. (7.187) in a new form:

$$\left(1+4\pi\chi_0^{(1)}+4\pi\chi_2^{(1)}(i\partial_x)^{-2}\right)\frac{\partial}{\partial t}\Pi_1-c\left(\sqrt{1+4\pi\chi_0}+\frac{8\pi^3\chi_2}{\sqrt{1+4\pi\chi_0}}(i\partial_x)^{-2}\right)\frac{\partial}{\partial x}\Pi_1$$
$$=-2\pi\frac{\partial}{\partial t}\left(\chi_{yyyy}\Pi_1(3\Pi_1^2+\Pi_2^2)\right). \tag{7.196}$$

With an analogy to Ref. [1] we can substitute in (7.196) a multiple scales ansatz. The ansatz, by construction, accounts for only one of the directed waves; hence, we simply do not take the opposite waves into account, regarding excitation of the correspondent mode,

$$\Pi_1(x,t)=\left(\kappa A_0(\phi,x_1,x_2)+\kappa^2A_1(\phi,x_1,x_2)+\cdots\right),$$

$$\Pi_2(x,t)=\left(\kappa B_0(\phi,x_1,x_2)+\kappa^2B_1(\phi,x_1,x_2)+\cdots\right),$$

where $\phi = \frac{ct-x}{\kappa}$ and $x_n = \kappa^n x$, then we get

$$\left(2+8\pi\chi_0^{(1)}-3\mu^{-\frac{1}{2}}\sqrt{1+4\pi\chi_0^{(1)}}\right)\frac{\partial^2 A_0}{\partial\phi\partial x_1}-\left(4\pi\chi_2^{(1)}-\frac{8\pi^3\chi_2}{\sqrt{1+4\pi\chi_0}}\right)A_0$$
$$=2\pi\left(\frac{\partial^2}{\partial\phi^2}\right)\chi^{(3)}A_0\left(3A_0^2+B_0^2\right), \tag{7.197}$$
$$\left(2+8\pi\chi_0^{(1)}-3\mu^{-\frac{1}{2}}\sqrt{1+4\pi\chi_0^{(1)}}\right)\frac{\partial^2 B_0}{\partial\phi\partial x_1}-\left(4\pi\chi_2^{(1)}-\frac{8\pi^3\chi_2}{\sqrt{1+4\pi\chi_0}}\right)B_0$$
$$=2\pi\left(\frac{\partial^2}{\partial\phi^2}\right)\chi^{(3)}B_0\left(3B_0^2+A_0^2\right). \tag{7.198}$$

Comparing our equations to the one obtained by Pietrzyk et al. [24]

$$\frac{\partial^2}{\partial x_1\partial\phi}A_0 = A_0+\frac{1}{6}\frac{\partial^2}{\partial\phi^2}\left(A_0^3+3A_0B_0^2\right), \tag{7.199}$$

we conclude the coincidence of the results.

7.6 PROJECTION METHOD FOR BOUNDARY REGIME PROPAGATION

In a theory of waveguide propagation, the conventional mathematical problem is formulated as a boundary one. An application of the projection operator method to a time-dependent boundary problem needs significant modification. From the point of view of the boundary problem to be considered, we use a version of antisymmetric in t basic functions E_y, H_y, E_z, H_z, defined on $t \in (-\infty,\infty)$, compare to Ref. [2], which guarantees the zero value of the field at $t = 0$. Such continuation of the fields to the range $t < 0$ gives us the opportunity to define its Fourier transforms as

$$E_y = \frac{1}{\sqrt{2\pi}}\int_{-\infty}^{\infty}\tilde{E}_y(\omega)e^{i\omega t}d\omega. \tag{7.200}$$

The consequent procedure is supposed to be provided for the variable $\omega \in (-\infty,\infty)$ with corresponding continuation.

It is essential to stress that $\hat{\varepsilon}$ and $\hat{\mu}$ are integral operators, whose forms differ from ones of Section 2:

$$D_y = \int_{-\infty}^{\infty}\int_{-\infty}^{\infty}\varepsilon(\omega)e^{-i\omega(t-t')}d\omega E_y(t')dt' \quad = \quad \hat{\varepsilon}E_y, \tag{7.201}$$

$$B_y = \int_{-\infty}^{\infty}\int_{-\infty}^{\infty}\mu(\omega)e^{-i\omega(t-t')}d\omega H_y(t')dt' \quad = \quad \hat{\mu}H_y, \tag{7.202}$$

here

$$D_y = \frac{1}{\sqrt{2\pi}} \int_{-\infty}^{\infty} \tilde{D}_y(\omega) e^{i\omega t} d\omega. \tag{7.203}$$

which gives a simple link between transforms

$$\tilde{D}_y(\omega) = \varepsilon(\omega) \tilde{E}_y(\omega), \tag{7.204}$$

which describes the material properties of a medium of propagation. The dielectric susceptibility coefficient $\varepsilon(\omega)$ either originated from a quantum version of the Lorentz formula (see e.g. [2,27]) or directly from phenomenology [1,28], for example, approximated as Taylor expansion at $\omega \Rightarrow \infty$ point by Pietrzyk et al. [24]

$$\varepsilon(\omega) \approx 1 + 4\pi\chi(\omega) \approx 1 - 4\pi\chi_0\omega^{-2}. \tag{7.205}$$

The problem for the Eq. (7.146) is accomplished by the boundary conditions

$$\begin{aligned} E_y(0,t) &= \phi_y(t), \quad E_z(0,t) = \phi_z(t), \\ H_y(0,t) &= \theta_y(t), \quad H_z(0,t) = \theta_z(t), \end{aligned} \tag{7.206}$$

at the time half-axis $t \geq 0$.

Exchanging the places of derivatives by x and t and taking a bit different interpretation, we present a new operator, which is similar to the evolution one L (7.155)

$$\frac{\partial \psi'}{\partial x} = L'\psi'. \tag{7.207}$$

To construct the x-shift generator L', it is necessary to use the integral operators $\hat{\varepsilon}$ and $\hat{\varepsilon}^{-1}$ in time representation (7.201), and, similarly treat the operator $\hat{\mu}$

$$\frac{\partial}{\partial x} \begin{pmatrix} E_z \\ B_y \\ E_y \\ B_z \end{pmatrix} = \begin{pmatrix} 0 & \frac{1}{c}\frac{\partial}{\partial t} & 0 & 0 \\ \frac{\hat{\mu}\hat{\varepsilon}}{c}\frac{\partial}{\partial t} & 0 & 0 & 0 \\ 0 & 0 & 0 & -\frac{1}{c}\frac{\partial}{\partial t} \\ 0 & 0 & -\frac{\hat{\mu}\hat{\varepsilon}}{c}\frac{\partial}{\partial t} & 0 \end{pmatrix} \begin{pmatrix} E_z \\ B_y \\ E_y \\ B_z \end{pmatrix}. \tag{7.208}$$

System (7.208) can be treated with a procedure similar to the one presented in (7.156) – (7.159). This leads us to a dispersion relation that is similar to (7.160)

$$k^2 = \mu(\omega)\varepsilon(\omega)\frac{\omega^2}{c^2}, \tag{7.209}$$

but ε and μ are functions of ω.

Strictly from those equations one can construct projection operators. Hence, we can present a direct relation between $\breve{E}_z$ and $\breve{H}_y$ as well as between $\breve{E}_y$ and $\breve{H}_z$,

$$\varepsilon^{\frac{1}{2}}\mu^{\frac{1}{2}}\breve{E}_z = -\breve{B}_y, \quad \varepsilon^{\frac{1}{2}}\mu^{\frac{1}{2}}\breve{E}_y = \breve{B}_z, \tag{7.210}$$

as $k = \frac{1}{c}\omega\mu^{\frac{1}{2}}\varepsilon^{\frac{1}{2}}$. For one of propagation direction for both polarizations, otherwise -

$$\varepsilon^{\frac{1}{2}}\mu^{\frac{1}{2}}\breve{E}_z = \breve{B}_y, \quad \varepsilon^{\frac{1}{2}}\mu^{\frac{1}{2}}\breve{E}_y = -\breve{B}_z, \tag{7.211}$$

as $k = -\frac{1}{c}\omega\mu^{\frac{1}{2}}\varepsilon^{\frac{1}{2}}$, with eigenvectors

$$\begin{aligned} \Psi_{11} &= \begin{pmatrix} \tilde{E}_z \\ \varepsilon^{\frac{1}{2}}\mu^{\frac{1}{2}}\tilde{E}_z \\ 0 \\ 0 \end{pmatrix}, \quad \Psi_{12} = \begin{pmatrix} 0 \\ 0 \\ -\tilde{E}_y \\ -\varepsilon^{\frac{1}{2}}\mu^{\frac{1}{2}}\tilde{E}_y \end{pmatrix}, \\ \Psi_{21} &= \begin{pmatrix} -\tilde{E}_z \\ -\varepsilon^{\frac{1}{2}}\mu^{\frac{1}{2}}\tilde{E}_z \\ 0 \\ 0 \end{pmatrix}, \quad \Psi_{22} = \begin{pmatrix} 0 \\ 0 \\ \tilde{E}_y \\ \varepsilon^{\frac{1}{2}}\mu^{\frac{1}{2}}\tilde{E}_y \end{pmatrix} \end{aligned} \tag{7.212}$$

that specify four modes by the projecting operators in ω-representation. As a starting point, we know that they have to fulfill the standard properties of orthogonal projecting operators: (7.153).

Let us show the form of two corresponding projection operators:

$$\mathscr{P}_{11} = \frac{1}{2}\begin{pmatrix} 1 & -\mu^{-\frac{1}{2}}\varepsilon^{-\frac{1}{2}} & 0 & 0 \\ -\varepsilon^{\frac{1}{2}}\mu^{\frac{1}{2}} & 1 & 0 & 0 \\ 0 & 0 & 0 & 0 \\ 0 & 0 & 0 & 0 \end{pmatrix}, \quad \mathscr{P}_{12} = \frac{1}{2}\begin{pmatrix} 0 & 0 & 0 & 0 \\ 0 & 0 & 0 & 0 \\ 0 & 0 & 1 & \mu^{-\frac{1}{2}}\varepsilon^{-\frac{1}{2}} \\ 0 & 0 & \varepsilon^{\frac{1}{2}}\mu^{\frac{1}{2}} & 1 \end{pmatrix}. \tag{7.213}$$

Projection operators for this model have the same form as projection operators generated in the Cauchy problem; however, with one significant difference hidden in form of $\hat{\varepsilon}$. After an inverse Fourier transformation, one contains its explicit form. The application of the first one yields the transition to new variables in the space of solutions

$$\hat{\mathscr{P}}_{11}\psi' = \frac{1}{2}\begin{pmatrix} 1 & -\hat{\mu}^{-\frac{1}{2}}\hat{\varepsilon}^{-\frac{1}{2}} & 0 & 0 \\ -\hat{\varepsilon}^{\frac{1}{2}}\mu^{\frac{1}{2}} & 1 & 0 & 0 \\ 0 & 0 & 0 & 0 \\ 0 & 0 & 0 & 0 \end{pmatrix}\begin{pmatrix} E_z \\ B_y \\ E_y \\ B_z \end{pmatrix} = \frac{1}{2}\begin{pmatrix} E_z - \hat{\mu}^{-\frac{1}{2}}\hat{\varepsilon}^{-\frac{1}{2}}B_y \\ -\hat{\varepsilon}^{\frac{1}{2}}\hat{\mu}^{\frac{1}{2}}E_z + B_y \\ 0 \\ 0 \end{pmatrix}, \tag{7.214}$$

and matrix operator action gives

$$\frac{1}{2}\begin{pmatrix} E_z - \mu^{-\frac{1}{2}}\hat{\varepsilon}^{-\frac{1}{2}}B_y \\ -\hat{\varepsilon}^{\frac{1}{2}}\mu^{\frac{1}{2}}E_z + B_y \\ 0 \\ 0 \end{pmatrix} = \begin{pmatrix} \Pi_1 \\ -\mu^{\frac{1}{2}}\hat{\varepsilon}^{\frac{1}{2}}\Pi_1 \\ 0 \\ 0 \end{pmatrix}. \tag{7.215}$$

The analogous actions allow us to define a system of new variables

$$\begin{aligned}\Pi_1' &= E_z - \hat{\mu}^{-\frac{1}{2}}\hat{\varepsilon}^{-\frac{1}{2}}B_y,\\ \Lambda_1' &= E_z + \hat{\mu}^{-\frac{1}{2}}\hat{\varepsilon}^{-\frac{1}{2}}B_y,\\ \Pi_2' &= E_y - \hat{\mu}^{-\frac{1}{2}}\hat{\varepsilon}^{-\frac{1}{2}}B_z,\\ \Lambda_2' &= E_y + \hat{\mu}^{-\frac{1}{2}}\hat{\varepsilon}^{-\frac{1}{2}}B_z,\end{aligned} \tag{7.216}$$

which compare with (7.168) and other modes' definitions for the Cauchy problem reformulation. Taking nonlinearity into consideration, the system is described by the equation

$$\partial_x \psi' - L'\psi' = \mathbb{N}(\psi'). \tag{7.217}$$

With the use of the new projection operators at Eq. (7.217) we get

$$\partial_x \hat{\mathscr{P}}_{ij}\psi' - L'\hat{\mathscr{P}}_{ij}\psi' = \hat{\mathscr{P}}_{ij}\mathbb{N}(\psi'). \tag{7.218}$$

This action generates the system of equations that describes the pulse propagation to the right and left directions

$$\partial_x \hat{\mathscr{P}}_{11}\psi' - L'\hat{\mathscr{P}}_{11}\psi' = \hat{\mathscr{P}}_{11}\mathbb{N}(\psi'), \tag{7.219}$$
$$\partial_x \hat{\mathscr{P}}_{12}\psi' - L'\hat{\mathscr{P}}_{12}\psi' = \hat{\mathscr{P}}_{12}\mathbb{N}(\psi'), \tag{7.220}$$
$$\partial_x \hat{\mathscr{P}}_{21}\psi' - L'\hat{\mathscr{P}}_{21}\psi' = \hat{\mathscr{P}}_{21}\mathbb{N}(\psi'), \tag{7.221}$$
$$\partial_x \hat{\mathscr{P}}_{22}\psi' - L'\hat{\mathscr{P}}_{22}\psi' = \hat{\mathscr{P}}_{22}\mathbb{N}(\psi'), \tag{7.222}$$

with an account for both polarizations. Let's take a closer look at Eq. (7.219), omitting primes,

$$\frac{\partial}{\partial x}\begin{pmatrix}\Pi_1\\ -\mu^{\frac{1}{2}}\hat{\varepsilon}^{\frac{1}{2}}\Pi_1\\ 0\\ 0\end{pmatrix} - \begin{pmatrix}0 & \frac{1}{c}\frac{\partial}{\partial t} & 0 & 0\\ \frac{\hat{\mu}\hat{\varepsilon}}{c}\frac{\partial}{\partial t} & 0 & 0 & 0\\ 0 & 0 & 0 & -\frac{1}{c}\frac{\partial}{\partial t}\\ 0 & 0 & -\frac{\hat{\mu}\hat{\varepsilon}}{c}\frac{\partial}{\partial t} & 0\end{pmatrix}\begin{pmatrix}\Pi_1\\ -\hat{\mu}^{\frac{1}{2}}\hat{\varepsilon}^{\frac{1}{2}}\Pi_1\\ 0\\ 0\end{pmatrix} = \hat{\mathscr{P}}_{11}\mathbb{N}(\psi'). \tag{7.223}$$

As the projection operators have similar form, the nonlinear term given by Eq. (7.184) can be used. The first line of the matrix system (7.223) reads

$$\frac{\partial}{\partial x}\hat{\varepsilon}\Pi_1 + \frac{1}{c}\frac{\partial}{\partial t}\hat{\mu}^{\frac{1}{2}}\hat{\varepsilon}^{\frac{3}{2}}\Pi_1 = -2\pi\frac{\partial}{\partial t}\chi_{yyyy}\left(3(\Pi_1-\Lambda_1)^3 + (\Pi_1+\Lambda_1)(\Pi_2-\Lambda_2)^2\right). \tag{7.224}$$

To compare the result with those from the works of others for nonmagnetic media, put $\hat{\mu} = \mu$. After a Taylor series expansion of the necessary powers of $\varepsilon(\omega)$ and an inverse Fourier transformation of (9.127), we get the formulas for them:

$$\varepsilon^{\frac{1}{2}}(\omega) \approx 1 - 2(\pi\chi_0)(i\partial_t)^{-2}, \tag{7.225}$$

$$\varepsilon^{\frac{3}{2}}(\omega) \approx 1 - 6(\pi\chi_0)(i\partial_t)^{-2}, \tag{7.226}$$

$$\varepsilon^{-1}(\omega) \approx 1 + 4\pi\chi_0(i\partial_t)^{-2}, \tag{7.227}$$

which we apply to (7.224).

Differentiating two times, both sides with respect to time, we obtain the equation

$$\begin{aligned} &\frac{\partial}{\partial x}\frac{\partial^2}{\partial t^2}\Pi_1 - 4\pi\chi_0\frac{\partial}{\partial x}\Pi_1 + \frac{1}{c}\mu^{\frac{1}{2}}\frac{\partial^3}{\partial t^3}\Pi_1 + \frac{1}{c}\mu^{\frac{1}{2}}6\pi\chi_0\frac{\partial}{\partial t}\Pi_1 \\ &= 2\pi\frac{\partial^3}{\partial t^3}\chi_{yyyy}\left(3(\Pi_1 - \Lambda_1)^3 + (\Pi_1 + \Lambda_1)(\Pi_2 - \Lambda_2)^2\right). \end{aligned} \tag{7.228}$$

With an analogy to [30], we can substitute a multiple scales ansatz, as we have done in Sections 7.2.1 and 7.5 3. The ansatz, by construction, accounts to only one of the directed waves; hence, we simply do not take the opposite waves into account, regarding the excitation of the corresponding mode, which is specified by the projection operators' application to the boundary conditions vector. Both right waves amplitudes are expressed similarly:

$$\Pi_1(x,t) = \left(\kappa A_0(\phi, x_1, x_2) + \kappa^2 A_1(\phi, x_1, x_2) + \cdots\right),$$

$$\Pi_2(x,t) = \left(\kappa B_0(\phi, x_1, x_2) + \kappa^2 B_1(\phi, x_1, x_2) + \cdots\right),$$

where $\phi = \frac{ct-x}{\kappa}$ and $x_n = \kappa^n x$, then we get $\Lambda_i = 0$,

$$\begin{aligned} \left(\frac{\partial^2}{\partial x_1 \partial\phi}\right)A_0 &= \frac{4\pi\chi_0}{c^2}\left(1 - \frac{3}{2}\mu^{\frac{1}{2}}\right)A_0 + 2\pi\chi_{yyyy}\frac{\partial^2}{\partial\phi^2}\left(3A_0^3 + A_0B_0^2\right), \\ \left(\frac{\partial^2}{\partial x_1 \partial\phi}\right)B_0 &= \frac{4\pi\chi_0}{c^2}\left(1 - \frac{3}{2}\mu^{\frac{1}{2}}\right)B_0 + 2\pi\chi_{yyyy}\frac{\partial^2}{\partial\phi^2}\left(3B_0^3 + B_0A_0^2\right). \end{aligned} \tag{7.229}$$

Comparing our result to the one obtained by Pietrzyk et al. [24]

$$\frac{\partial^2}{\partial x_1 \partial\phi}A_0 = A_0 + \frac{1}{6}\frac{\partial^2}{\partial\phi^2}\left(A_0^3 + 3A_0B_0^2\right), \tag{7.230}$$

we state that the system (7.229) contains the result of Refs [24,31] in the frame of the accepted approximations. The system describes collisions of oppositely propagated short pulses as in Ref. [20].

REFERENCES

1. Schäfer, T., and G.E.Wayne. 2004. Propagation of ultra-short optical pulses in cubic nonlinear media. *Physica D: Nonlinear Phenomena* 196(1–2): 90–105.
2. Boyd, R.W. 1992. *Nonlinear Optics*. Academic Press, Boston.
3. Leble, S.B. 1991. *Nonlinear Waves in Waveguides with Stratification*. Springer-Verlag, Berlin.
4. Kuszner, M., and S. Leble. 2011. Directed electromagnetic pulse dynamics: Projecting operators method. *Journal of the Physical Society of Japan* 80: 024002.
5. Kinsler, P., B.P. Radnor, and G.C. New. 2005. Theory of direction pulse propagation. *Physical Review* A 72: 063807.
6. Kinsler, P. 2010. Unidirectional optical pulse propagation equation for materials with both electric and magnetic responses. *Physical Reviews*. A 81: 023808.
7. Leble, S. 2003. Nonlinear waves in optical waveguides and soliton theory applications. In: Porsezian, K. and Kuriakose, V.C. (Eds), *Optical Solitons: Theoretical and Experimental Challenges*. Springer, Berlin, pp. 71–104.
8. Mlejnek, M., M. Kolesik, and J.V. Moloney. 2002. Unidirectional optical pulse propagation equation. *Physical Review Letters* 89(28): 283902; Moloney, J. V., and M. Kolesik. 2004. Nonlinear optical pulse propagation simulation: From maxwells to unidirectional equations. *Physical Review* E 70(3 Pt 2): 036604.
9. Perelomova, A. 2006. Development of linear projecting in studies of non-linear flow. Acoustic heating induced by non-periodic sound. *Physics Letters* A 357: 42–47.
10. Agarwal, G.P. 2001. *Nonlinear Fiber Optics*, 3rd ed., Optics and Photonics Series. Academic Press, San Diego, CA.
11. Zakharov, V.E. 1968. *Journal of Applied Mechanics and Technical Physics* 9(2): 190, Springer New York.
12. Fleck, J.A. 1970. Ultrashort-Pulse Generation by Q-swiched Lasers. *Physical Review* B 1: 84.
13. Mamyshev, P.V., and S.V. Chenikov. 1990. Ultrashort-pulse propagation in optical fibers. *Optics Letters* 15(19): 1076–1078.
14. Caputo, J-G., and A.I. Maimistov. 2002. Unidirectional propagation of an ultra-short electromagnetic pulse in a resonant medium with high frequency Stark shift. *Physics Letters* A 296: 34–42.
15. Ranka, J.K., and A.L. Gaeta. 1998. Breakdown of the slowly varying envelope approximation in the self-focusing of ultrashort pulses. *Optical Letters* 23:534–536.
16. Rothenberg, J.E. 1992. Space-time focusing: Breakdown of the slowly varying envelope approximation in the self-focusing of femtosecond pulses. *Optical Letters* 17:1340–1342.
17. Brabec, T., and F. Krausz. 1997. Nonlinear optical pulse propagation in the single-cycle regime. *Physical Review Letters* 78: 3282.
18. Voronin, A.A., and A.M. Zheltikov. 2008. Soliton-number analysis of soliton-effect pulse compression to single-cycle pulse widths. *Physical Review* A 78: 063834.
19. Kinsler, P. 2007. Limits of the unidirectional pulse propagation approximation. *Journal of the Optical Society of America* B 24: 2363.
20. Pitois, S., G. Millot, and S. Wabnitz. 2001. Nonlinear polarization dynamics of counterpropagating waves in an isotropic optical fiber: Theory and experiments. *Journal of the Optical Society of America* B 18(4): 432.
21. Montes, C., et al. 2013. *Without Bounds: A Scientific Canvas of Nonlinearity and Complex Dynamics*, Springer.

22. Crepin, T., et al. 2005. Experimental evidence of backward waves on terahertz left-handed transmission lines. *Applied Physics Letters* 87: 104105.
23. Kuszner, M., and S. Leble. 2014. Ultrashort opposite directed pulses dynamics with Kerr effect and polarization account. *Journal of the Physical Society of Japan* 83: 034005.
24. Pietrzyk, M., I. Kanattsikov, and U. Bandelow. 2008. On the propagation of vector ultra-short pulses. *Journal of Nonlinear Mathematical Physics* 15.2: 162–170.
25. Galitskii, V.M., and V.M. Ermachenko. 1988. *Macroscopic Electrodynamics*. Moscow High School.
26. Kielich, S. 1977. *Nonlinear Molecular Optics*. PWN, Warsaw.
27. Fock, V.A. 1978. *Fundamentals of Quantum Mechanics*. Mir Publishers, Moscow.
28. Chung, Y., C.K.R.T. Jones, T. Schäfer, and C.E. Wayne. 2005. Ultra-short pulses in linear and nonlinear media. *Nonlinearity* 18: 1351–1374. nlin.SI/0408020.
29. Brabec, T., and F. Krausz. 2000. Intense few-cycle laser fields: Frontiers of nonlinear optics. *Reviews of Modern Physics* 72: 545.
30. Chung, Y., and T. Schäfer. 2008. Stabilization of ultra-short pulses in cubic nonlinear media.
31. Sakovich, S. 2008. Integrability of the vector short pulse equation. *Journal of the Physical Society of Japan* 77: 123001.
32. Menyuk, C.R., and B.S. Marks. 2006. Interaction of polarization mode dispersion and nonlinearity in optical fiber transmission systems. *Journal of Lightwave Technology* 24(7): 2806–2826.
33. Remoissenet, M. 1996. *Waves Called Solitons: Concepts and Experiments.* Springer, Berlin.
34. Wegener, M. 2005. *Extreme Nonlinear Optics:* An Introduction, Advanced Texts in Physics. Springer, Berlin.
35. Kinsler, P., S.B.P. Radnor, and G.H.C. New. 2005. Theory of direction pulse propagation. *Physical Reviews* A 72: 063807.
36. Amiranashvili, Sh., and A. Demircan. 2011. Ultrashort Optical Pulse Propagation in terms of Analytic Signal. *Advances in Optical Technologies* 2011, Article ID 989515 http://dx.doi.org/10.1155/2011/989515.
37. Chernikov, S.V., and P.V. Mamyshev. 1990. Ultrashort-pulse propagation in optical fibers. *Optics Letters* 15(19): 1076.
38. C.R. Menyuk. 1987. Nonlinear pulse propagation in birefringent optical fibers. *IEEE Journal of Quantum of Electron* QE-23(2): 174.
39. Sakovich, A., and S. Sakovich. 2006. Solitary wave solutions of the short pulse equation. *Journal of Physics A: Mathematical and General* 39(22): L361–L367.
40. Wen, S., Y. Xiang, X. Dai, Z. Tang, W. Su, and D. Fan. 2007. Theoretical models for ultrashort electromagnetic pulse propagation in nonlinear metamaterials. *Physical Review* A 75: 033815.
41. Reichel, B., and S. Leble. On convergence and stability of a numerical scheme of coupled nonlinear Schrödinger rquations. *Computers and Mathematics with Applications* 55: 745–759.
42. Zhao, Y., C. Argyropoulos, and Y. Hao. 2008. Full-wave finite-difference time-domain simulation of electromagnetic cloaking structures. *Optical Society of America* 16: 6717–6730.

8 Metamaterials

8.1 STATEMENT OF PROBLEM FOR METAMATERIALS

8.1.1 TWO WORDS ON METAMATERIALS

One of the most interesting modern problems for researchers is the theory of metamaterials with simultaneously negative dielectric permittivity and magnetic permeability, recently experimentally verified [1,2]. The ideas connected with negative refraction appeared in the 1940s–1950s [3,4] and were actively developed until today [5–7]. Typically, the refractive index n is determined using $n = \pm\sqrt{\varepsilon\mu}$, where by convention the positive square root is chosen for n [8]. However, in negative phase velocity (NPV) materials, we reverse that convention and pick the negative sign to mimic the fact that the wave vector (and hence phase velocity) are likewise reversed. Speaking in the language of this book, the refractive index is a derived quantity telling us how the wave vector is related to the optical frequency and propagation direction of the light (directed mode); thus, the sign of n must be chosen to match the physical situation. For more details about the language and metamaterials see the introduction of this book. The directed waves, its phase and pointing vector direction, play a fundamental role in metamaterial analysis [9]. The next chapter touches on a waveguide propagation that also could be filled with metamaterials [10,11]. In this chapter, we show how in conditions of 1D propagation the direction of propagation and general dispersion properties appear. The next subsection is based mainly on publication [12].

8.1.2 MAXWELL'S EQUATIONS. OPERATORS OF DIELECTRIC PERMITTIVITY AND MAGNETIC PERMEABILITY

Our starting point is again the Maxwell equations for linear isotropic dispersive dielectric media; unlike the Lorentz units choice in Section 7.1, for a reader convenience, we write the system in the SI unit system:

$$\mathrm{div}\vec{D}(\vec{r},t) = 0, \tag{8.1}$$

$$\mathrm{div}\vec{B}(\vec{r},t) = 0, \tag{8.2}$$

$$\mathrm{rot}\vec{E}(\vec{r},t) = -\frac{\partial \vec{B}(\vec{r},t)}{\partial t}, \tag{8.3}$$

$$\mathrm{rot}\vec{H}(\vec{r},t) = \frac{\partial \vec{D}(\vec{r},t)}{\partial t}. \tag{8.4}$$

We restrict ourselves to a one-dimensional model, similar to Schäfer and Wayne [13], Kuszner and Leble [14], in which the x-axis is chosen as the direction of wave propagation. As mentioned earlier in this section, we assume $D_x = 0$ and $B_x = 0$, taking into account the only polarization of electromagnetic waves. This allows us to write the Maxwell equations as

$$\frac{\partial D_y}{\partial t} = -\frac{\partial H_z}{\partial x}, \qquad (8.5)$$
$$\frac{\partial B_z}{\partial t} = -\frac{\partial E_y}{\partial x}.$$

Such form of the equations with t-derivatives on the L.H.S. implies the Cauchy problem formulation identically, as in Section 7.1. Further indices will be omitted for conciseness. We introduce four variables $\mathscr{E}, \mathscr{B}, \mathscr{D}, \mathscr{H}$ as Fourier images of E, B, D and H that are connected by inverse Fourier transformations

$$E(x,t) = \frac{1}{\sqrt{2\pi}} \int_{-\infty}^{\infty} \mathscr{E}(x,\omega) \exp(i\omega t) d\omega, \qquad (8.6)$$

$$B(x,t) = \frac{1}{\sqrt{2\pi}} \int_{-\infty}^{\infty} \mathscr{B}(x,\omega) \exp(i\omega t) d\omega, \qquad (8.7)$$

$$D(x,t) = \frac{1}{\sqrt{2\pi}} \int_{-\infty}^{\infty} \mathscr{D}(x,\omega) \exp(i\omega t) d\omega, \qquad (8.8)$$

$$H(x,t) = \frac{1}{\sqrt{2\pi}} \int_{-\infty}^{\infty} \mathscr{H}(x,\omega) \exp(i\omega t) d\omega. \qquad (8.9)$$

The domain of Fourier images we call ω−representation or a frequency domain. The functions E, B, D, H are in t-representation or in a time domain. Linear material equations in ω-representation in SI units we take as

$$\mathscr{D} = \varepsilon_0 \varepsilon(\omega) \mathscr{E}, \qquad (8.10)$$

$$\mathscr{B} = \mu_0 \mu(\omega) \mathscr{H}. \qquad (8.11)$$

Here:
$\varepsilon(\omega)$ - dielectric permittivity of a medium, ε_0 - dielectric permittivity of the vacuum. $\mu(\omega)$ - magnetic permeability of a medium and μ_0 - magnetic permeability of the vacuum. Recall that $\mathscr{B}$ is an image of the field B in ω-representation. For calculation

purposes of physical realization we need to use t-representation. In this representation ε and μ become integral operators. Then

$$D(x,t) = \frac{1}{\sqrt{2\pi}} \int_{-\infty}^{\infty} \mathscr{D}(x,\omega) \exp(i\omega t) d\omega$$

$$= \frac{\varepsilon_0}{\sqrt{2\pi}} \int_{-\infty}^{\infty} \varepsilon(\omega) \mathscr{E}(x,\omega) \exp(i\omega t) d\omega, \quad (8.12)$$

$$B(x,t) = \frac{1}{\sqrt{2\pi}} \int_{-\infty}^{\infty} \mathscr{B}(x,\omega) \exp(i\omega t) d\omega$$

$$= \frac{\mu_0}{\sqrt{2\pi}} \int_{-\infty}^{\infty} \mu(\omega) \mathscr{H}(x,\omega) \exp(i\omega t) d\omega. \quad (8.13)$$

Plugging the inverse transform

$$\mathscr{E}(x,\omega) = \frac{1}{\sqrt{2\pi}} \int_{-\infty}^{\infty} E(x,s) \exp(-i\omega s) ds \quad (8.14)$$

into (8.12), we obtain the expression that contains a double integral:

$$D(x,t) = \frac{\varepsilon_0}{2\pi} \int_{-\infty}^{\infty} \varepsilon(\omega) \int_{-\infty}^{\infty} E(x,s) \exp(-i\omega s) ds \exp(i\omega t) d\omega. \quad (8.15)$$

If the function $E(x,s)\exp(-i\omega s)$ satisfies the basic conditions of Fubini's theorem, we rewrite:

$$D(x,t) = \frac{\varepsilon_0}{2\pi} \int_{-\infty}^{\infty} \int_{-\infty}^{\infty} \varepsilon(\omega) \exp(i\omega(t-s)) d\omega E(x,s) ds = \int_{-\infty}^{\infty} \tilde{\varepsilon}(t-s) E(x,s) ds, \quad (8.16)$$

where the kernel

$$\tilde{\varepsilon}(t-s) = \frac{\varepsilon_0}{2\pi} \int_{-\infty}^{\infty} \varepsilon(\omega) \exp(i\omega(t-s)) d\omega, \quad (8.17)$$

defines the integral operator of convolution type

$$\widehat{\varepsilon}\psi(x,t) = \int_{-\infty}^{\infty} \tilde{\varepsilon}(t-s) \psi(x,s) ds, \quad (8.18)$$

or

$$D(x,t) = \widehat{\varepsilon} E(x,t), \quad (8.19)$$

which expresses the material relation for the definition (8.10). Doing operations for E and magnetic components of the field, we continue:

$$E(x,t) = \int_{-\infty}^{\infty} \tilde{e}(t-s)D(x,s)ds, \tag{8.20}$$

$$B(x,t) = \int_{-\infty}^{\infty} \tilde{\mu}(t-s)H(x,s)ds = \widehat{\mu}H(x,t), \tag{8.21}$$

$$H(x,t) = \int_{-\infty}^{\infty} \tilde{m}(t-s)\mathscr{B}(x,s)ds = \widehat{\mu}^{-1}B(x,t), \tag{8.22}$$

having integral operators with kernels

$$\begin{aligned} \tilde{e}(t-s) &= \frac{1}{2\pi}\int_{-\infty}^{\infty} \varepsilon^{-1}(\omega)\exp(i\omega(t-s))d\omega, \\ \tilde{\mu}(t-s) &= \frac{\mu_0}{2\pi}\int_{-\infty}^{\infty} \mu(\omega)\exp(i\omega(t-s))d\omega, \\ \tilde{m}(t-s) &= \frac{1}{2\pi\mu_0}\int_{-\infty}^{\infty} \mu^{-1}(\omega)\exp(i\omega(t-s))d\omega. \end{aligned} \tag{8.23}$$

The transforms define the fields in the time domain, using the conventional continuation of the fields to the half space $t < 0$ and causality condition [15].

8.1.3 BOUNDARY REGIME PROBLEM

As we see in the time domain, for the E and B equation, the system (8.5) may be written in operator form:

$$\begin{aligned} \frac{\partial}{\partial t}(\widehat{\varepsilon}E) &= -\frac{\partial}{\partial x}(\widehat{\mu}^{-1}B), \\ \frac{\partial B}{\partial t} &= -\frac{\partial E}{\partial x}. \end{aligned} \tag{8.24}$$

Action of operators $\widehat{\varepsilon}$ and $\widehat{\mu}$ was defined by (8.18 and 8.22).

Going to a boundary regime problem we, quite as in the preceding section, could rewrite the system (8.24), exchanging the left and right sides. The details we shall show in the next subsection. We also must add the boundary regime conditions:

$$E(0,t) = j(t), \; B(0,t) = k(t), \tag{8.25}$$

j and k are arbitrary functions, continued to the half space $t < 0$ antisymmetricaly:

$$j(-t) = -j(t), k(-t) = -k(t). \tag{8.26}$$

8.2 DYNAMIC PROJECTING OPERATORS

The algorithm and applications of the projecting technique are similar to those developed for a Cauchy problem, but when one interchanges x and t variables, the Fourier transform and evolution also interchange so attention should be paid. Let us start with the transformations (8.12) and (8.13). We plug them into the system of equations (8.24) and have the closed system

$$\frac{\partial}{\partial t}\left(\int_{-\infty}^{\infty} \varepsilon(\omega)\mathscr{E}(x,\omega)\exp(i\omega t)d\omega\right) = -\frac{1}{\mu_0\varepsilon_0}\frac{\partial}{\partial x}\left(\int_{-\infty}^{\infty} \frac{\mathscr{B}(x,\omega)}{\mu(\omega)}\exp(i\omega t)d\omega\right) \tag{8.27}$$

The inverse Fourier transformation yields in the first equation of (8.24)

$$\frac{\partial \mathscr{B}}{\partial x} = -i\omega\mu_0\varepsilon_0\mu(\omega)\varepsilon(\omega)\mathscr{E}. \tag{8.28}$$

The second equation gives similarly

$$\frac{\partial \mathscr{E}}{\partial x} = -i\omega\mathscr{B}. \tag{8.29}$$

The natural notation reads

$$\mu_0\varepsilon_0\varepsilon(\omega)\mu(\omega) \equiv c^{-2}\varepsilon(\omega)\mu(\omega) \equiv a^2(\omega), \tag{8.30}$$

where $c^2 = \frac{1}{\sqrt{\varepsilon_0\mu_0}}$ is the velocity of light in vacuum. Hence, the system (8.28 and 8.29) simplifies as

$$\frac{\partial \mathscr{B}}{\partial x} = -i\omega a^2(\omega)\mathscr{E}, \tag{8.31}$$

$$\frac{\partial \mathscr{E}}{\partial x} = -i\omega\mathscr{B}. \tag{8.32}$$

The system (8.24) may be written in matrix form. For this purpose, we introduce matrices $\mathscr{L}$ and $\tilde{\Psi}$ as

$$\tilde{\Psi} = \begin{pmatrix} \mathscr{B} \\ \mathscr{E} \end{pmatrix} \tag{8.33}$$

$$\mathscr{L} = \begin{pmatrix} 0 & -i\omega a^2(\omega) \\ -i\omega & 0 \end{pmatrix}. \tag{8.34}$$

Hence, the matrix form of (8.28, 8.29) is

$$\frac{\partial \tilde{\Psi}}{\partial x} = \mathscr{L}\tilde{\Psi}. \tag{8.35}$$

(8.32) is a system of ordinary differential equations with constant coefficients that have exponential-type solutions. Following the technique described in [14], we rely upon a 2×2 eigenvalue problem for the evolution operator $\mathscr{L}$. Look for such matrices as $P^{(i)}, i = \overline{1,2}$, for which $P^{(i)}\Psi = \Psi_i$ would be eigenvectors of the evolution matrix (8.125) with an unit upper element. The standard properties of orthogonal projecting operators

$$P^{(1)}P^{(2)} = 0, i \neq j,$$
$$P^{(i)} \cdot P^{(i)} = P^{(i)}, \tag{8.36}$$
$$P^{(1)} + P^{(2)} = I,$$

are implied. It's easy to show: the projectors $P^{(i)}$ in ω-representation have the form

$$P^{(1)}(\omega) = \frac{1}{2}\begin{pmatrix} 1 & -a(\omega) \\ -\frac{1}{a(\omega)} & 1 \end{pmatrix}, \tag{8.37}$$

$$P^{(2)}(\omega) = \frac{1}{2}\begin{pmatrix} 1 & a(\omega) \\ \frac{1}{a(\omega)} & 1 \end{pmatrix}. \tag{8.38}$$

The inverse Fourier transformation $\widehat{\mathbf{P}}^{(i)} = \mathscr{F}P^{(i)}\mathscr{F}^{-1}$, where the $\mathscr{F}$ - operator of the Fourier transformation leads to projectors in t-representation:

$$\widehat{\mathbf{P}}^{(1,2)}(t) = \frac{1}{2}\begin{pmatrix} 1 & \widehat{P}_{12}^{(1,2)} \\ \widehat{P}_{21}^{(1,2)} & 1 \end{pmatrix}, \tag{8.39}$$

The diagonal elements of projectors are proportional to the identity

$$\widehat{P}_{11}^{(1,2)}\xi(x,t) = \frac{1}{4\pi}\int_{-\infty}^{\infty} \exp(i\omega t - i\omega\tau)\xi(x,\tau)d\omega$$
$$= \frac{1}{2}\int_{-\infty}^{\infty} \delta(t-\tau)\xi(x,\tau)d\tau = \frac{1}{2}\xi(x,t),$$
$$\widehat{P}_{22}^{(1,2)}\eta(t) = \frac{1}{4\pi}\int_{-\infty}^{\infty} \exp(i\omega t - i\omega\tau)\xi(x,\tau)d\omega$$
$$= \frac{1}{2}\int_{-\infty}^{\infty} \delta(t-\tau)\xi(x,\tau)d\tau = \frac{1}{2}\eta(x,t). \tag{8.40}$$

Non-diagonal elements act as integral operators:

$$\widehat{P}_{12}^{(1,2)}\eta(x,t) = \int_{-\infty}^{\infty} p_{12}^{(1,2)}(t,\tau)\eta(x,\tau)d\tau, \tag{8.41}$$

$$\widehat{P}_{21}^{(1,2)}\xi(x,t) = \int_{-\infty}^{\infty} p_{21}^{(1,2)}(t,\tau)\xi(x,\tau)d\tau, \tag{8.42}$$

Integral operator kernels of $\widehat{P}_{12}^{(1,2)}(t,\tau)$ and $\widehat{P}_{21}^{(1,2)}(t,\tau)$ are

$$p_{12}^{(1,2)}(t,\tau) = \mp\frac{1}{4\pi}\int_{-\infty}^{\infty} a(\omega)\exp(i\omega t - i\omega\tau)d\omega, \tag{8.43}$$

$$p_{21}^{(1,2)}(t,\tau) = \mp\frac{1}{4\pi}\int_{-\infty}^{\infty} \frac{1}{a(\omega)}\exp(i\omega t - i\omega\tau)d\omega, \tag{8.44}$$

So, we can define them via the operator $\widehat{a}$:

$$\widehat{P}_{12}^{(1,2)}\eta(x,t) = \mp\widehat{a}\eta(x,t), \tag{8.45}$$

$$\widehat{P}_{21}^{(1,2)}\xi(x,t) = \mp\widehat{a}^{-1}\xi(x,t), \tag{8.46}$$

whose action is defined by means of the function (8.30):

$$\begin{aligned}\widehat{a}\eta(x,t) &= \frac{1}{2\pi}\int_{-\infty}^{\infty}\left[\eta(x,\tau)\int_{-\infty}^{\infty} a(\omega)\exp(i\omega(t-\tau))d\omega\right]d\tau,\\ \widehat{a}^{-1}\xi(x,t) &= \frac{1}{2\pi}\int_{-\infty}^{\infty}\left[\xi(x,\tau)\int_{-\infty}^{\infty} \frac{1}{a(\omega)}\exp(i\omega(t-\tau))d\omega\right]d\tau.\end{aligned} \tag{8.47}$$

8.3 SEPARATED EQUATIONS AND DEFINITIONS FOR HYBRID WAVES

We again introduce the shorthands

$$\partial_t \equiv \frac{\partial}{\partial t}, \partial_x \equiv \frac{\partial}{\partial x}. \tag{8.48}$$

In t-representation, matrix equation (8.32) takes the form

$$\partial_x\Psi = \widehat{L}\Psi, \tag{8.49}$$

where

$$\Psi = \begin{pmatrix} B \\ E \end{pmatrix}, \tag{8.50}$$

$$\widehat{L} = \begin{pmatrix} 0 & -\partial_t\widehat{a^2} \\ -\partial_t & 0 \end{pmatrix}. \tag{8.51}$$

One can check that the operator $\widehat{a^2}$ defined as

$$\widehat{a^2}\psi(x,t) = \frac{1}{2\pi}\int_{-\infty}^{\infty} a^2(\omega)\exp(i\omega t - i\omega\tau)\psi(x,\tau)d\omega d\tau, \tag{8.52}$$

acts as the square of $\hat{a}$, defined by (8.47).

Making similar calculations, we find that $\widehat{a^2}$ is expressed as the product of permittivity operators (see, e.g. 8.18, 8.22):

$$\widehat{a^2} = \widehat{\varepsilon}\widehat{\mu},$$

that commute

$$\widehat{\varepsilon}\widehat{\mu}\psi(x,t) = \widehat{\mu}\widehat{\varepsilon}\psi(x,t). \tag{8.53}$$

We note that this relation is true only if operators $\widehat{\varepsilon}$ and $\widehat{\mu}$ are convolution type integrals. For the further operations, we also prove the commutation of operators ∂_t and $\widehat{a^2}$.

Acting by the operator $\widehat{\mathbf{P}}^{(1)}$ (8.39) on Eq. (8.49) we find

$$\widehat{\mathbf{P}}^{(1)}\partial_x\Psi = \widehat{\mathbf{P}}^{(1)}\widehat{L}\tilde{\Psi}. \tag{8.54}$$

We can commute $\widehat{\mathbf{P}}^{(1)}$ and ∂_x, because projectors don't depend on x. Using also the proven relations, we write

$$\partial_x\widehat{\mathbf{P}}^{(1)}(t)\Psi = \widehat{\mathbf{P}}^{(1)}(t)\widehat{L}\Psi = \widehat{L}\widehat{\mathbf{P}}^{(1)}(t)\Psi. \tag{8.55}$$

Plugging Ψ and $\widehat{L}$ (8.50 and 8.51) as well as $\widehat{\mathbf{P}}^{(2)}$ (8.39) into Eq. (8.55), we find two copies of the x-evolution equation of one of the directed waves:

$$\partial_x\begin{pmatrix} \frac{1}{2}B + \frac{1}{2}\widehat{a}E \\ \frac{1}{2}\widehat{a}^{-1}B + \frac{1}{2}E \end{pmatrix} = \begin{pmatrix} -\frac{1}{2}\widehat{a}\partial_t B - \frac{1}{2}\widehat{a^2}\partial_t E \\ -\frac{1}{2}\partial_t B - \frac{1}{2}\widehat{a}^{-1}\widehat{a^2}\partial_t E \end{pmatrix}. \tag{8.56}$$

Applying the projection operators to the vector Ψ (8.50), we introduce new (hybrid) variables Π and Λ as

$$\Lambda \equiv \frac{1}{2}(B - \widehat{a}E), \tag{8.57}$$

$$\Pi \equiv \frac{1}{2}(B + \widehat{a}E). \tag{8.58}$$

These are the left and right waves variables. From (8.56) and by similar action with $\widehat{\mathbf{P}}^{(1)}(t)$ we get two equations that determine evolution with respect to the variable x of the boundary regime (8.25):

$$\begin{aligned} \partial_x\Pi(x,t) &= -\widehat{a}\partial_t\Pi, \\ \partial_x\Lambda(x,t) &= \widehat{a}\partial_t\Lambda. \end{aligned} \tag{8.59}$$

Using relations (8.58 and 8.57) from (8.25) we derive boundary regime conditions for the left and right waves:

$$\begin{aligned} \Lambda(0,t) &= \frac{1}{2}(B(0,t) - \widehat{a}E(0,t)) = \frac{1}{2}(k(t) - \widehat{a}j(t)), \\ \Pi(0,t) &= \frac{1}{2}(B(0,t) + \widehat{a}E(0,t)) = \frac{1}{2}(k(t) + \widehat{a}j(t)) \end{aligned} \tag{8.60}$$

So, we obtain two boundary regime problems as the system of independent operator equations for time-domain dispersion of the hybrid, or left and right waves in the linear case.

8.4 NONLINEARITY ACCOUNT

Let's consider a nonlinear problem. We start again from the Maxwell's equations (8.5) with generalized material relations:

$$\begin{aligned} D &= \widehat{\varepsilon}E + P_{NL}, \\ B &= \widehat{\mu}H + M_{NL}, \end{aligned} \tag{8.61}$$

P_{NL} - nonlinear part of polarization (M_{NL} - one for magnetization). For our purposes, the linear parts of polarization and magnetization have already been taken into account. In the time-domain, a closed nonlinear version of (8.24) is

$$\begin{aligned} \partial_t(\widehat{\varepsilon}E) + \partial_t P_{NL} &= -\partial_x \widehat{\mu}^{-1} B + \partial_x \widehat{\mu}^{-1} M_{NL}, \\ \frac{\partial B}{\partial t} &= -\frac{\partial E}{\partial x}. \end{aligned} \tag{8.62}$$

The action of operator $\widehat{\mu}$ on the first system of equations (8.163) and the use of the same notations Ψ and $\widehat{L}$ from (8.50 and 8.51) once more, produce a nonlinear analogue of the matrix equation (8.49):

$$\partial_x \Psi - \widehat{L}\Psi = \begin{pmatrix} \partial_x M_{NL} \\ 0 \end{pmatrix} - \begin{pmatrix} \widehat{\mu}\partial_t P_{NL} \\ 0 \end{pmatrix}. \tag{8.63}$$

We introduce a vector of nonlinearity:

$$\mathbb{N}(E,B) = \begin{pmatrix} \partial_x M_{NL} - \partial_t \widehat{\mu} P_{NL} \\ 0 \end{pmatrix}. \tag{8.64}$$

Then we get the nonlinear analogue of matrix equation (8.49):

$$\partial_x \Psi - \widehat{L}\Psi = \mathbb{N}(E,B). \tag{8.65}$$

Next, acting by operators $\widehat{\mathbf{P}}^{(1,2)}$ (8.39) on Eq. (8.65) we find

$$\begin{aligned} \partial_x \Pi + \widehat{a}\partial_t \Pi &= -\mathbb{N}_1(\widehat{a}^{-1}(\Pi - \Lambda), \Pi + \Lambda), \\ \partial_x \Lambda - \widehat{a}\partial_t \Lambda &= \mathbb{N}_1(\widehat{a}^{-1}(\Pi - \Lambda), \Pi + \Lambda), \end{aligned} \tag{8.66}$$

where

$$\mathbb{N}_1(E,B) \equiv \frac{1}{2}(\partial_x M_{NL} - \partial_t \widehat{\mu} P_{NL}). \tag{8.67}$$

Equation (8.66) is the system of equations of left and right waves interacting due to the arbitrary nonlinearity with the general temporal dispersion account. It is the principal result of this section.

8.5 GENERAL EQUATIONS OF 1D WAVE PROPAGATION IN A METAMATERIAL THAT IS DESCRIBED BY THE LOSSLESS DRUDE MODEL

Since a metamaterial is inherently a lossy material, it appears that the electrons are not bonded to atoms but free to move around, colliding with one another, as in a conducting material. The Drude model represents this situation. For this case, we use relations that are known (see e.g. [16]):

$$\varepsilon(\omega) = \left(1 - \frac{\omega_{pe}^2}{\omega^2}\right), \tag{8.68}$$

$$\mu(\omega) = \left(1 - \frac{\omega_{pm}^2}{\omega^2}\right). \tag{8.69}$$

This model is used by many authors (e.g. [17,18]) to describe the material properties of metamaterial. Energy density is positive at ω range, for which (8.68 and 8.69) are valid:

$$W = \frac{d(\omega\varepsilon(\omega))}{d\omega}E^2 + \frac{d(\omega\mu(\omega))}{d\omega}H^2 = \left(1 + \frac{\omega_{pe}^2}{\omega^2}\right)E^2 + \left(\frac{\omega_{pm}^2}{\omega^2} + 1\right)H^2 > 0, \tag{8.70}$$

where ω_{pe} and ω_{pm} - parameters, dependent on the density, charge, and mass of the charge carriers. These parameters are commonly known as the electric and magnetic plasma frequencies [16]. The function $a(\omega)$ that enters the kernel of the integral operator $\widehat{a}$, see (8.47), is

$$a(\omega) = c^{-1}\sqrt{\left(1 - \frac{\omega_{pe}^2}{\omega^2}\right)\left(1 - \frac{\omega_{pm}^2}{\omega^2}\right)}. \tag{8.71}$$

A three-term Laurent expansion approximation of $a(\omega)$ gives for the operator $\widehat{a}$ in the t-domain the following expression:

$$\widehat{a} \approx c^{-1}\left[\omega_{pe}\omega_{pm}\partial_t^{-2} - \frac{1}{2}\frac{\omega_{pe}^2 + \omega_{pm}^2}{\omega_{pe}\omega_{pm}} + \left(\frac{1}{2\omega_{pe}\omega_{pm}} - \frac{1}{8}\frac{(\omega_{pe}^2 + \omega_{pm}^2)^2}{\omega_{pe}^3\omega_{pm}^3}\right)\partial_t^2\right]. \tag{8.72}$$

This approximation for $a(\omega)$ is good in a range of frequencies $\omega < 0.9\omega_{pe}$; the relative error for such frequencies is less than 1%. The first term that we use in

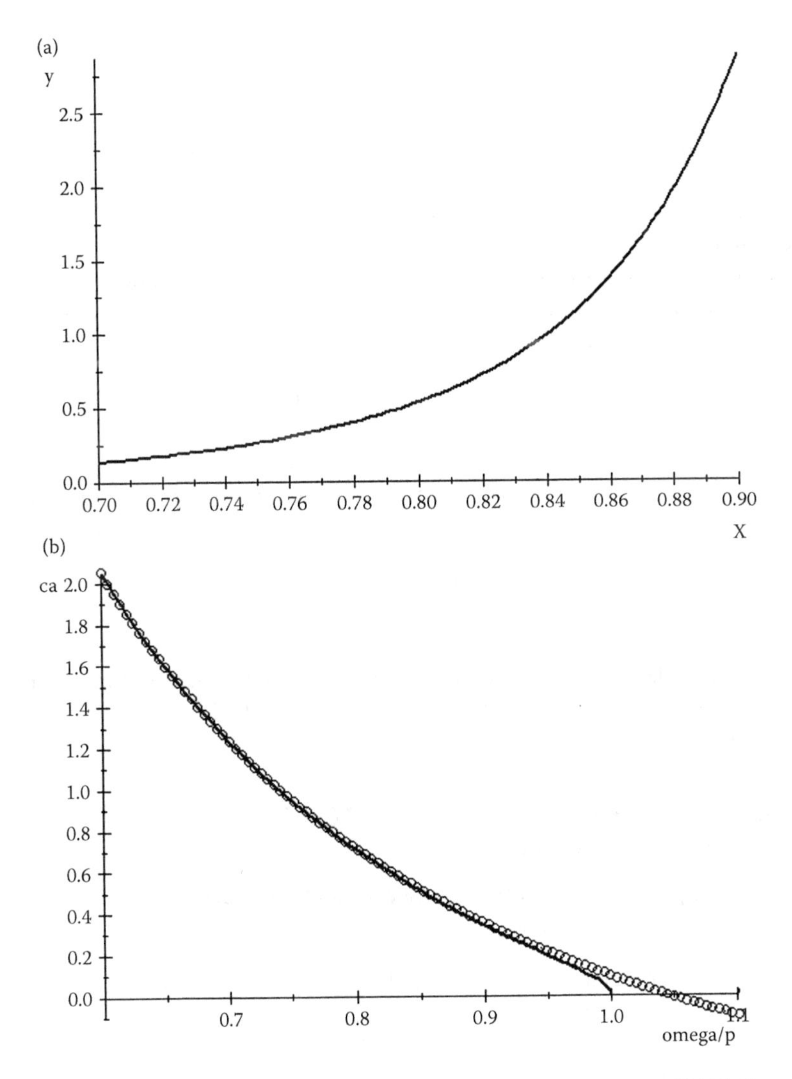

Figure 8.1 Relative error (in percent) of the expansion (8.72) at ω range $[0.7, 0.9]$ ω_{pe} (a), and comparison of expressions for *ca*, (8.72) — solid line and (8.71) — dotdotdash at ω range $[0.6, 1.1]$ ω_{pe} (b), $\omega_{pm} = 1.1\omega_{pe}$ [19]

further simplifications is the leading one within the 1% relative error as it is seen from the numerical analysis up to $0.1\omega_{pe}$ (see the Figure 8.1b). The operator ∂_α^{-1} is defined as the primitive integral

$$\partial_\alpha^{-1} f(\alpha) = \int_0^\alpha f(\beta) d\beta. \tag{8.73}$$

In the case of $\omega_{pm} = \omega_{pe}$ we find

$$ca(\omega) = \sqrt{\left(1 - \frac{\omega_{pe}^2}{\omega^2}\right)\left(1 - \frac{\omega_{pe}^2}{\omega^2}\right)} = \sqrt{\left(1 - \frac{\omega_{pe}^2}{\omega^2}\right)^2} = \left(1 - \frac{\omega_{pe}^2}{\omega^2}\right), \quad (8.74)$$

which already has algebraic form and goes to differential operators in coordinate representation.

Accounting for all estimations, we left the only term in the relation (8.72). Next, for conciseness, we mark ω_{pe} as p and ω_{pm} as q. Plugging this minimal version of (8.72) into the system (8.59) we arrive at

$$\begin{aligned} \partial_x \Pi &= -c^{-1} pq \partial_t^{-1} \Pi, \\ \partial_x \Lambda &= c^{-1} pq \partial_t^{-1} \Lambda. \end{aligned} \quad (8.75)$$

Differentiating this system with respect to t once more, we write the resulting system in which the right and left wave amplitudes are completely separated:

$$\begin{aligned} \partial_{xt} \Pi &= -c^{-1} pq \Pi, \\ \partial_{xt} \Lambda &= c^{-1} pq \Lambda. \end{aligned} \quad (8.76)$$

Both equations describe the wave dispersion and are equivalent to the 1+1 Klein-Fock equation in oblique variables $\Box\phi_\pm = m_\pm \phi_\pm$ with the corresponding mass parameters $m_\pm = \pm c^{-1} pq$.

8.6 KERR NONLINEARITY ACCOUNT FOR LOSSLESS DRUDE METAMATERIALS

8.6.1 EQUATIONS OF INTERACTION OF LEFT AND RIGHT WAVES WITH KERR EFFECT

For nonlinear Kerr metamaterials [20,21], the third-order nonlinear part of polarization [14,21] has the form

$$P_{NL} = \chi^{(3)} E^3.$$

From (8.67) we find the term $\mathbb{N}_1$:

$$\mathbb{N}_1 \equiv \frac{1}{2} \widehat{\mu} (\widehat{\mu}^{-1} \partial_x M_{NL} - \partial_t P_{NL}) = -\frac{\chi^{(3)}}{2} \widehat{\mu} \partial_t E^3. \quad (8.77)$$

Operator $\widehat{\mu}$ for the chosen model is just $\mu_0(1 - q^2 \partial_t^{-2})$. Moreover, the effect of negative permeability was demonstrated at THz range [22]. Hence, the $q^2 \partial_t^{-2}$ contribution prevails. Then, from (8.66) we obtain

$$\begin{aligned} c\partial_x \Pi - pq\partial_t^{-1} \Pi &= -\frac{\chi^{(3)}}{2} \mu_0 q^2 \partial_t^{-1} [\widehat{a}^{-1} (\Pi - \Lambda)]^3, \\ c\partial_x \Lambda + pq\partial_t^{-1} \Lambda &= \frac{\chi^{(3)}}{2} \mu_0 q^2 \partial_t^{-1} [\widehat{a}^{-1} (\Pi - \Lambda)]^3, \end{aligned} \quad (8.78)$$

The same approximation for the operator $\widehat{a}^{-1}$ reads as

$$\widehat{a}^{-1}\eta(x,t) \approx \frac{c}{pq}\partial_t^2\eta(x,t). \tag{8.79}$$

We substitute it into the system (8.78) and differentiate by time, denoting derivatives by indices for more conciseness:

$$\begin{aligned} c\partial_{xt}\Pi + pq\Pi &= -\frac{\mu_0\chi^{(3)}c^3}{2p^3q}\left[(\Pi-\Lambda)_{tt}\right]^3, \\ c\partial_{xt}\Lambda - pq\Lambda &= \frac{\mu_0\chi^{(3)}c^3}{2p^3q}\left[(\Pi-\Lambda)_{tt}\right]^3, \end{aligned} \tag{8.80}$$

We consider this system the **main result** of the study within the frame of made approximations. This system we will name the Short Pulse system (SPS); it describes the opposite pulses interaction, including such phenomena as reflection. The equivalent system is obtained by triple differentiating both equations of the system with respect to time and rescaling $\Pi_{tt} = \gamma\pi$, $\Lambda_{tt} = \gamma\lambda$, $x = \beta\zeta$ with the choice $\gamma = \sqrt{\frac{2p^4q^2}{\mu_0\chi^{(3)}c^3}}$, $\beta = \frac{c}{pq}$. Then

$$\begin{aligned} \pi_{\zeta t} + \pi &= -\partial_t^2(\pi-\lambda)^3, \\ \lambda_{\zeta t} - \lambda &= \partial_t^2(\pi-\lambda)^3, \end{aligned} \tag{8.81}$$

with extra boundary conditions to maintain the equivalence.

Consider the unidirectional case of (8.80) with $\Lambda = 0$, which corresponds to special initial conditions from (8.60) : $(k(t) - \widehat{a}j(t)) = 0$ and is valid till the effect of the left left-wave generation would be noticeable. After the rescaling $\Pi = \gamma\tilde{\Pi}$, $x = \beta\zeta$, it reads as

$$\tilde{\Pi}_{\zeta t} + \tilde{\Pi} = -\partial_{tt}(\tilde{\Pi})^3. \tag{8.82}$$

Comparing with Schafer-Wayne equation [13]

$$u_{xt} - u = \frac{1}{6}\partial_{tt}u^3, \tag{8.83}$$

we see that left and right waves for ordinary materials interchange their direction compared to metamaterials. The system (8.81)

$$\pi_{\zeta t} + \pi = -\partial_t^2\pi^3. \tag{8.84}$$

Similarly for $\pi = 0$

$$\lambda_{\zeta t} - \lambda = -\partial_t^2\lambda^3. \tag{8.85}$$

8.6.2 STATIONARY SOLUTION

Now we would write a particular solution of the Short Pulse system, taking into account only the directed wave. It is, as shown, equivalent to SPE (Shafer-Wayne) and seeks a stationary solution of it. z Starting from the original system (8.80) with $\Pi = 0$, we change the variables so as to be convenient to investigate the wave propagation in its rest frame moving with velocity v:

$$x = \eta, \xi = x - vt. \tag{8.86}$$

Then, the derivatives in (8.80) are transformed as

$$\begin{aligned} \partial_t &= -v\partial_\xi, \partial_x = \partial_\eta + \partial_\xi, \\ \partial_{xt} &= -v[\partial_{\eta\xi} + \partial_\xi^2]. \end{aligned} \tag{8.87}$$

We declare the independence of a solution Λ on η as a definition of stationary state:

$$\partial_\eta \Lambda = 0.$$

Equation (8.80) (with $\Pi = 0$) after the transition to variables (8.86) and use of the stationarity condition simplifies as ordinary differential one:

$$vc\Lambda_{\xi\xi} + pq\Lambda = -\alpha \left[\Lambda_{\xi\xi}\right]^3, \tag{8.88}$$

where

$$\alpha \equiv \frac{\mu_0 \chi^{(3)} c^3 v^6}{2p^3 q}. \tag{8.89}$$

The Equation (8.88) is the cubic algebraic one with respect to $\Lambda_{\xi\xi}$; rearranging terms, write:

$$\left[\Lambda_{\xi\xi}\right]^3 + v\frac{c}{\alpha}\Lambda_{\xi\xi} + \frac{pq}{\alpha}\Lambda = 0. \tag{8.90}$$

It can be solved by the Cardano formula, using obviously a positive discriminant:

$$Q = \left(\frac{vc}{3\alpha}\right)^3 + \left(\frac{pq}{2\alpha}\Lambda\right)^2 > 0. \tag{8.91}$$

For physical reason, for a solution search, we use only the real root $(\Lambda_{\xi\xi})_1$ of the cubic equation (8.90):

$$(\Lambda_{\xi\xi})_1 = \sqrt[3]{-\tfrac{pq}{2\alpha}\Lambda + \sqrt{Q}} + \sqrt[3]{-\tfrac{pq}{2\alpha}\Lambda - \sqrt{Q}}. \tag{8.92}$$

An analysis of dimensionless amplitude of a wave to be small in conditions of the previously mentioned experiments allows us to expand the root in Taylor series, arriving at

$$(\Lambda_{\xi\xi})_1 \approx -\frac{pq}{vc}\Lambda + \frac{p^3 q^3 \alpha}{v^4 c^4}\Lambda^3 + \cdots. \tag{8.93}$$

Accounting only for the first term, we obtain the linear equation

$$\Lambda_{\xi\xi} = -\frac{pq}{vc}\Lambda, \tag{8.94}$$

which has a solution oscillating along ξ with the wavelength $2\pi/\mathsf{k}, k = \sqrt{\frac{pq}{vc}}$. Accounting for the second term of expansion (8.93) yields the equation of the cubic oscillator form

$$\Lambda_{\xi\xi} + \frac{pq}{vc}\Lambda - \frac{p^3q^3\alpha}{v^4c^4}\Lambda^3 = 0. \tag{8.95}$$

Accounting for (8.89) yields

$$\frac{p^3q^3}{v^4c^4}\frac{\mu_0\chi^{(3)}c^3v^6}{2p^3q} = \frac{q^2}{c}\frac{\mu_0\chi^{(3)}v^2}{2},$$

which results in

$$\Lambda_{\xi\xi} + \frac{pq}{vc}\Lambda - \frac{q^2}{c}\frac{\mu_0\chi^{(3)}v^2}{2}\Lambda^3 = 0.$$

The parameter of expansion is defined via the last terms ratio: $\frac{1}{2p}qv^3\Lambda^2\chi^{(3)}\mu_0$. Going to dimensionless variables:

$$\begin{aligned} f\Lambda &= y, \\ g\xi &= z, \end{aligned} \tag{8.96}$$

where $f, g \neq 0$ are parameters, results in

$$y_{zz} + \frac{\mathsf{k}^2}{g^2}y - \frac{\mu_0\chi^{(3)}q^2v^2}{2cf^2g^2}y^3 = 0. \tag{8.97}$$

Comparing it with the differential equation for Jacobi elliptic *sn* functions with modulus m

$$y_{zz} + (1+m^2)y - 2m^2y^3 = 0, \tag{8.98}$$

we find

$$\begin{aligned} 1+m^2 &= \frac{\mathsf{k}^2}{g^2}, \\ m^2 &= \frac{\mu_0\chi^{(3)}q^2v^2}{4cf^2g^2}. \end{aligned} \tag{8.99}$$

Combining these relations yields

$$f^2g^2v - \frac{pf^2q}{c} + \frac{\mu_0q^2\chi^{(3)}}{4c}v^3 = 0, \tag{8.100}$$

that is a cubic equation for v with a ratio of the last terms $\frac{\frac{\mu_0q^2\chi^{(3)}}{4c}v^3}{\frac{pf^2q}{c}} = \frac{qv^3\chi^3\mu_0}{4f^2p} = r.$

Denoting,

$$\frac{\mu_0\chi^{(3)}q^2}{4c} = \chi,$$

Equation (8.100) simplifies as $g^2 v - \frac{pq}{c} + \frac{\chi}{f^2} v^3 = 0$. If we go to the frequency ω of a periodic solution by $v = \frac{\omega}{g}$, we have

$$g^4 \omega - \frac{1}{c} pq g^3 + \frac{1}{f^2} \chi \omega^3 = 0, \tag{8.101}$$

which is the nonlinear dispersion equation for the Cauchy problem. As seen from the solution arguments, the frequency ω and the amplitude $1/f$ are prescribed by the boundary regime. The dispersion relation $v(g,f)$ or $\omega(g,f)$ is expressed again by means of Cardano formula. For a boundary regime problem in similar way one obtains $g(\omega,f)$, see Figure 8.2b.

The expansion (8.93) is valid for $\max_z \frac{qv^3}{2p} \Lambda^2 \chi^{(3)} \mu_0 = \frac{qv^3 \chi^{(3)} \mu_0}{2pf^2} < 1$. The three-term part Taylor series is rigoriosly valid for

$$\frac{q}{p} \frac{\mu_0 \chi^{(3)}}{2f^2} v^3 \ll 1, \tag{8.102}$$

which includes $\frac{q}{p} \frac{\mu_0 \chi^{(3)}}{2f^2} v^3 = 2r$ in the l.h.s..

Having $\frac{\mu_0 q \chi^{(3)}}{4f^2 p} v^3 = 1 - \frac{cg^2 v}{pq}$, the strong inequality (8.102) may be rewritten as $\frac{1}{2} \ll \frac{c\omega g(\omega,f)}{pq}$, or

$$\frac{pq}{2c\omega} \ll g(\omega,f).$$

The leading term is defined by $g = \frac{pq}{c\omega}, v = \frac{\omega}{g} = \frac{c}{pq}\omega^2$, as in linear approximation. In this approximation, the amplitude square is restricted as

$$\frac{1}{f^2} \ll \frac{2p^4 q^2}{c^3 \chi^{(3)} \omega^6 \mu_0},$$

or rewriting,

$$\frac{\mu_0 \chi^{(3)} q^2 v^2}{4cf^2 g^2} = \frac{\mu_0 \chi^{(3)} q v^3}{2f^2 p} \frac{qp}{2cg^2 v} \ll 1,$$

we obtain the restriction for the modulus (8.99):

$$m^2 = \frac{\mu_0 \chi^{(3)} q v^3}{2f^2 p} \ll \frac{2cg\omega}{qp} \sim \frac{2c \frac{pq}{c\omega} \omega}{qp} = 2.$$

In this range of amplitudes

$$g^2 v - \frac{pq}{c} = 0, \; v = \frac{pq}{cg^2},$$

that is, it should be

$$\frac{1}{pq}\frac{2c\chi}{f^2}v^3 = \frac{1}{pq}\frac{2c\chi}{f^2}\left(\frac{pq}{cg^2}\right)^3 = \frac{2}{c^2 f^2 g^6}p^2q^2\chi \ll 1.$$

Denoting the frequency at a boundary as

$$vg = \frac{pq}{cg} = \omega, g = \frac{pq}{\omega c},$$

we estimate the parameters for the case of [19], taking

$$q = 1.1p, \omega = 0.7q,$$

$$g = \left[\frac{pq}{\omega c}\right]_{\omega=0.7q,q=1.1p} = \frac{1.4286}{c}p.$$

Solving the first equation of the system (8.99) with respect to g^2 we obtain

$$g^2 = \frac{k^2}{1+m^2}. \tag{8.103}$$

Then, the solution of equation (8.98) is

$$y = sn(z,m) = sn\left(z, \frac{qv}{2fg}\sqrt{\frac{\mu_0\chi^{(3)}}{c}}\right), \tag{8.104}$$

or, returning to the physical origin, we write the hybrid wave amplitude (8.57)

$$\Lambda = \frac{1}{f}sn\left[g(x-vt), \frac{qv}{2fg}\sqrt{\frac{\mu_0\chi^{(3)}}{c}}\right]. \tag{8.105}$$

It is the second argument of the function with the name “elliptic sinus” that is marked “sn”, the modulus is defined by a period of boundary regime. Its value determines the velocity of propagation and the wavelength of a nonlinear wave taking into account the value of amplitude $\frac{1}{f}$.

Such a representation of the solution is convenient from the point of a comparison with the linear case. Namely, if $\chi^{(3)} \to 0$, or amplitude $\frac{1}{f} \to 0$, the limit gives

$$m^2 = 0, g = \mathrm{k},$$
$$\Lambda_L = \frac{1}{f}\sin\left(\sqrt{k^2}\xi\right) = \frac{1}{f}\sin(k\xi) = \frac{1}{f}\sin(k(x-v_0t)). \tag{8.106}$$

This limit explains the notations and the link between frequency and the k. Certainly in nonlinear case there is the dependence of velocity of propagation, period, and amplitude expressed by (8.99) and (8.103).

The effect of negative permittivity was demonstrated at GHz region [22]. In such conditions and big amplitudes, the squared modulus (8.103) simplifies to

$$m^2 \approx \frac{\mu_0 \chi^{(3)}}{2vkf^2}. \tag{8.107}$$

The evaluation of m by the general expression (8.99) and the velocity is found as the real root of (8.100) as a function of amplitude $1/f$. This dependence is presented in Figure 8.2c.

In Figure 8.2c the evolution of the wave profile of the solution expressed by (8.106), which is represented by Figure 8.2a.

The results of the simulations of the solutions with different amplitudes but with boundary regimes of the same frequency show the peculiarities of the Λ wave behavior as a function of time. The corresponding boundary regime at $x = 0$ is directly seen from Figure 8.2 and the formulas for Λ in the nonlinear case (8.105) or, in the particular linear case (8.106). It is defined by the parameter k in the function *sn* argument.

A problem that could be solved on the basis of the derived systems (8.80) or its version (8.81), as well as, more generally, by (8.66) with general nonlinearity and dispersion may be formulated as a wave reflection by a wall or slab with auxiliary boundary conditions. There are versions of this general equation written for the special choice of the Drude dispersion (8.76) and the Kerr nonlinearity (8.77). The estimation of the parameters shows that the frequency at an almost linear regime of generation is very close to that of nonlinear oscillations. In the realm of nonlinearity, the parameters of oscillations ω and a frequency connected with the Jacobi function modulus and velocity of propagation deviate one from another. Hence, the question of the approximate equations' validity needs further investigation. There are efforts in investigations of analytical schemes for such a system's integrability [23]. It gives solitonic solutions [24], but with specific loops. For our elliptic solution, a solitonic one can be also extracted from (8.105) with $m = 1$. A natural application of such a simulation or analytics gives the parameters of metamaterial dispersion $\omega_{pe} = p, \omega_{pm} = q$, see (8.68), and of a nonlinearity one, α, see (8.89). A detail modeling of the process of a nonlinear evolution describing propagation and reflection of arbitrary wave needs mathematical investigation in the spirit of [25,26] and elaboration of special numerical codes.

8.7 STATEMENT OF PROBLEM FOR WAVES WITH TWO POLARIZATIONS

8.7.1 MAXWELL'S EQUATIONS. BOUNDARY REGIME PROBLEM

We restrict ourselves to a one-dimensional model with two possible polarizations, similar to Kuszner, Leble [27], where the x-axis is chosen as the direction of a wave propagation for wave propagation in the opposite direction. As mentioned authors,

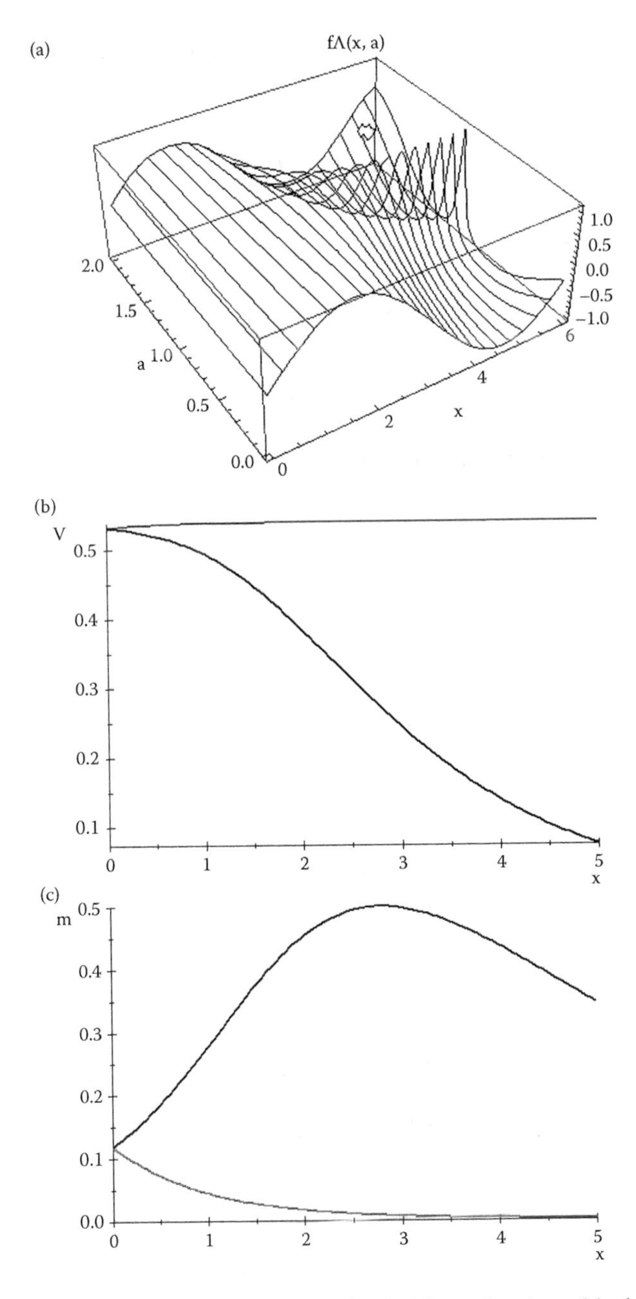

Figure 8.2 (a) The solution (multiplied on f) (8.105) as function of both variables for $p = 1GHz, q = 1.1p, I = 1GW/cm^2 = 10^{13}W/m^2, \chi^{(3)} = 10^{-20}m^2/V^2$ [21], (b) velocity of propagation as function of amplitude in logarithmic scale $x = ln(f)$ — lower line and for $x = ln(\frac{1}{f})$ — upper line, (c) the elliptic sinus modulus is plotted as the function of the same variables

we assume $D_x = 0$ and $B_x = 0$. This allows us to write the Maxwell equations in SI as

$$\frac{\partial D_y}{\partial t} = -\frac{\partial H_z}{\partial x},\qquad (8.108)$$
$$\frac{\partial D_z}{\partial t} = \frac{\partial H_y}{\partial x},$$
$$\frac{\partial B_y}{\partial t} = \frac{\partial E_z}{\partial x},$$
$$\frac{\partial B_z}{\partial t} = -\frac{\partial E_y}{\partial x}.$$

Further, it is convenient to mark these projections as

$$D_y \equiv D_1, D_z \equiv D_2, B_y \equiv B_1, B_z \equiv B_2, \qquad (8.109)$$

with additional material relations

$$\begin{aligned} D_i(x,t) &= \widehat{\varepsilon} E_i(x,t), i = 1,2, \\ B_i(x,t) &= \widehat{\mu} H_i(x,t), i = 1,2, \end{aligned} \qquad (8.110)$$

where $\widehat{\mu}$ and $\widehat{\varepsilon}$ are integral convolution-type operators as in the preceding section, see (8.21) and [12]:

$$\begin{aligned} \widehat{\varepsilon}\psi(x,t) &= \int_{-\infty}^{\infty} \tilde{\varepsilon}(t-s)\psi(x,s)ds, \\ \widehat{\mu}\psi(x,t) &= \int_{-\infty}^{\infty} \tilde{\mu}(t-s)\psi(x,s)ds, \end{aligned} \qquad (8.111)$$

with kernels

$$\begin{aligned} \tilde{\varepsilon}(t-s) &= \frac{\varepsilon_0}{2\pi}\int_{-\infty}^{\infty} \varepsilon(\omega)\exp(i\omega(t-s))d\omega, \\ \tilde{\mu}(t-s) &= \frac{\mu_0}{2\pi}\int_{-\infty}^{\infty} \mu(\omega)\exp(i\omega(t-s))d\omega, \end{aligned} \qquad (8.112)$$

Plugging operator links (8.110) into Eq. (8.108), we write it down in short form:

$$\begin{aligned} \partial_t(\widehat{\varepsilon}E_{1,2}) &= \mp\partial_x(\widehat{\mu}^{-1}B_{2,1}), \\ \partial_t B_{1,2} &= \pm\partial_x E_{2,1}. \end{aligned} \qquad (8.113)$$

Here we marked again the derivatives as

$$\frac{\partial}{\partial x} \equiv \partial_x, \frac{\partial}{\partial t} \equiv \partial_t. \qquad (8.114)$$

We also must add the time dependent boundary conditions to finalize the boundary regime problem:

$$E_{1,2}(0,t) = j_{1,2}(t),\ B_{1,2}(0,t) = \ell_{1,2}(t), \tag{8.115}$$

j_i and ℓ_i are arbitrary functions, continued to the half space $t < 0$ anti-symmetrically:

$$j_i(-t) = -j_i(t), \ell_i(-t) = -\ell_i(t), i = 1,2. \tag{8.116}$$

8.8 DYNAMIC PROJECTING OPERATORS

Doing the Fourier transformations like in previous chapter, see also [12], and plugging them into the system of equations (8.113), we have the closed system

$$\partial_t \left(\int_{-\infty}^{\infty} \varepsilon(\omega)\mathscr{E}_{1,2}(x,\omega)\exp(i\omega t)d\omega \right) = \frac{\mp\partial_x}{\mu_0\varepsilon_0} \left(\int_{-\infty}^{\infty} \frac{\mathscr{B}_{2,1}(x,\omega)}{\mu(\omega)} \exp(i\omega t)d\omega \right). \tag{8.117}$$

The inverse Fourier transformation yields in the first pair of equations of (8.113)

$$\partial_x \mathscr{B}_{2,1} = \mp i\omega\mu_0\varepsilon_0\mu(\omega)\varepsilon(\omega)\mathscr{E}_{1,2}. \tag{8.118}$$

The second pair of equations looks similar

$$\partial_x \mathscr{E}_{2,1} = \pm i\omega\mathscr{B}_{1,2}. \tag{8.119}$$

As in the previous sections, we introduce auxiliary function $a(\omega)$ as

$$\mu_0\varepsilon_0\varepsilon(\omega)\mu(\omega) \equiv c^{-2}\varepsilon(\omega)\mu(\omega) \equiv a^2(\omega), \tag{8.120}$$

where $c = \frac{1}{\sqrt{\varepsilon_0\mu_0}}$ is the velocity of light in vacuum. In the time domain there is an operator [12]:

$$\widehat{a}^2 = \widehat{a^2} = c^{-2}\widehat{\varepsilon}\widehat{\mu} = c^{-2}\widehat{\mu}\widehat{\varepsilon}, \tag{8.121}$$

defined as an integral one:

$$\widehat{a^2}\psi(x,t) = \frac{1}{2\pi} \int_{-\infty}^{\infty} a^2(\omega)\exp(i\omega t - i\omega\tau)\psi(x,\tau)d\omega d\tau, \tag{8.122}$$

Hence, the system (8.118 and 8.119) simplifies as

$$\partial_x \mathscr{B}_{2,1} = \mp i\omega a^2(\omega)\mathscr{E}_{1,2}, \tag{8.123}$$

$$\pm\partial_x \mathscr{E}_{2,1} = i\omega\mathscr{B}_{1,2}. \tag{8.124}$$

The system (8.113) may be written also in matrix form, convenient for projectors construction. For this purpose we introduce matrices $\mathscr{L}$ and $\tilde{\Psi}$ as

$$\tilde{\Psi} = \begin{pmatrix} \mathscr{B}_2 \\ \mathscr{B}_1 \\ \mathscr{E}_2 \\ \mathscr{E}_1 \end{pmatrix}, \tag{8.125}$$

or, in the short block columns form

$$\tilde{\Psi} = \begin{pmatrix} \mathscr{B}_{2,1} \\ \mathscr{E}_{2,1} \end{pmatrix}, \tag{8.126}$$

$$\mathscr{L} = \begin{pmatrix} 0 & 0 & 0 & -i\omega a^2 \\ 0 & 0 & i\omega a^2 & 0 \\ 0 & i\omega & 0 & 0 \\ -i\omega & 0 & 0 & 0 \end{pmatrix}. \tag{8.127}$$

In the block form

$$\mathscr{L} = \begin{pmatrix} \widehat{0} & \mathscr{L}_1 \\ \mathscr{L}_2 & \widehat{0} \end{pmatrix}, \tag{8.128}$$

where

$$\mathscr{L}_1 = \begin{pmatrix} 0 & -i\omega a^2 \\ i\omega a^2 & 0 \end{pmatrix}, \mathscr{L}_2 = \begin{pmatrix} 0 & i\omega \\ -i\omega & 0 \end{pmatrix}. \tag{8.129}$$

Hence, the matrix form of (8.118) and (8.119) is

$$\partial_x \tilde{\Psi} = \mathscr{L}\tilde{\Psi}. \tag{8.130}$$

Consider the eigenvalue problem for matrix $\mathscr{L}$

$$\begin{pmatrix} 0 & 0 & 0 & -i\omega a^2 \\ 0 & 0 & i\omega a^2 & 0 \\ 0 & i\omega & 0 & 0 \\ -i\omega & 0 & 0 & 0 \end{pmatrix} \begin{pmatrix} A_1 \\ A_2 \\ A_3 \\ A_4 \end{pmatrix} = \lambda \begin{pmatrix} A_1 \\ A_2 \\ A_3 \\ A_4 \end{pmatrix} \tag{8.131}$$

λ we find from equation

$$\det(\mathscr{L} - \lambda I) = 0, \tag{8.132}$$

where I - identity matrix. We obtain

$$\det \begin{pmatrix} -\lambda & 0 & 0 & -i\omega a^2 \\ 0 & -\lambda & i\omega a^2 & 0 \\ 0 & i\omega & -\lambda & 0 \\ -i\omega & 0 & 0 & -\lambda \end{pmatrix} = 0, \tag{8.133}$$

or

$$\lambda^4 + 2\omega^2 a^2 \lambda^2 + \omega^4 a^4 = 0 \rightarrow (\lambda^2 + \omega^2 a^2)^2 = 0. \tag{8.134}$$

The result shows that the eigenvalues

$$\lambda^2 = -\omega^2 a^2 \rightarrow \lambda_{1,2} = \pm i a \omega \tag{8.135}$$

are degenerate, the subspaces are two-dimensional, and, from (8.131), we derive

$$\begin{aligned} -i\omega a^2 A_4 &= ia\omega A_1, \\ i\omega a^2 A_3 &= -ia\omega A_2, \\ i\omega A_2 &= ia\omega A_3, \\ -i\omega A_1 &= -ia\omega A_4. \end{aligned} \tag{8.136}$$

After simplifications we have

$$\begin{aligned} -aA_4 &= A_1, \\ aA_3 &= -A_2, \\ A_2 &= aA_3, \\ A_1 &= aA_4. \end{aligned} \tag{8.137}$$

For $\lambda_{1,2}$ we choose two set of the elements; the first one

$$A_{11} = 1, A_{12} = 0, A_{13} = 0, A_{14} = -\frac{1}{a}. \tag{8.138}$$

and, second, orthogonal to the first

$$A_{21} = 0, A_{22} = 1, A_{23} = -\frac{1}{a}, A_{24} = 0. \tag{8.139}$$

for $\lambda_{3,4}$, similarly

$$A_{31} = 0, A_{32} = a, A_{33} = 1, A_{34} = 0, \tag{8.140}$$

and, again, the orthogonal one

$$A_{41} = a, A_{42} = 0, A_{43} = 0, A_{44} = 1. \tag{8.141}$$

We construct from these elements an auxiliary matrix of eigenvectors Φ as

$$\Phi = \begin{pmatrix} A_{11} & A_{21} & A_{31} & A_{41} \\ A_{12} & A_{22} & A_{32} & A_{42} \\ A_{13} & A_{23} & A_{33} & A_{43} \\ A_{14} & A_{24} & A_{34} & A_{44} \end{pmatrix} \rightarrow \Phi = \begin{pmatrix} 1 & 0 & 0 & a \\ 0 & 1 & a & 0 \\ 0 & -\frac{1}{a} & 1 & 0 \\ -\frac{1}{a} & 0 & 0 & 1 \end{pmatrix} \tag{8.142}$$

The inverse matrix Φ^{-1} is

$$\Phi^{-1} = \frac{1}{2}\begin{pmatrix} 1 & 0 & 0 & -a \\ 0 & 1 & -a & 0 \\ 0 & \frac{1}{a} & 1 & 0 \\ \frac{1}{a} & 0 & 0 & 1 \end{pmatrix} \tag{8.143}$$

Corresponding matrix projecting operators we find from the basic formula (2.13) from Section 2.1:

$$P_{ij}^{(n)}(\omega) = \Phi_{in}\Phi_{nj}^{-1}. \tag{8.144}$$

Using it, we get four projectors:

$$P^{(1)} = \frac{1}{2}\begin{pmatrix} 1 & 0 & 0 & -a \\ 0 & 0 & 0 & 0 \\ 0 & 0 & 0 & 0 \\ -\frac{1}{a} & 0 & 0 & 1 \end{pmatrix},$$

$$P^{(2)} = \frac{1}{2}\begin{pmatrix} 0 & 0 & 0 & 0 \\ 0 & 1 & -a & 0 \\ 0 & -\frac{1}{a} & 1 & 0 \\ 0 & 0 & 0 & 0 \end{pmatrix},$$

$$P^{(3)} = \frac{1}{2}\begin{pmatrix} 0 & 0 & 0 & 0 \\ 0 & 1 & a & 0 \\ 0 & \frac{1}{a} & 1 & 0 \\ 0 & 0 & 0 & 0 \end{pmatrix},$$

$$P^{(4)} = \frac{1}{2}\begin{pmatrix} 1 & 0 & 0 & a \\ 0 & 0 & 0 & 0 \\ 0 & 0 & 0 & 0 \\ \frac{1}{a} & 0 & 0 & 1 \end{pmatrix}. \tag{8.145}$$

Recall the standard properties of orthogonal projecting operators:

$$P^{(i)}P^{k)} = P^{(i)}\delta_{ik}, \sum_{i=1}^{4} P^{(i)} = I, \tag{8.146}$$

(see Section 2.1). The inverse Fourier transformation $\mathbf{P}^{(i)} = \mathscr{F} P^{(i)} \mathscr{F}^{-1}$, where $\mathscr{F}$ - operator of Fourier transformation, leads to the dynamic projectors in t-representation:

$$\mathbf{P}^{(1)} = \frac{1}{2}\begin{pmatrix} 1 & 0 & 0 & -\widehat{a} \\ 0 & 0 & 0 & 0 \\ 0 & 0 & 0 & 0 \\ -\widehat{a}^{-1} & 0 & 0 & 1 \end{pmatrix},$$

$$\mathbf{P}^{(2)} = \frac{1}{2}\begin{pmatrix} 0 & 0 & 0 & 0 \\ 0 & 1 & \widehat{a} & 0 \\ 0 & \widehat{a}^{-1} & 1 & 0 \\ 0 & 0 & 0 & 0 \end{pmatrix},$$

$$\mathbf{P}^{(3)} = \frac{1}{2}\begin{pmatrix} 1 & 0 & 0 & \widehat{a} \\ 0 & 0 & 0 & 0 \\ 0 & 0 & 0 & 0 \\ \widehat{a}^{-1} & 0 & 0 & 1 \end{pmatrix},$$

$$\mathbf{P}^{(4)} = \frac{1}{2}\begin{pmatrix} 0 & 0 & 0 & 0 \\ 0 & 1 & \widehat{a} & 0 \\ 0 & \widehat{a}^{-1} & 1 & 0 \\ 0 & 0 & 0 & 0 \end{pmatrix} \tag{8.147}$$

Here are the definitions of the integral operators $\widehat{a}$, $\widehat{a}^{-1}$ [12]:

$$\widehat{a}\eta(x,t) = \frac{1}{2\pi}\int_{-\infty}^{\infty}\left[\eta(x,\tau)\int_{-\infty}^{\infty} a(\omega)\exp(i\omega(t-\tau))d\omega\right]d\tau,$$
$$\widehat{a}^{-1}\xi(x,t) = \frac{1}{2\pi}\int_{-\infty}^{\infty}\left[\xi(x,\tau)\int_{-\infty}^{\infty}\frac{1}{a(\omega)}\exp(i\omega(t-\tau))d\omega\right]d\tau. \tag{8.148}$$

8.9 SEPARATED EQUATIONS AND DEFINITION FOR LEFT AND RIGHT WAVES

We return to the time domain. Let's write the matrix equation (8.130) in this representation:

$$\partial_x \Psi = \widehat{L}\Psi, \tag{8.149}$$

where the wave vector is defined by

$$\Psi = \begin{pmatrix} B_2 \\ B_1 \\ E_2 \\ E_1 \end{pmatrix}, \tag{8.150}$$

and the x-evolution operator has the matrix form

$$\widehat{L} = \begin{pmatrix} 0 & 0 & 0 & -\partial_t \widehat{a}^2 \\ 0 & 0 & \partial_t \widehat{a}^2 & 0 \\ 0 & \partial_t & 0 & 0 \\ -\partial_t & 0 & 0 & 0 \end{pmatrix}. \tag{8.151}$$

Acting by projectors $\mathbf{P}^{(i)}$ on (8.149), we introduce four wave variables that differ by direction of propagation and polarization. Let us take the case of $\mathbf{P}^{(1)}$, the action on the L.H.S. of (8.149) gives

$$\frac{1}{2}\partial_x \begin{pmatrix} 1 & 0 & 0 & -\widehat{a} \\ 0 & 0 & 0 & 0 \\ 0 & 0 & 0 & 0 \\ -\widehat{a}^{-1} & 0 & 0 & 1 \end{pmatrix} \begin{pmatrix} B_2 \\ B_1 \\ E_2 \\ E_1 \end{pmatrix} = \partial_x \begin{pmatrix} \frac{1}{2}B_2 - \frac{1}{2}aE_1 \\ 0 \\ 0 \\ \frac{1}{2}E_1 - \frac{1}{2a}B_2 \end{pmatrix},$$

and on the R.H.S., one obtains

$$\frac{1}{2}\begin{pmatrix} 0 & 0 & 0 & -\partial_t \widehat{a}^2 \\ 0 & 0 & \partial_t \widehat{a}^2 & 0 \\ 0 & \partial_t & 0 & 0 \\ -\partial_t & 0 & 0 & 0 \end{pmatrix} \begin{pmatrix} 1 & 0 & 0 & -\widehat{a} \\ 0 & 0 & 0 & 0 \\ 0 & 0 & 0 & 0 \\ -\widehat{a}^{-1} & 0 & 0 & 1 \end{pmatrix} \begin{pmatrix} B_2 \\ B_1 \\ E_2 \\ E_1 \end{pmatrix}$$

$$= \begin{pmatrix} \frac{1}{2}\widehat{a}\partial_t B_2 - \frac{1}{2}\widehat{a}^2 \partial_t E_1 \\ 0 \\ 0 \\ \frac{1}{2}\widehat{a}\partial_t E_1 - \frac{1}{2}\partial_t B_2 \end{pmatrix}.$$

Reading the first line of the equality yields the equation

$$\partial_x \left(\frac{1}{2}B_2 - \frac{1}{2}\widehat{a}E_1 \right) = \frac{1}{2}\widehat{a}\partial_t B_2 - \frac{1}{2}\partial_t \widehat{a}^2 E_1.$$

If one fixes a direction of propagation, for example by $i = 1, 2$, the following combinations Π_1, Π_2

$$\Pi_1 = \frac{1}{2}(B_2 - \widehat{a}E_1), \tag{8.152}$$

$$\Pi_2 = \frac{1}{2}(B_1 + \widehat{a}E_2) \tag{8.153}$$

satisfy the equations

$$\partial_x \Pi_1 + \partial_t \widehat{a}\Pi_1 = 0, \partial_x \Pi_2 + \partial_t \widehat{a}\Pi_2 = 0. \tag{8.154}$$

Correspondingly,

$$\Lambda_1 = \frac{1}{2}(B_2 + \widehat{a}E_1), \tag{8.155}$$

$$\Lambda_2 = -\frac{1}{2}(B_1 - \widehat{a}E_2), \tag{8.156}$$

describe the opposite directed waves; by the way, Λ_1 is Λ-wave, obtained for the case of one polarization [12]. The functions solve the equations:

$$\begin{aligned} \partial_x \Lambda_1 &= \partial_t \widehat{a} \Lambda_1, \\ \partial_x \Lambda_2 &= \partial_t \widehat{a} \Lambda_2. \end{aligned} \tag{8.157}$$

The indexes in both cases mark the polarizations. The inverse transformation gives the electric and magnetic fields

$$B_2 = \Pi_1 + \Lambda_1, B_1 = \Pi_2 - \Lambda_2, \tag{8.158}$$

$$E_2 = \widehat{a}^{-1}(\Lambda_2 + \Pi_2), E_1 = \widehat{a}^{-1}(\Lambda_1 - \Pi_1). \tag{8.159}$$

Note that the real direction of propagation depends on the sign of $\hat{a}$. It shows the metamaterial peculiarities, determined by the type of the material ($\varepsilon > 0, \mu < 0$, or opposite.)

Using definitions of (8.152, 8.153, 8.155, 8.156) and (8.115) we derive boundary regime conditions for left and right waves with both polarizations:

$$\begin{aligned} \Lambda_1(0,t) &= \frac{1}{2}(B_2(0,t) - \widehat{a}E_1(0,t)) = \frac{1}{2}(k_2(t) - \widehat{a}j_1(t)), \\ \Lambda_2(0,t) &= \frac{1}{2}(B_1(0,t) + \widehat{a}E_2(0,t)) = \frac{1}{2}(k_1(t) + \widehat{a}j_2(t)), \\ \Pi_1(0,t) &= \frac{1}{2}(B_2(0,t) + \widehat{a}E_1(0,t)) = \frac{1}{2}(\ell_2(t) + \widehat{a}j_1(t)), \\ \Pi_2(0,t) &= \frac{1}{2}(B_1(0,t) - \widehat{a}E_2(0,t)) = \frac{1}{2}(\ell_1(t) - \widehat{a}j_2(t)). \end{aligned} \tag{8.160}$$

8.10 GENERAL NONLINEARITY ACCOUNT

Let's consider a nonlinear problem. We start again from the Maxwell's equations (8.108) for two components of electric and two of magnetic fields with generalized nonlinear material relations:

$$\begin{aligned} D_i &= \widehat{\varepsilon}E_i + P_i^{(NL)}, \\ B_i &= \widehat{\mu}H_i + M_i^{(NL)}, i = 1,2 \end{aligned} \tag{8.161}$$

$\vec{P}^{NL}$ - nonlinear part of polarization vector ($\vec{M}^{NL}$ - one for magnetization). For our purposes, the linear parts of polarization and magnetization have already been taken into account. In the time domain, a closed version of (8.108) with the nonlinear terms inserted is

$$\frac{\partial(\widehat{\varepsilon}E_1 + P_1^{(NL)})}{\partial t} = -\frac{\partial\widehat{\mu}^{-1}(B_2 - M_2^{(NL)})}{\partial x},$$
$$\frac{\partial(\widehat{\varepsilon}E_2 + P_2^{(NL)})}{\partial t} = \frac{\partial\widehat{\mu}^{-1}(B_1 - M_1^{(NL)})}{\partial x},$$
$$\frac{\partial B_y}{\partial t} = \frac{\partial E_z}{\partial x},$$
$$\frac{\partial B_z}{\partial t} = -\frac{\partial E_y}{\partial x}. \tag{8.162}$$

It terms of (8.109) it is rewritten as

$$\partial_t\widehat{\varepsilon}E_{2,1} + \partial_t P_{2,1}^{(NL)} = \pm(\partial_x\widehat{\mu}^{-1}B_{1,2} - \partial_x\widehat{\mu}^{-1}M_{1,2}^{(NL)}),$$
$$\partial_t B_{1,2} = \mp\partial_x E_{2,1}. \tag{8.163}$$

The action of operator $\widehat{\mu}$ on the first pair of equations in system (8.163) and the use of the same notations Ψ and $\widehat{L}$ from (8.150 and 8.151) once more, produce a nonlinear analogue of the matrix equation (8.149):

$$\partial_x\Psi - \widehat{L}\Psi = \partial_x \begin{pmatrix} M_2^{(NL)} \\ M_1^{(NL)} \\ 0 \\ 0 \end{pmatrix} - \widehat{\mu}\partial_t \begin{pmatrix} P_1^{(NL)} \\ -P_2^{(NL)} \\ 0 \\ 0 \end{pmatrix}. \tag{8.164}$$

On the R.H.S., there is a vector of nonlinearity for the case of the opposite directed 1D-waves:

$$\mathbb{N} = \begin{pmatrix} \partial_x M_2^{(NL)} - \widehat{\mu}\partial_t P_1^{(NL)} \\ \partial_x M_1^{(NL)} + \widehat{\mu}\partial_t P_2^{(NL)} \\ 0 \\ 0 \end{pmatrix}. \tag{8.165}$$

Next, acting by the operators $\widehat{\mathbf{P}}^{(1,2,3,4)}$ (8.147) on Eq. (8.164) we find

$$\partial_x\Pi_1 + \partial_t\widehat{a}\Pi_1 = \frac{1}{2}\partial_x M_2^{(NL)} - \frac{1}{2}\widehat{\mu}\partial_t P_2^{(NL)},$$
$$\partial_x\Pi_2 + \partial_t\widehat{a}\Pi_2 = \frac{1}{2}\partial_x M_2^{(NL)} + \frac{1}{2}\widehat{\mu}\partial_t P_1^{(NL)},$$
$$\partial_x\Lambda_1 - \partial_t\widehat{a}\Lambda_1 = \frac{1}{2}\partial_x M_1^{(NL)} - \frac{1}{2}\widehat{\mu}\partial_t P_2^{(NL)},$$
$$\partial_x\Lambda_2 - \partial_t\widehat{a}\Lambda_2 = \frac{1}{2}\partial_x M_2^{(NL)} + \frac{1}{2}\widehat{\mu}\partial_t P_1^{(NL)}. \tag{8.166}$$

Generally, the R.H.S. of each equation (8.170) depends on the field vectors $\vec{E}, \vec{B}$, that should be presented in terms of the fields $\vec{\Pi}, \vec{\Lambda}$ to re-express the system in hybrid waves language. The vectors components are expressed by means of the inverse transformation of (8.152, 8.153, 8.155, 8.156).

8.11 KERR NONLINEARITY ACCOUNT FOR LOSSLESS DRUDE METAMATERIALS

8.11.1 EQUATIONS OF INTERACTION OF THE WAVES VIA KERR EFFECT

For nonlinear Kerr materials [21], the third-order nonlinear part of polarization [14,21] has the form

$$P_{1,2}^{(NL)} = \varepsilon_0 \chi_e^{(3)} (E_{1,2}^3 + E_{1,2} E_{2,1}^2).$$

From (8.165), deleting magnetic nonlinearity, one can find the vector N:

$$\mathbb{N} = -\varepsilon_0 \widehat{\mu} \chi_e^{(3)} \partial_t \begin{pmatrix} -E_2^3 - E_2 E_1^2 \\ E_1^3 + E_1 E_2^2 \\ 0 \\ 0 \end{pmatrix}. \tag{8.167}$$

The system for left and right waves with two polarization equations in a medium with Kerr nonlinearity is

$$\begin{aligned} \partial_x \Pi_1 + \partial_t \widehat{a} \Pi_1 &= -\frac{1}{2} \widehat{\mu} \varepsilon_0 \chi_e^{(3)} \partial_t (E_1^3 + E_1 E_2^2), \\ \partial_x \Pi_2 + \partial_t \widehat{a} \Pi_2 &= \frac{1}{2} \widehat{\mu} \varepsilon_0 \chi_e^{(3)} \partial_t (E_2^3 + E_2 E_1^2), \\ \partial_x \Lambda_1 - \partial_t \widehat{a} \Lambda_1 &= -\frac{1}{2} \widehat{\mu} \varepsilon_0 \chi_e^{(3)} \partial_t (E_1^3 + E_1 E_2^2), \\ \partial_x \Lambda_2 - \partial_t \widehat{a} \Lambda_2 &= \frac{1}{2} \widehat{\mu} \varepsilon_0 \chi_e^{(3)} \partial_t (E_2^3 + E_2 E_1^2). \end{aligned} \tag{8.168}$$

The definitions of the hybrid fields as (8.155), or (8.169) gives

$$E_2 = \widehat{a}^{-1} (\Lambda_2 + \Pi_2), E_1 = \widehat{a}^{-1} (\Lambda_1 - \Pi_1). \tag{8.169}$$

In the unidirectional case with $\Lambda_i = 0$ we obtain the system that describes the interaction between hybrid fields with different polarizations:

$$\begin{aligned} \partial_x \Pi_1 + \partial_t \widehat{a} \Pi_1 &= -\frac{1}{2} \widehat{\mu} \varepsilon_0 \chi_e^{(3)} \partial_t ((\widehat{a}^{-1}(-\Pi_1))^3 + (\widehat{a}^{-1}(-\Pi_1))(\widehat{a}^{-1}(\Pi_2))^2), \\ \partial_x \Pi_2 + \partial_t \widehat{a} \Pi_2 &= \frac{1}{2} \widehat{\mu} \varepsilon_0 \chi_e^{(3)} \partial_t ((\widehat{a}^{-1}(\Pi_2))^3 + (\widehat{a}^{-1}(\Pi_2)(\widehat{a}^{-1}(-\Pi_1))^2). \end{aligned} \tag{8.170}$$

Applying the Drude model (see Section 7.6.1 for details, Eq. (8.79), $\widehat{a}^{-1} \approx \frac{c}{pq}\partial_t^2$) and neglecting the derivatives terms in the nonlinear part of the equations ($\widehat{\mu} \approx -\mu_0 q^2 \partial_t^{-2}$) finally we get the SPE system (SPES) for metamaterials:

$$\begin{aligned}\partial_x \Pi_1 + \frac{pq}{c}\partial_t^{-1}\Pi_1 &= \frac{1}{2}\mu_0\varepsilon_0\chi_e^{(3)}\frac{c^3}{p^3 q}\partial_t^{-1}\left((\partial_t^2\Pi_1)^3 + (\partial_t^2\Pi_1)(\partial_t^2\Pi_2)^2\right),\\ \partial_x \Pi_2 + \frac{pq}{c}\partial_t^{-1}\Pi_2 &= \frac{1}{2}\mu_0\varepsilon_0\chi_e^{(3)}\frac{c^3}{p^3 q}\partial_t^{-1}\left((\partial_t^2\Pi_2)^3 + (\partial_t^2\Pi_2)(\partial_t^2\Pi_1)^2\right).\end{aligned} \tag{8.171}$$

Repeating the arguments from Section 8.6 about the effect of negative permeability (it was demonstrated at THz range [22]), we again use the expression $-\mu_0 q^2 \partial_t^{-2}$ which contribution in $\widehat{\mu}$ prevails. Using $\varepsilon_0\mu_0 = c^{-2}$ and differentiating by t simplifies the equations

$$\begin{aligned}\partial_{xt}\Pi_1 + \frac{pq}{c}\Pi_1 &= \frac{c\chi_e^{(3)}}{2p^3 q}((\partial_t^2\Pi_1)^3 + (\partial_t^2\Pi_1)(\partial_t^2\Pi_2)^2),\\ \partial_{xt}\Pi_2 + \frac{pq}{c}\Pi_2 &= \frac{c\chi_e^{(3)}}{2p^3 q}((\partial_t^2\Pi_2)^3 + (\partial_t^2\Pi_2)(\partial_t^2\Pi_1)^2).\end{aligned} \tag{8.172}$$

Introducing new field functions π_1 and π_2 and variable χ as

$$\begin{aligned}\pi_1 &\equiv \partial_t^2\gamma\Pi_1,\\ \pi_2 &\equiv \partial_t^2\gamma\Pi_2,\\ \gamma &= \sqrt{\frac{c^2\chi_e^{(3)}}{2p^4q^2}}\\ \partial_x &= \frac{pq}{c}\partial_\chi,\end{aligned} \tag{8.173}$$

we obtain the system

$$\begin{aligned}\partial_{\chi t}\pi_1 + \pi_1 &= \partial_t^2(\pi_1^3 + \pi_1\pi_2^2),\\ \partial_{\chi t}\pi_2 + \pi_2 &= \partial_t^2(\pi_2^3 + \pi_2\pi_1^2),\end{aligned} \tag{8.174}$$

or, in matrix form

$$\partial_{\chi t}\mathbb{U} + \mathbb{U} = \partial_t^2(\mathbb{U}(\mathbb{U}^2 + C\mathbb{U}^2 C)), \tag{8.175}$$

here

$$\mathbb{U} = \begin{pmatrix} \pi_1 & 0, \\ 0 & \pi_2 \end{pmatrix}, \mathbb{C} = \begin{pmatrix} 0 & 1 \\ 1 & 0 \end{pmatrix}. \tag{8.176}$$

This is the analog of vector ultrashort SPE [28], see also [18]

$$\partial_{\chi t}\mathbb{U} - \mathbb{U} - \frac{1}{6}\partial_t^2\mathbb{U}^3 = 0, \tag{8.177}$$

modified for the case of wave propagation in metamaterials with nonlinearity terms as in (8.167) but with different $\widehat{\mu}$. In our case, the nonlinear part is written via $\mathbb{U}$, that is, the matrix of the fields π_i and the matrix of constants $\mathbb{C}$. The difference seems to be connected with the properties of the metamaterial. We also notice the propagation direction change of a sole Π wave.

8.12 WAVE PACKETS

8.12.1 LINEAR WAVE PACKETS FOR THE RIGHT WAVES

To explain the elementary properties of electromagnetic waves in a Drude metamaterial, we linearize the basic system. To do that, we start from Eq. (8.172) with zero on the RHS which are identical. Hence, take one of them

$$c\partial_{xt}\Pi + pq\Pi = 0, \tag{8.178}$$

and plug it into the conventional wavetrain solution, which we prepare for a comparison with the nonlinear case:

$$\Pi = A(x,t)\exp[i(kx-\omega t)] + c.c.. \tag{8.179}$$

Differentiating

$$\partial_{xt}\Pi = A_{xt}\exp[i(kx-\omega t)] + \\ ikA_t\exp[i(kx-\omega t)] - i\omega A_x\exp[i(kx-\omega t)] - i\omega(ik)A\exp(kx-\omega t) + c.c., \tag{8.180}$$

we put the result in Eq. (8.178), assuming slow varying amplitude:

$$A_x \ll kA, A_t \ll \omega A, \tag{8.181}$$

arriving at

$$A_t - \frac{\omega}{k}A_x = 0. \tag{8.182}$$

We neglect the mixed derivative and cut the term proportional to A, getting the dispersion relation, which originated from Drude formula within the approximation we use in Section 7.6.1, see also [12]

$$k(\omega) = -\frac{pq}{c\omega}. \tag{8.183}$$

For this case, $\frac{\omega}{k} = v_g$ is a group velocity that is negative for a metamaterial. To fix a unique solution of Eq. (8.182), we add the boundary condition:

$$A(0,t) = A_0\exp\left[-\left(\frac{t}{\tau}\right)^2\right]. \tag{8.184}$$

Here, τ characterizes the duration of the wave packet. Next, we conventionally change the variables:

$$\eta = t - \frac{x}{v_g}, \xi = t + \frac{x}{v_g}, \tag{8.185}$$

writing for the operators and amplitude function

$$\partial_t = \partial_\xi + \partial_\eta, \partial_x = \frac{1}{v_g}\partial_\xi - \frac{1}{v_g}\partial_\eta, A(x,t) \to \mathbb{A}(\eta,\xi). \tag{8.186}$$

Equation (8.182) trivializes as

$$2\mathbb{A}_\eta = 0. \tag{8.187}$$

As we've shown, $\mathbb{A}$ is independent on η, hence

$$\mathbb{A} = f(\xi); A(x,t) = f\left(t + \frac{x}{v_g}\right).$$

Accounting for the boundary regime (8.184), we obtain the explicit formula for the Π-wave, propagated to the right,

$$\Pi(x,t) = A_0 \exp\left\{-\left(\frac{t+\frac{x}{v_g}}{\tau}\right)^2 + i(-\frac{pq}{c\omega}x - \omega t)\right\} + c.c., \tag{8.188}$$

which leads to conventional wave packets for typical values for the frequency range, which lies within the admissible terahertz (THz) range of applicability of the approximations we had made.

The wavetrains with other polarization differ only by electric and magnetic fields components numbers as they are prescribed by (8.153). The opposite directed waves are defined by (8.155) and (8.156); its formulas differ from (8.188) only by signs by $\frac{x}{v_g}$.

8.12.2 UNIDIRECTIONAL WAVETRAINS INTERACTION

Let us plug (8.179) into Eq. (8.172) as Π_1 together with a similar expression for Π_2, but change A by B with an account for (8.181) and similar strong inequalities for B. The R.H.S.'s are proportional to

$$\begin{gathered}((A\exp[ikx - i\omega t] + c.c.)^3 + (A\exp[i(kx - \omega t)] + c.c.)(B\exp[ikx - i\omega t] + c.c.)^2),\\ ((B\exp[ikx - i\omega t] + c.c.)^3 + (B\exp[ikx - i\omega t] + c.c.)(A\exp[ikx - i\omega t] + c.c.)^2).\end{gathered} \tag{8.189}$$

Differentiations act only on exponents in the leading order because of strong inequalities (8.181) for A, B. The complete equations read

$$\begin{aligned} ikA_t - i\omega A_x &= \frac{c\omega^6\chi_e^{(3)}}{2p^3q}(3A^2A^* + 2ABB^* + A^*B^2),\\ ikB_t - i\omega B_x &= \frac{c\omega^6\chi_e^{(3)}}{2p^3q}(3B^2B^* + 2BAA^* + B^*A^2). \end{aligned} \tag{8.190}$$

The zero-order term is equal to zero if the frequency ω and the wavenumber k of the carrier is such that (8.183) is still valid: $k(\omega) = -\frac{pq}{c\omega}$. We also conventionally denote

$$v_g = \frac{\omega}{k} = -\frac{c\omega^2}{pq}, \tag{8.191}$$

and leave the only leading nonlinear resonant term in both R.H.S.'s, using the linear independence of exponential functions, having

$$\begin{aligned} iA_t - iv_g A_x &= -\frac{c^2\chi_e^{(3)}\omega^7}{2p^4q^2}(3A^2A^* + 2ABB^* + A^*B^2), \\ iB_t - iv_g B_x &= -\frac{c^2\chi_e^{(3)}\omega^7}{2p^4q^2}(3B^2B^* + 2BAA^* + B^*A^2). \end{aligned} \tag{8.192}$$

This system of equations describes the interaction between orthogonal polarization modes. After transition (8.220) to characteristics, one arrives at a system of nonlinear ODE:

$$\eta = t - \frac{x}{v_g}, \xi = t + \frac{x}{v_g}, \tag{8.193}$$

which can be written for reader convenience as

$$\partial_t = \partial_\xi + \partial_\eta, \partial_x = \frac{1}{v_g}\partial_\xi - \frac{1}{v_g}\partial_\eta, \tag{8.194}$$

arriving at

$$\begin{aligned} A_\eta &= i\frac{c^2\chi_e^{(3)}\omega^7}{4p^4q^2}(3A^2A^* + 2ABB^* + A^*B^2), \\ B_\eta &= i\frac{c^2\chi_e^{(3)}\omega^7}{4p^4q^2}(3B^2B^* + 2BAA^* + B^*A^2). \end{aligned} \tag{8.195}$$

Rescaling variables

$$\mathbb{A} = fa, \mathbb{B} = fb,$$

it is convenient to choose

$$f^2\frac{c^2\chi_e^{(3)}\omega^7}{4p^4q^2} = 1.$$

Finally, we simplify the system (8.195) as

$$\begin{aligned} a_\eta &= i\left(3a^2a^* + 2abb^* + a^*b^2\right), \\ b_\eta &= i\left(3b^2b^* + 2baa^* + b^*a^2\right). \end{aligned} \tag{8.196}$$

Going to new variables $\phi(\eta), \psi(\eta)$, we plug in the first of (8.196)

$$a = a_0 \exp[i\phi(\eta)], \tag{8.197}$$

where a_0 is constant. Similarly, for the second equation of (8.196), we put

$$b = b_0 \exp[i\psi(\eta)], \tag{8.198}$$

with constant b_0. Plugging the results into the original system (8.196), we obtain

$$\begin{aligned} \phi_\eta &= 3a_0a_0^* + 2b_0b_0^* + \tfrac{a_0^*}{a_0}b_0^2\exp[i2(\psi(\eta) - \phi(\eta))], \\ \psi_\eta &= 3b_0b_0^* + 2a_0a_0^* + \tfrac{b_0^*}{b_0}a_0^2\exp[i2(\phi(\eta) - \psi(\eta))]. \end{aligned} \tag{8.199}$$

Subtracting the equations gives

$$\psi_\eta - \phi_\eta = 3a_0a_0^* + 2b_0b_0^* + \frac{a_0^*}{a_0}b_0^2\exp[i2(\psi(\eta) - \phi(\eta))] - 3b_0b_0^* - 2a_0a_0^* - \frac{b_0^*}{b_0}a_0^2\exp[i2(\phi(\eta) - \psi(\eta))] = a_0a_0^* - b_0b_0^* + \frac{1}{a_0}b_0^2a_0^*\exp[i2(\psi(\eta) - \phi(\eta))] - \frac{a_0^2}{b_0}b_0^* \exp[i2(\phi(\eta) - \psi(\eta))].$$

If we denote,

$$2(\psi(\eta) - \phi(\eta)) = s$$

and differentiate, it yields

$$s_\eta = 2a_0a_0^* - b_0b_0^* + \frac{b_0^2}{a_0}a_0^*\exp[is] - \frac{a_0^2}{b_0}b_0^*\exp[-is],$$

introducing

$$\exp[is] = y,$$

one has

$$(\exp[is])_\eta = i\exp[is]s_\eta = y_\eta,$$

$$s_\eta = -i\frac{y_\eta}{y},$$

$$-i\frac{y_\eta}{y} = 2a_0a_0^* - b_0b_0^* + \frac{b_0^2}{a_0}a_0^*y - \frac{a_0^2}{b_0}b_0^*y^{-1},$$

Finally, one arrives at

$$y_\eta = i(2a_0a_0^* - b_0b_0^*)y + i\frac{b_0^2}{a_0}a_0^*y^2 - i\frac{a_0^2}{b_0}b_0^*. \tag{8.200}$$

which is solved in elliptic integrals [29]. The function $\phi(\eta)$ is obtained from the first equation of (8.200) by direct integration

$$\phi = \int [3a_0a_0^* + 2b_0b_0^* + \frac{a_0^*}{a_0}b_0^2y(\eta)]d\eta, \tag{8.201}$$

whereas the second variable is evaluated as

$$\psi(\eta) = \frac{1}{2}[-i\ln(y) + \phi(\eta)]. \tag{8.202}$$

A simple solution may be obtained by choosing the difference of $\psi(\eta)$ and $\phi(\eta)$ as a constant.

Plugging $\psi(\eta) - \phi(\eta) = c/2i$ into both R.H.S.'s of Eq. (8.200) should equalize them, having

$$a_0a_0^* - b_0b_0^* + \frac{b_0^2}{a_0}a_0^*e^c - \frac{a_0^2}{b_0}b_0^*e^{-c} = 0.$$

Let us mark $e^c = z$, rewriting the equation as

$$a_0 a_0^* - b_0 b_0^* + \frac{z}{a_0} b_0^2 a_0^* - \frac{1}{z} \frac{a_0^2}{b_0} b_0^* = \left(a_0^2 + z b_0^2\right) \frac{-a_0 b_0^* + z b_0 a_0^*}{z a_0 b_0} = 0,$$

which we have happily factorized, giving two roots

1. $a_0^2 + z b_0^2 = 0$, the solution is: $z_1 = -\frac{a_0^2}{b_0^2}$
2. $-a_0 b_0^* + z b_0 a_0^* = 0$, the solution is: $z_2 = \frac{a_0}{b_0 a_0^*} b_0^*$.

Continuing as

$$\phi_\eta = 3a_0 a_0^* + 2b_0 b_0^* + \frac{a_0^*}{a_0} b_0^2 z,$$

and integrating

$$\phi = \left(3a_0 a_0^* + 2b_0 b_0^* + \frac{a_0^*}{a_0} b_0^2 z\right) \eta + \phi_0,$$

we write for two cases:

1. $\phi = \phi_0 + 2\eta \left(a_0 a_0^* + b_0 b_0^*\right)$,
2. $\phi = \phi_0 + \eta \left(3a_0 a_0^* + 3b_0 b_0^*\right)$.

Recall that $\psi(\eta) = \phi(\eta) + c/2i$, calculate the function b as

$$b = b_0 \exp[i\psi(\eta)] = b_0 \exp[i(\phi(\eta) + c/2i)] = b_0 \exp[i\phi(\eta)] \exp[\frac{c}{2}]$$
$$= b_0 \sqrt{z} \exp[i\phi(\eta)]$$

and, again for the two cases

$$\begin{aligned} a = a_0 \exp[i\phi(\eta)] = a_0 \exp[i\left(\phi_0 + 2\eta \left(a_0 a_0^* + b_0 b_0^*\right)\right)], \\ b = b_0 \exp[i\psi(\eta)] = b_0 \sqrt{z} \exp[i\phi(\eta)] = \\ a_0 \exp[i(\phi(\eta) + \pi)] = a_0 \exp[i(\phi_0 + 2\eta \left(a_0 a_0^* + b_0 b_0^*\right) + \pi)]. \end{aligned} \tag{8.203}$$

$$b = b_0 \exp[i\psi(\eta)] = \sqrt{\frac{a_0}{a_0^*} b_0 b_0^*} \exp[i\left(\phi_0 + 3\eta \left(a_0 a_0^* + b_0 b_0^*\right)\right)]$$

$$a_0 = r \exp[i\varphi], \sqrt{\frac{a_0}{a_0^*} b_0 b_0^*} = \pm |b_0| \exp[i\varphi]$$

Returning to the original variables, finally, the right waves with orthogonal polarizations are expressed by

$$\begin{aligned} \Pi_1 = f a_0 \exp[i\phi(t - \tfrac{x}{v_g}) \exp[ikx - i\omega t] + c.c, \\ \Pi_2 = f b_0 \exp[i\psi(t - \tfrac{x}{v_g})] \exp[ikx - i\omega t] + c.c. \end{aligned} \tag{8.204}$$

So, the system (8.172) has an approximate explicit solution in terms of elliptic functions (8.201) and (8.202) with appropriate boundary conditions for electric and magnetic fields, as defined by (8.158 and 8.169).

For the simplest solution (case 1) we plug the formulas (8.203) into the envelopes A, B, having

$$\begin{aligned} A &= f a_0 \exp[i(2\,(a_0 a_0^* + b_0 b_0^*)\,(t - \tfrac{x}{v_g})], \\ B &= f b_0 \exp[i(2\,(a_0 a_0^* + b_0 b_0^*)\,(t - \tfrac{x}{v_g}) + \pi)]. \end{aligned} \tag{8.205}$$

The initial phase ϕ_0 is dropped. Next, inserting in Π_i, we arrive at

$$\begin{aligned} \Pi_1 &= f a_0 \exp[i(2\,(a_0 a_0^* + b_0 b_0^*))(t - \tfrac{x}{v_g})] \exp[ikx - i\omega t] + c.c, \\ \Pi_2 &= f b_0 \exp[i(2\,(a_0 a_0^* + b_0 b_0^*)\,(t - \tfrac{x}{v_g}) + \pi)] \exp[ikx - i\omega t] + c.c. \end{aligned} \tag{8.206}$$

Collecting terms proportional to t and x, we can find the frequency and wave vector shifts.

The frequency parameter ω is defined by the initiating boundary frequencies $\omega_{a,b}$ by the relations

$$2\,(a_0 a_0^* + b_0 b_0^*) - \omega = \omega_b, \tag{8.207}$$

with explicit dependence on amplitudes a_0, b_0. The choice of the boundary frequency ω_b defines the wave vector inside the material; hence, both polarizations are defined by the following expressions:

$$k - \frac{2\left(a_0 a_0^* + b_0 b_0^*\right)}{v_g} = k_b \tag{8.208}$$

Let us illustrate the results, plotting both functions a, b as functions of η for both solutions (Figure 8.3).

A comparison and eventual applications of the solutions of the linear (8.182) and the nonlinear (8.192) problems can be provided by a special choice of the boundary conditions. We take the simplest harmonic wave solution of (8.182) with amplitude a_0 and frequency ω' to prescribe the equality of amplitudes at the boundary point $x = 0$. The result points out the differences in phase and group velocities in the metamaterial as functions of the material parameters p, q and the incident waves amplitude.

8.12.3 COUPLED NONLINEAR SCHRODINGER EQUATIONS

We accounted for the next term of expansion by the dispersion parameter, and, in the same approximation, we left an identical nonlinear resonant term on the R.H.S. of both equations of the system (8.192). The nonlinear system for amplitude function A, B introduced by (8.179) for $\Pi_{1,2}$ takes the form

$$\begin{aligned} iA_t - iv_g A_x + \tfrac{k}{\omega} A_{tt} &= -\frac{c^2 \chi_e^{(3)} \omega^7}{2p^4 q^2} (3A^2 A^* + 2ABB^* + A^* B^2), \\ iB_t - iv_g B_x + \tfrac{k}{\omega} B_{tt} &= -\frac{c^2 \chi_e^{(3)} \omega^7}{2p^4 q^2} (3B^2 B^* + 2BAA^* + B^* A^2). \end{aligned} \tag{8.209}$$

This is the typical coupled nonlinear Schrödinger equation (CNSE) [30] with the same L.H.S.. We choose the variables as

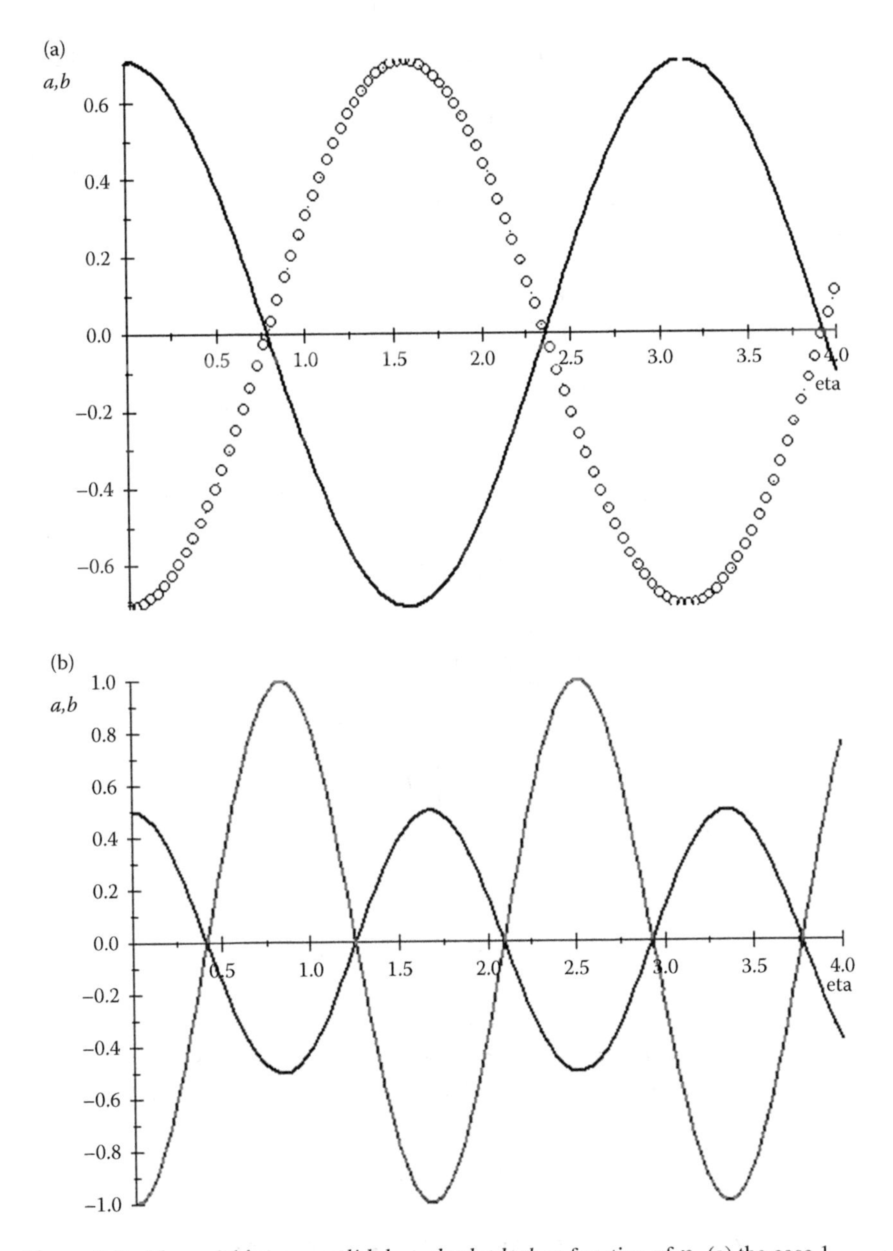

Figure 8.3 The variables a — *solid*, b — *dotdotdash* as function of η. (a) the case 1., (b) the case 2

$$\eta = x, \xi = t - \frac{x}{v_g}, \tag{8.210}$$

$$\partial_x = \partial_\eta + \frac{1}{v_g}\partial_\xi, \partial_t = \partial_\xi, \partial_{tt} = \partial_{\xi\xi}$$
$$A(x,t) \to \mathbb{A}(\eta,\xi), B(x,t) \to \mathbb{B}(\eta,\xi), \tag{8.211}$$

which gives

$$\frac{k}{\omega}\mathbb{A}_{\xi\xi} + i(k\mathbb{A}_\xi - \omega\mathbb{A}_\eta - \frac{\omega}{v_g}\mathbb{A}_\xi) = -\frac{c^2\chi_e^{(3)}\omega^7}{2p^4q^2}(3\mathbb{A}^2\mathbb{A}^* + 2\mathbb{A}\mathbb{B}\mathbb{B}^* + \mathbb{A}^*\mathbb{B}^2),$$

$$\frac{k}{\omega}\mathbb{B}_{\xi\xi} + i(k\mathbb{B}_\xi - \omega\mathbb{B}_\eta - \frac{\omega}{v_g}\mathbb{B}_\xi) = -\frac{c^2\chi_e^{(3)}\omega^7}{2p^4q^2}(3\mathbb{B}^2\mathbb{B}^* + 2\mathbb{B}\mathbb{A}\mathbb{A}^* + \mathbb{B}^*\mathbb{A}^2). \tag{8.212}$$

From this system, we derive by taking v_g, as in the linear case $\frac{\omega}{k}$, such that

$$\frac{k}{\omega}\mathbb{A}_{\xi\xi} - i\omega\mathbb{A}_\eta = -\frac{c^2\chi_e^{(3)}\omega^7}{2p^4q^2}(3\mathbb{A}^2\mathbb{A}^* + 2\mathbb{A}\mathbb{B}\mathbb{B}^* + \mathbb{A}^*\mathbb{B}^2),$$
$$\frac{k}{\omega}\mathbb{B}_{\xi\xi} - i\omega\mathbb{B}_\eta = -\frac{c^2\chi_e^{(3)}\omega^7}{2p^4q^2}(3\mathbb{B}^2\mathbb{B}^* + 2\mathbb{B}\mathbb{A}\mathbb{A}^* + \mathbb{B}^*\mathbb{A}^2). \tag{8.213}$$

Rescaling as follows

$$z = -\frac{k}{\omega^2}\eta,$$

with $k = -\frac{c\omega}{pq}$, leads to

$$ia_z + a_{\xi\xi} = a(3aa^* + 2bb^*) + b^*a^2,$$
$$ib_z + b_{\xi\xi} = b(3bb^* + 2aa^*) + a^*b^2 \tag{8.214}$$

Here, $\gamma = \frac{c^2\chi_e^{(3)}\omega^7}{2p^3q}, a = \frac{\mathbb{A}}{\sqrt{\gamma}}, b = \frac{\mathbb{B}}{\sqrt{\gamma}}$. The system is the particular case of general NLSE. The effective stable and convergent numerical scheme of its integration was written by Leble, Reichel in [30]. The reductions $b = a = u$ results in the celebrated nonlinear Schrödinger equation [31]

$$iu_z + u_{\xi\xi} = 6u|u|^2. \tag{8.215}$$

8.13 STATIONARY SOLUTIONS OF SPE SYSTEM FOR UNIDIRECTIONAL WAVES

Let us return to the SPE system (8.174) and, after a bit, rearrange

$$\pi_1^{-1}\partial^{t-2}(\partial_{\chi t}\pi_1 + \pi_1) = \pi_1^2 + \pi_2^2,$$
$$\pi_2^{-1}\partial^{t-2}(\partial_{\chi t}\pi_2 + \pi_2) = \pi_2^2 + \pi_1^2, \tag{8.216}$$

with the obvious bilinear corollary

$$\pi_1^{-1}\partial_t^{-2}(\partial_{\chi t}\pi_1+\pi_1)=\pi_2^{-1}\partial_t^{-2}(\partial_{\chi t}\pi_2+\pi_2)=u(x,t). \tag{8.217}$$

It reads as two equivalent linear Laplace-like equations:

$$\begin{aligned}\partial_{\chi t}\pi_1+\pi_1&=\partial_t^2 u(x,t)\pi_1,\\ \partial_{\chi t}\pi_2+\pi_2&=\partial_t^2 u(x,t)\pi_2,\end{aligned} \tag{8.218}$$

with the same potential $u(x,t)$. It opens a possibility to apply a dressing method [32] with reductions that follow from (8.216).

Returning back to (8.172), we repeat

$$\begin{aligned}\frac{c}{pq}\partial_{xt}\Pi_1+\Pi_1&=\varepsilon((\partial_t^2\Pi_1)^3+(\partial_t^2\Pi_1)(\partial_t^2\Pi_2)^2),\\ \partial_{xt}\Pi_2+\frac{c}{pq}\Pi_2&=\varepsilon((\partial_t^2\Pi_2)^3+(\partial_t^2\Pi_2)(\partial_t^2\Pi_1)^2).\end{aligned} \tag{8.219}$$

with notation

$$\frac{c^2\chi_e^{(3)}}{2p^4q^2}=\varepsilon$$

Next, we change the variables

$$\eta=x,\xi=t+x/v, \tag{8.220}$$

$v=\frac{c}{pq}$, and the derivatives operators

$$\begin{aligned}\partial_t&=\partial_\xi,\partial_\chi=\partial_\eta+\frac{1}{v}\partial_\xi,\\ \partial_{xt}&=\partial_\xi(\partial_\eta+\frac{1}{v}\partial_\xi),\end{aligned} \tag{8.221}$$

preserving the same notations for Π_i. The structure of the system (8.174), which is similar to one for a case of one polarization. Then, declaring the independence of Π_i on η, denoting $\partial_{\xi\xi}\Pi_i=\Pi_i''$ for the traveling wave solutions, we write

$$\begin{aligned}\Pi_1''+\Pi_1&=\varepsilon((\Pi_1)^3+(\Pi_1)(\Pi_2)^2),\\ \Pi_2''+\Pi_2&=\varepsilon((\Pi_2)^3+(\Pi_2)(\Pi_1)^2).\end{aligned} \tag{8.222}$$

A reduction $\Pi_2=a\Pi_1$ with a real a sends both equations to

$$\Pi_1''+\Pi_1=\varepsilon((\Pi_1)^3+(\Pi_1)(a\Pi_1)^2), \tag{8.223}$$

Rescaling again by amplitude factor $\Pi_1 = \frac{f}{\sqrt{\varepsilon}} z$ and $\xi = d\zeta$, and use the small amlitude approximation in the R.H.S. $z'' \approx -d^2 z$ yields

$$z'' + d^2 z + d^2(1+a^2) f^2 z^3 = 0. \tag{8.224}$$

Comparison with the equation of elliptic cosinus $cn(x,m)$:

$$z'' + (1+m^2)z + 2m^2 z^3 = 0, \tag{8.225}$$

leads to an overdetermined system relative to m.

$$\begin{aligned} 1+m^2 &= d^2, \\ 2m^2 &= d^2(1+a^2) f^2. \end{aligned} \tag{8.226}$$

Further, it links the amplitude f and scale d parameters

$$2 + d^2[(1+a^2) f^2 - 2] = 0. \tag{8.227}$$

Finally, the wave field π_1 with polarization along y is expressed via Jacobi ellipticsn(ζ, m):

$$\Pi_1 = \frac{f}{\sqrt{\varepsilon}} \mathrm{sn}\left(\frac{\xi}{d}, m\right) \tag{8.228}$$

with $m = df\sqrt{\frac{1+a^2}{2}}$, $d = \sqrt{\frac{2}{2-(1+a^2)f^2}}$ in the amplitude range $f^2 < \frac{2}{(1+a^2)}$. The second field within the reduction under consideration is proportional to the first one $\Pi_2 = a\Pi_1$.

REFERENCES

1. Smith, D.R., W.J. Padilla, D.C. Vier, S.C. Nemat-Nasser, and S. Schultz. 2000. Composite medium with simultaneously negative permeability and permittivity. *Physical Review Letters* 84: 41844187.
2. Shelby, R.A., D.R. Smith, and S. Schultz. 2001. Experimental verification of a negative index of refraction. *Science* 292(5514): 77–79.
3. Veselago, V.G. 1966. FTT 8: 3571; 1967. *Soviet Physics, Solid State* 8: 2853. Veselago, V.G. 1968. The electrodynamics of substances with simultaneously negative values of ε and μ. *Soviet Physics Uspekhi* 10: 509514.
4. Malyuzhinets, G.D. 1951. A note on the radiation principle. *Zhurnal Technicheskoi Fiziki* 21: 940–942 (in Russian. English translation in Soviet Physics Technical Physics).
5. Slusar, V. 2009. Metamaterials in antennas: History and basic principles. *Electonics: NTB* 7: 64–74 (in Russian).
6. Sihvola, A. 2007. Metamaterials in electromagnetics. *Metamaterials* 1: 2–11.
7. Lee, J., M. Tymchenko, C. Argyropoulos, P.Y. Chen, F. Lu, F. Demmerle, G. Boehm, M.C. Amann, A. Alu, and M.A. Belkin. 2014. Giant nonlinear response from plasmonic metasurfaces coupled to intersubband transitions. *Nature*. DOI:10.1038/nature13455.
8. Yoshizawa, T. ed. 2015. *Handbook of Optical Metrology: Principles and Applications*, 2nd ed., CRC Press, Boca Raton, FL.

9. Islam, R., F. Eleck, and G.V. Eleftheriades. 2004. Coupled-line metamaterial coupler having co-directional phase but contra-directional power flow. *Electronics Letters* 40(5): 315–317.
10. Collin, R.E. 1991. *Field Theory of Guided Waves*, 2nd ed., IEEE Press, New York.
11. Alu, A., and N. Engheta. 2004. Guided modes in a waveguide filled with a pair of single-negative (SNG), double-negative (DNG), and/or double-positive (DPS) layers. *IEEE Transactions on Microwave Theory and Techniques* 52(1): 199–210.
12. Leble, S. and D. Ampilogov. 2016. Directed electromagnetic wave propagation in 1D metamaterial: Projecting operators method. *Physics Letters A* 380(29–30): 2271–2278.
13. Schäfer, T., and G.E. Wayne. 2002. *Propagation of Ultra-Short Optical Pulses in Nonlinear Media*. Elsevier Science.
14. Kuszner, M. and S. Leble. 2011. Directed electromagnetic pulse dynamics: Projecting operators method. *Journal of the Physical Society of Japan* 80: 024002.
15. Boyd, R.W. 1992. *Nonlinear Optics*. Academic Press, Boston.
16. Wen, S., Y. Xiang, X. Dai, Z. Tang, W. Su, and D. Fan. 2007. Theoretical models for ultrashort electromagnetic pulse propagation in nonlinear metamaterials. *Physical Review A* 75: 033815.
17. Engheta, N., and R.W. Ziolkowski. 2005. A positive future for double-negative metamaterials. *IEEE Transactions on Microwave Theory and Techniques* 53(4): 1535–1556.
18. Pietrzyk, M., and I. Kanattsikov. 2015. Generalized short pulse equation for propagation of few-cycle pulses in metamaterials: 1–4. arXiv:1512.09322.
19. Lagarkov, A.N., V.N. Kisel, A.K. Sarychev, and V.N. Semenenko. 2010. Electrophysics and electrodynamics of metamaterials. *High Temperature* 48: 983. https://doi.org/10.1134/S0018151X10060258. Original Russian Text Lagarkov, A.N., V.N. Kisel, A.K. Sarychev, and V.N. Semenenko. 2010, published in Teplofizika Vysokikh Temperatur 48(6): 1031–1048.
20. Chen, L., W. Ding, X.-J. Dang, and C.-H. Liang. 2007. Counter-propagating energy-flows in nonlinear left-handed metamaterials. *Progress in Electromagnetics Research, PIER* 70: 257–267.
21. Argyropoulos, C. et al. Enhanced nonlinear effects in metamaterials and plasmonics. *Advanced Electromagnetics* 1(1): 46–51.
22. Ipatov, M., V. Zhukova, L.V. Panina, and A. Zhukov. Ferromagnetic microwires composite metamaterials with tuneable microwave electromagnetic parameters. *PIERS Online* 5(6): 586–590.
23. Sakovich, S. 2008. Integrability of the vector short pulse equation. *Journal of the Physical Society of Japan* 77: 123001.
24. Sakovich, A., and S. Sakovich. 2006. Solitary wave solutions of the short pulse equation. *Journal of Physics A: Mathematical and General* 39: L361–L367.
25. Liu, Y., D. Pelinovski, and A. Sakovich. 2009. Wave breaking in the short-pulse equation. *Dynamics of PDE* 6: 291310.
26. Pelinovski, D., and A. Sakovich. 2010. Global well-posedness of the short-pulse and Sine–Gordon equations in energy space. *Communications in PDE* 35: 613–629.
27. Kuszner, M., and S. Leble. 2014. Ultrashort opposite directed pulses dynamics with Kerr effect and polarization account. *Journal of the Physical Society of Japan* 83: 034005.
28. Pietrzyk, M., I. Kanattsikov, and U. Bandelow. 2008. On the propagation of vector ultra-short pulses. *Journal of Nonlinear Mathematical Physics* 15: 2.
29. Baker, A. 2014. *Elliptic Functions: An Elementary Text-Book for Students of Mathematics*. John Wiley & Sons, New York.

30. Leble, S., and B. Reichel. 2008. On convergence and stability of a numerical scheme of coupled nonlinear Schrodinger equations. *Computers and Mathematics with Applications* 55: 745–759.
31. Zakharov, V.E. 1968. Stability of periodic waves of finite amplitude on the surface of a deep fluid. *Journal of Applied Mechanics and Technical Physics* 9: 2, Springer, New York, 190.
32. Doktorov, E., and S.B. Leble. 2007. *Dressing Method in Mathematical Physics.* Springer-Verlag, Berlin.

9 Waves in Waveguides

The case of a waveguide propagation, which is generally a 3D problem, differs from propagation in open space by the presence of boundaries, which restrict the wave presence by some domain. The mathematical description, then, includes boundary conditions that could be satisfied by some elementary or special functions depending on waveguide geometry and the character of the boundaries. "Therefore, as a subdivision of longitudinal and transversal coordinates is used." The conventional approach to the problem's solution consists of an expansion of a state vector on the basis built from the eigenfunction of a transversal coordinate operator and a Fourier transformation in longitudinal coordinate or time [1].

9.1 ELECTROMAGNETIC WAVES IN METAL RECTANGULAR WAVEGUIDE FILLED WITH A MATERIAL: PROJECTING OPERATORS METHOD

The general wave equations for an electromagnetic field in a metal rectangular waveguide with a metamaterial inside are derived by the projecting operators method. The equations form a system for wave modes with the opposite direction of propagation, in which interaction arises from nonlinearity [3,4]. Examples of Drude models for both permittivities and Kerr nonlinearity are studied as examples.

9.1.1 MAXWELL'S EQUATIONS FOR A WAVEGUIDE. BOUNDARY CONDITIONS

For reader convenience, we reproduce here the system of Maxwell equations for a medium inside a waveguide; it is now written in SI units:

$$\mathrm{div}\vec{D} = 0, \tag{9.1}$$

$$\mathrm{div}\vec{B} = 0, \tag{9.2}$$

$$\mathrm{rot}\vec{E} = -\frac{\partial \vec{B}}{\partial t}, \tag{9.3}$$

$$\mathrm{rot}\vec{H} = \frac{\partial \vec{D}}{\partial t}. \tag{9.4}$$

The material relation we write as in Section 7.3.2, taking the isotropic case for simplicity:

$$\vec{D} = \varepsilon_0 \hat{\varepsilon} \vec{E}, \tag{9.5}$$

$$\vec{H} = \frac{1}{\mu_0} \hat{\mu}^{-1} \vec{B}.$$

Here, the operators $\hat{\varepsilon}$ and $\hat{\mu}$ act as integral ones

$$\hat{\varepsilon}\vec{E} = \int_{-\infty}^{\infty} \tilde{\varepsilon}(t-s)\vec{E}(s)ds,\ \hat{\mu}^{-1}\vec{B} = \frac{1}{\sqrt{2\pi}}\int_{-\infty}^{\infty} \tilde{\mu}^{-1}(t-\tau)\vec{B}(\tau)d\tau, \tag{9.6}$$

where

$$\tilde{\varepsilon}(t-s) = \frac{1}{2\pi}\int_{-\infty}^{\infty} \varepsilon(\omega)\exp(i\omega(t-s))d\omega, \tag{9.7}$$

$$\tilde{\mu}(t-\tau) = \frac{1}{\sqrt{2\pi}}\int_{-\infty}^{\infty} \mu(\omega)\exp(i\omega(t-s))d\omega, \tag{9.8}$$

Here, ε_0, μ_0 are vacuum SI constants.

The Ampere equation of the Maxwell system is transformed as

$$\mathrm{rot}\vec{B} = \frac{\hat{\varepsilon}\hat{\mu}}{c^2}\frac{\partial \vec{E}}{\partial t}. \tag{9.9}$$

On the walls of the waveguide, we suppose that the following boundary conditions hold

$$E_{x,y}\big|_{z=0,b} = 0, \qquad E_{x,z}\big|_{y=0,b} = 0, \tag{9.10}$$

$$H_z\big|_{z=0,a} = 0, \qquad H_y\big|_{y=0,a} = 0. \tag{9.11}$$

The complete problem formulation needs either initial or boundary regime conditions. These are formulated in the next section.

9.1.2 THE TRANSVERSAL WAVEGUIDE MODES EVOLUTION

The transversal waveguide modes we use have the standard form [1]. The expansion for an x-component of the electric field vector looks like

$$E_x = \sum_{n,m} \varphi_{n,m}(x,t)\sin\frac{\pi m}{a}y\sin\frac{\pi n}{b}z. \tag{9.12}$$

The rest of the components of the electric field should be chosen as

$$E_y = \sum_{n,m} \chi_{n,m}(x,t)\cos\frac{\pi m}{a}y\sin\frac{\pi n}{b}z, \tag{9.13}$$

$$E_z = \sum_{n,m} \psi_{n,m}(x,t)\sin\frac{\pi m}{a}y\cos\frac{\pi n}{b}z, \tag{9.14}$$

to adjust them to the boundary conditions (9.10). The magnetic field expansions should account for the Faraday equation (9.3) and should naturally fit the boundaries (9.11).

$$B_x = \sum_{n,m} \alpha_{n,m}(x,t) \cos\frac{\pi m}{a} y \cos\frac{\pi n}{b} z,$$
$$B_y = \sum_{n,m} \beta_{n,m}(x,t) \sin\frac{\pi m}{a} y \cos\frac{\pi n}{b} z, \quad (9.15)$$
$$B_z = \sum_{n,m} \gamma_{n,m}(x,t) \cos\frac{\pi m}{a} y \sin\frac{\pi n}{b} z.$$

Plugging the expansions (9.12–9.15) into the Maxwell system (9.1–9.4), omitting indexes, yields

$$\frac{\partial \varphi}{\partial x} - \frac{\pi m}{a}\chi - \frac{\pi n}{b}\psi = 0, \quad (9.16)$$

$$\frac{\partial \alpha}{\partial x} + \frac{\pi m}{a}\beta + \frac{\pi n}{b}\gamma = 0. \quad (9.17)$$

For the dynamic part of magnetic field, we obtain

$$\frac{\pi m}{a}\psi - \frac{\pi n}{b}\chi + \frac{\partial \alpha}{\partial t} = 0,$$
$$\frac{\pi n}{b}\varphi - \frac{\partial \psi}{\partial x} + \frac{\partial \beta}{\partial t} = 0, \quad (9.18)$$
$$\frac{\partial \chi}{\partial x} - \frac{\pi m}{a}\varphi + \frac{\partial \gamma}{\partial t} = 0.$$

The electric field evolution is described by

$$-\frac{\pi m}{a}\gamma + \frac{\pi n}{b}\beta - \frac{\hat{\varepsilon}\hat{\mu}}{c^2}\frac{\partial \varphi}{\partial t} = 0,$$
$$-\frac{\pi n}{b}\alpha - \frac{\partial \gamma}{\partial x} - \frac{\hat{\varepsilon}\hat{\mu}}{c^2}\frac{\partial \chi}{\partial t} = 0, \quad (9.19)$$
$$\frac{\partial \beta}{\partial x} - \frac{\pi m}{a}\alpha - \frac{\hat{\varepsilon}\hat{\mu}}{c^2}\frac{\partial \psi}{\partial t} = 0.$$

Equation (9.19) differ from the ones from [2] only by the presence of $\widehat{\mu}$, notations, units, and hats over ε and μ because we account for a dispersion now.

Excluding ψ, γ by the relations (9.16) and (9.17), we arrive at equations similar to the ones in [2], Eq. (3.18):

$$\frac{\partial \alpha}{\partial t} = q\chi - p\frac{\partial \varphi}{\partial x}, \tag{9.20}$$

$$\frac{\partial \beta}{\partial t} = \frac{1}{rp}\frac{\partial^2 \varphi}{\partial x^2} - \frac{r}{p}\varphi - p\frac{\partial \chi}{\partial x}, \tag{9.21}$$

$$\frac{\hat{\varepsilon}\hat{\mu}}{c^2}\frac{\partial \varphi}{\partial t} = p\frac{\partial \alpha}{\partial x} + q\beta, \tag{9.22}$$

$$\frac{\hat{\varepsilon}\hat{\mu}}{c^2}\frac{\partial \chi}{\partial t} = -r\alpha + r^{-1}\frac{\partial^2 \alpha}{\partial x^2} + p\frac{\partial \beta}{\partial x}, \tag{9.23}$$

where the constants are expressed in terms of the waveguide dimension parameters and the transverse mode numbers m, n:

$$p \equiv \frac{bm}{an}, \; q \equiv \pi\left(\frac{m^2 b}{a^2 n} + \frac{\pi n}{b}\right), \; r \equiv \frac{\pi n}{b}.$$

Recall the notation (8.121) from Section 8.8:

$$\frac{\hat{\varepsilon}\hat{\mu}}{c^2} = \hat{a}^2,$$

writing the last equations of the set (9.20) in standard form

$$\frac{\partial \varphi}{\partial t} = \hat{a}^{-2}\left(p\frac{\partial \alpha}{\partial x} + q\beta\right), \tag{9.24}$$

$$\frac{\partial \chi}{\partial t} = \hat{a}^{-2}\left(r\alpha + r^{-1}\frac{\partial^2 \alpha}{\partial x^2} + s\frac{\partial \beta}{\partial x}\right). \tag{9.25}$$

The equations are rewritten to accomplish the system for the Cauchy problem formulation. The standard actions in the spirit of this book produce the projecting operators to split the initial problem.

The geometry of a waveguide and physics imply a boundary regime propagation rather than a Cauchy problem: suppose the excitation is realized in a vicinity of one or both ends of the waveguides. Therefore, we need a reformulation of the problem in the spirit of Section 2.1.3. Equations (9.20)–(9.23) we would transform as follows. Start with (9.20)–(9.23) with constant ε, μ; one can easily put hats over the material parameters.

Transforming as

$$\frac{\partial \alpha}{\partial x} = \frac{\widehat{a}^2}{p}\frac{\partial \varphi}{\partial t} - \frac{q}{p}\beta, \tag{9.26}$$

$$\frac{\partial \chi}{\partial x} = -\frac{r}{p}\varphi + \frac{1}{rp}\frac{\partial^2 \varphi}{\partial x^2} - \frac{1}{p}\frac{\partial \beta}{\partial t}, \tag{9.27}$$

$$\frac{\partial \beta}{\partial x} = \frac{r}{p}\alpha - \frac{1}{rp}\frac{\partial^2 \alpha}{\partial x^2} + \frac{\widehat{a}^2}{p}\frac{\partial \chi}{\partial t}, \tag{9.28}$$

$$\frac{\partial \varphi}{\partial x} = \frac{q}{p}\chi - \frac{1}{p}\frac{\partial \alpha}{\partial t}, \tag{9.29}$$

and combining last and first equations yields

$$\frac{\partial^2 \varphi}{\partial x^2} = \frac{\partial}{\partial x}\left(\frac{\partial \varphi}{\partial x}\right) = \frac{q}{p}\frac{\partial \chi}{\partial x} - \frac{1}{p}\frac{\partial}{\partial t}\left(\frac{\widehat{a}^2}{p}\frac{\partial \varphi}{\partial t} - \frac{q}{p}\beta\right) \tag{9.30}$$

Let us use again the shorthands as

$$\partial_x \equiv \frac{\partial}{\partial x}, \partial_t \equiv \frac{\partial}{\partial t}, \partial_{tt} \equiv \frac{\partial^2}{\partial t^2}, \partial_{xx} \equiv \frac{\partial^2}{\partial x^2},$$

and, also, go to notations of the partial derivatives by the indices, as, for example, $U_x = \frac{\partial U}{\partial x}$, rewriting (9.30) as

$$\varphi_{xx} = \frac{q}{p}\chi_x - \frac{\widehat{a}^2}{p^2}\varphi_{tt} + \frac{q}{p^2}\beta.$$

Plugging in (9.27)

$$\chi_x = -\frac{r}{p}\varphi + \frac{q}{rp^2}\chi_x - \frac{\widehat{a}^2}{rp^3}\varphi_{tt} + \frac{q}{rp^3}\beta - \frac{1}{p}\beta_t,$$

which immediately, with extra notation

$$rp = \frac{\pi m}{a} = s,$$

gives

$$\chi_x\left(1 - \frac{q}{sp}\right) = -\frac{r}{p}\varphi - \frac{\widehat{a}^2}{sp^2}\varphi_{tt} + \frac{q}{sp^2}\beta - \frac{1}{p}\beta_t \tag{9.31}$$

Similarly, from (9.26) and (9.28),

$$\alpha_{xx} = \partial_x\alpha_x = \partial_x\left(\frac{\widehat{a}^2}{p}\frac{\partial \varphi}{\partial t} - \frac{q}{p}\beta\right) = \frac{\widehat{a}^2}{p}\varphi_{tx} - \frac{q}{p}\beta_x. \tag{9.32}$$

Then, using Eq. (9.29)

$$\varphi_{xt} = \frac{q}{p}\chi_t - \frac{1}{p}\alpha_{tt},$$

and plugging into (9.32)

$$\alpha_{xx} = \frac{\hat{a}^2}{p}\left(\frac{q}{p}\chi_t - \frac{1}{p}\alpha_{tt}\right) - \frac{q}{p}\beta_x,$$

next, in (9.28)

$$\beta_x = \frac{r}{p}\alpha - \frac{1}{s}\left(\frac{\hat{a}^2}{p}\left(\frac{q}{p}\chi_t - \frac{1}{p}\alpha_{tt}\right) - \frac{q}{p}\beta_x\right) + \frac{\hat{a}^2}{p}\chi_t$$

further,

$$\left(1 - \frac{q}{rp^2}\right)\beta_x = \frac{r}{p}\alpha - \frac{1}{s}\left(\frac{\hat{a}^2}{p}\frac{q}{p}\chi_t - \frac{1}{p}\alpha_{tt}\right) + \frac{\hat{a}^2}{p}\chi_t,$$

or

$$\beta_x = \frac{1}{1 - \frac{q}{rp^2}}\left(\frac{r}{p}\alpha + \frac{\hat{a}^2}{p}\chi_t - \frac{\hat{a}^2}{sp}\frac{q}{p}\chi_t - \frac{1}{sp}\alpha_{tt}\right).$$

This finalized the x-evolution system formation (the waveguide dispersion is observed at the R.H.S.'s):

$$\alpha_x = -\frac{q}{p}\beta + \frac{\hat{a}^2}{p}\partial_t\varphi, \tag{9.33}$$

$$\chi_x = \left(1 - \frac{q}{sp}\right)^{-1}[(\frac{q}{p^2} - \frac{1}{p}\partial_t)\beta + \left(-\frac{r}{p} - \frac{\hat{a}^2}{p^2}\partial_{tt}\right)\varphi], \tag{9.34}$$

$$\beta_x = \frac{1}{1 - \frac{q}{sp}}\left[\left(\frac{r}{p} - \frac{1}{sp}\partial_{tt}\right)\alpha + \frac{\hat{a}^2}{p}\left(1 - \frac{q}{sp}\right)\partial_t\chi\right], \tag{9.35}$$

$$\varphi_x = -\frac{1}{p}\partial_t\alpha + \frac{q}{p}\chi, \tag{9.36}$$

Going to matrix operators notations, as used through the book, we write

$$\frac{\partial\Psi}{\partial x} = L\Psi. \tag{9.37}$$

With the standard Ψ as the column

$$\Psi = \begin{pmatrix} \alpha \\ \chi \\ \beta \\ \varphi \end{pmatrix}, \tag{9.38}$$

and with the matrix operator L as

$$L = \begin{pmatrix} 0 & 0 & -\frac{q}{p} & \frac{\hat{a}^2}{p}\partial_t \\ 0 & 0 & \frac{q}{p^2\nu} - \frac{1}{p\nu}\partial_t & -\frac{r}{p\nu} - \frac{\hat{a}^2}{p^2\nu}\partial_{tt} \\ \frac{r}{p\nu} - \frac{1}{sp\nu}\partial_{tt} & \frac{\hat{a}^2}{p\nu}(1 - \frac{q}{sp})\partial_t & 0 & 0 \\ -\frac{1}{p}\partial_t & \frac{q}{p} & 0 & 0 \end{pmatrix}. \tag{9.39}$$

where

$$\nu = 1 - \frac{q}{sp},$$

Delivering the Fourier transformation for the four basic variables, $\varphi(x,t)$, $\psi(x,t)$, $\alpha(x,t)$ and $\beta(x,t)$, presents as

$$\varphi(x,t) = \frac{1}{\sqrt{2\pi}} \int \hat{\varphi}(\omega,x) \exp(i\omega t)\, dx, \tag{9.40}$$

$$\chi(x,t) = \frac{1}{\sqrt{2\pi}} \int \hat{\chi}(\omega,x) \exp(i\omega t)\, dx, \tag{9.41}$$

$$\alpha(x,t) = \frac{1}{\sqrt{2\pi}} \int \hat{\alpha}(\omega,x) \exp(i\omega t)\, dx, \tag{9.42}$$

$$\beta(x,t) = \frac{1}{\sqrt{2\pi}} \int \hat{\beta}(\omega,x) \exp(i\omega t)\, dx. \tag{9.43}$$

The Fourier transform of the vector-column (9.38), looks now as $\tilde{\Psi}$

$$\tilde{\Psi} = \begin{pmatrix} \hat{\alpha} \\ \hat{\chi} \\ \hat{\beta} \\ \hat{\varphi} \end{pmatrix}. \tag{9.44}$$

and the matrix $\hat{L}$ represents the operator (9.39):

$$\hat{L} = \begin{pmatrix} 0 & 0 & -\frac{q}{p} & \frac{\varepsilon\mu}{pi\omega} \\ 0 & 0 & \frac{q}{p^2\nu} - \frac{i\omega}{p\nu} & -\frac{r}{p\nu} + \frac{\varepsilon\mu\omega^2}{p^2\nu} \\ \frac{r}{p\nu} + \frac{\omega^2}{sp\nu} & \frac{\varepsilon\mu}{p\nu}(1 - \frac{i\omega q}{sp}) & 0 & 0 \\ -\frac{i\omega}{p} & \frac{q}{p} & 0 & 0 \end{pmatrix} \tag{9.45}$$

arguments in $\varepsilon(\omega)$ and in $\mu(\omega)$ are omitted. Along the general scheme, consider the spectral problem for the x-evolution matrix operator $\hat{L}$ (9.45), parametrized by the frequncy ω,

$$\hat{L}\Omega = \Omega\Lambda. \tag{9.46}$$

for the matrix of solutions Ω , and the matrix of eigenvalues:

$$\Lambda = diag\lambda_i \tag{9.47}$$

The structure of the matrix $\widehat{L}$ admits a division of the vector space into two block subspaces so as

$$\widehat{L} = \begin{pmatrix} 0 & B \\ C & 0 \end{pmatrix}. \tag{9.48}$$

The column vectors

$$\Xi_1 = \begin{pmatrix} 1 \\ \Omega_2 \end{pmatrix}, \Xi_2 = \begin{pmatrix} \Omega_3 \\ \Omega_4 \end{pmatrix}$$

satisfy the equation

$$\widehat{L}\Omega = \begin{pmatrix} B\Xi_2 \\ C\Xi_1 \end{pmatrix} = \lambda \begin{pmatrix} \Xi_1 \\ \Xi_2 \end{pmatrix}, \tag{9.49}$$

that is equivalent to

$$BC\Xi_1 = \lambda^2 \Xi_1. \tag{9.50}$$

The condition of a nonzero solution existence reads

$$\det(BC - \lambda^2 \widehat{I}) = 0. \tag{9.51}$$

It is equivalent to the bi-quadratic equation

$$\lambda^4 + b\lambda^2 + d = 0, \tag{9.52}$$

where the coefficients b and d are expressed via the Wiet theorem; see the Appendix 9.7. The vectors components we obtain, reading the first line of (9.185), which allows to express the only non-unit component Ω_2 as

$$\Omega_2 = (\lambda^2 - (BC)_{11})/(BC)_{12}. \tag{9.53}$$

Having Ξ_1, we calculate the components of Ξ_2, plugging for each vector $i = 1,2,3,4$ the corresponding eigenvalue λ_i as

$$\Xi_2 = \lambda^{-1} C\Xi_1. \tag{9.54}$$

To have a view of the whole eigenvalue problem for the matrix operator $\widehat{L}$ in explicit form, look at the equality:

$$\begin{pmatrix} 0 & 0 & -\frac{q}{p} & \frac{\varepsilon\mu}{pi\omega} \\ 0 & 0 & \frac{q}{p^2 v} - \frac{i\omega}{pv} & -\frac{r}{pv} + \frac{\varepsilon\mu\omega^2}{p^2 v} \\ \frac{r}{pv} + \frac{\omega^2}{spv} & \frac{\varepsilon\mu}{pv}(1 - \frac{i\omega q}{sp}) & 0 & 0 \\ -\frac{i\omega}{p} & \frac{q}{p} & 0 & 0 \end{pmatrix} \tag{9.55}$$
$$\Omega = \begin{pmatrix} 1 & 1 & 1 & 1 \\ \Omega_{21} & \Omega_{22} & \Omega_{23} & \Omega_{24} \\ \Omega_{31} & \Omega_{32} & \Omega_{33} & \Omega_{34} \\ \Omega_{41} & \Omega_{42} & \Omega_{43} & \Omega_{44} \end{pmatrix} \begin{pmatrix} \lambda_1 & 0 & 0 & 0 \\ 0 & \lambda_2 & 0 & 0 \\ 0 & 0 & \lambda_3 & 0 \\ 0 & 0 & 0 & \lambda_4 \end{pmatrix}$$

9.2 PROJECTING OPERATORS

The matrix Ω, calculated for four eigenvalues, gives the direct way to construct the corresponding projecting operators by the basic formula (2.13). In our notations, it reads

$$P_{ik}^{(j)} = \Omega_{ij}\Omega_{jk}^{-1}, \quad i,j,k = 1,2,3,4. \tag{9.56}$$

Such formalism may be used in practice the direct application of a program's symbolic computations. In the example calculations, we use the SWP program. We, however, in the Appendix 9.7, give explicit formulas for the operators, using the special form of the evolution matrix (9.39) and the normalization choice of the eigenvectors matrix Ω as in (9.55). The dynamic projection operators are obtained by transition to t-domain by Fourier transformation, prolongating the functions to the whole t- and ω-axis similar to previous sections, where 1+1 case was considered.

9.3 POLARIZATIONS AND DIRECTED MODES IN RECTANGULAR WAVEGUIDES

The roots of Eq. (9.52) classify all possible modes that can be exited in a rectangular waveguide. If, as in Appendix 9.7 fix $(\lambda^2)_i$, the magnetic (TM) or electric (TE) polarization mode is specified. For each such root we have two possible values of $\lambda_\pm$ that correspond to the opposite directions of propagation. Properties of the waves depend on the waveguide dispersion, which is explicitly, but cumbersome described by the functions of frequency ω. We however do not put it here, because the dependence is rather complicated. Note that the formula contains a waveguide dimensions. The material dispersion also included in the description via material parameters $\varepsilon(\omega)$ and $\mu(\omega)$, which may be chosen in the modeling of a wave excitation and propagation. In the preceded chapter, the Drude dispersion for a metamaterial was introduced [5]. It can be exploited here as well.

9.4 CYLINDRICAL DIELECTRIC WAVEGUIDES

9.4.1 ON TRANSVERSAL FIBER MODES

In this section, we consider a problem directly related to the fiber optics problem, applying the same algorithm as in the previous one. We use the complete field description [15], apart from the standard mode distribution of this field, producing the basis of separated transversal and longitudinal z variables, implying a standard Sturm-Liouville problem for the transverse variables. Starting from an expansion of the fields under consideration with respect to the transversal coordinates basis, going in the next step to dynamic projecting in time or longitudinal coordinate evolution within a transversal mode subspace. We limit our consideration to a dielectric cylindric waveguide with nonlinearities arising only from the third-order dielectric susceptibility. The dielectric susceptibility for silica as an isotropic medium is represented only by one component, $\chi_{xxxx}^{(3)}$, of the whole susceptibility tensor. This assumption restricts the area of time of the pulse to pulses longer than 0.1ps, which means an immediate response of the medium and it is a standard model for communication fibers in this range. We match solutions of the transversal Hondros-Debye problem [1] (in polar variables r, φ) inside and outside infinite waveguide at boundary $r = r_0$, getting the equation for the eigenvalues α_{ln}.

9.4.2 A FORMULATION AND SOLUTION OF LINEAR PROBLEM, A STEP TO DYNAMIC PROJECTING PROCEDURE

The cylindrical geometry of the problem implies a corresponding basis choice. A division of variables in terms of the longitudinal coordinate s and the transversal ones r, φ leads to a differential equation in transversal coordinates that, in the next step, introduces cylindrical (Bessel) functions $J_l(\alpha_{nl}r)$ and elementary exponential functions $e^{il\varphi}$. Bessel functions satisfy orthogonality conditions and hence form a convenient basis. To complete the statement of the problem, we define the boundary regime problem with the conditions at the boundary $z = 0$, which also fixes the transversal components (polarizations) of the electric field.

The procedure of the projecting is as prescribed by the division of variables procedure [6], firstly we expand all basic fields in series with respect to Hondros-Debye basis $N_{nl}^{-1}J_l(\alpha_{nl}r)e^{il\varphi}$. The form of the expansion is obvious for $E_z(r,\varphi,z,t)$, but we need an analysis of the result of substitution into the Maxwell equations to have a close expressions. The result is as follows

$$E_z(r,\varphi,z,t) = \sum_{l,n} \mathscr{A}_{ln}(z,t) J_l(\alpha_{nl}r) e^{il\varphi} + c.c., \tag{9.57a}$$

$$E_r(r,\varphi,z,t) = \sum_{l,n} \mathscr{B}_{ln}(z,t) \frac{il}{r} J_l(\alpha_{ln}r) e^{il\varphi} + \sum_{l,n} \mathscr{C}_{ln}(z,t) \frac{\partial J_l(\alpha_{ln}r)}{\partial r} e^{il\varphi} + c.c., \tag{9.57b}$$

$$E_\varphi(r,\varphi,z,t) = \sum_{l,n} \mathscr{D}_{ln}(z,t) \frac{il}{r} J_l(\alpha_{ln}r) e^{il\varphi} + \sum_{l,n} \mathscr{F}_{ln}(z,t) \frac{\partial J_l(\alpha_{ln}r)}{\partial r} e^{il\varphi} + c.c., \tag{9.57c}$$

$$B_z(r,\varphi,z,t) = \sum_{l,n} \mathscr{K}_{ln}(z,t) J_l(\alpha_{nl}r) e^{il\varphi} + c.c., \tag{9.57d}$$

$$B_r(r,\varphi,z,t) = \sum_{l,n} \mathscr{L}_{ln}(z,t) \frac{il}{r} J_l(\alpha_{ln}r) e^{il\varphi} + \sum_{l,n} \mathscr{M}_{ln}(z,t) \frac{\partial J_l(\alpha_{ln}r)}{\partial r} e^{il\varphi} + c.c., \tag{9.57e}$$

$$B_\varphi(r,\varphi,z,t) = \sum_{l,n} \mathscr{R}_{ln}(z,t) \frac{il}{r} J_l(\alpha_{ln}r) e^{il\varphi} + \sum_{l,n} \mathscr{S}_{ln}(z,t) \frac{\partial J_l(\alpha_{ln}r)}{\partial r} e^{il\varphi} + c.c., \tag{9.57f}$$

as a natural step, similar to the previous chapter. As it is said, we should substitute the expansion to the linearized Maxwell equations in cylindrical coordinates, that have the following form for Gauss units:

$$\frac{\partial}{r\partial r}(rD_r)+\frac{\partial D_\varphi}{r\partial\varphi}+\frac{\partial D_z}{\partial z}=0, \tag{9.58}$$

$$\frac{\partial}{r\partial r}(rB_r)+\frac{\partial B_\varphi}{r\partial\varphi}+\frac{\partial B_z}{\partial z}=0, \tag{9.59}$$

$$\left(\frac{\partial E_z}{r\partial\varphi}-\frac{\partial E_\varphi}{\partial z}\right)=-\frac{1}{c}\frac{\partial B_r}{\partial t}, \tag{9.60}$$

$$\left(\frac{\partial E_r}{\partial z}-\frac{\partial E_z}{\partial r}\right)=-\frac{1}{c}\frac{\partial B_\varphi}{\partial t}, \tag{9.61}$$

$$\frac{1}{r}\left(\frac{\partial(rE_\varphi)}{\partial r}-\frac{\partial E_r}{\partial\varphi}\right)=-\frac{1}{c}\frac{\partial B_z}{\partial t}, \tag{9.62}$$

$$\left(\frac{\partial H_z}{r\partial\varphi}-\frac{\partial H_\varphi}{\partial z}\right)=\frac{1}{c}\frac{\partial D_r}{\partial t}, \tag{9.63}$$

$$\left(\frac{\partial H_r}{\partial z}-\frac{\partial H_z}{\partial r}\right)=\frac{1}{c}\frac{\partial D_\varphi}{\partial t}, \tag{9.64}$$

$$\frac{1}{r}\left(\frac{\partial(rH_\varphi)}{\partial r}-\frac{\partial H_r}{\partial\varphi}\right)=\frac{1}{c}\frac{\partial D_z}{\partial t}. \tag{9.65}$$

For simplicity, at the first stage, we take an isotropic material relations case of constant dielectric and magnetc permeabilities, supposing

$$\vec{D}=\varepsilon\vec{E},\vec{B}=\mu\vec{H}. \tag{9.66}$$

9.4.3 TRANSITION TO BESSEL FUNCTIONS BASIS

In this subsection, we show the technical details omitted in publication [6], which show a transition of the first two Maxwell equations to the Bessel basis, plugging in the expansions (9.57), as follows

$$\begin{aligned}&\frac{\partial}{r\partial r}r\left(\sum_{l,n}\mathscr{B}_{ln}(z,t)\frac{il}{r}J_l(\alpha_{ln}r)e^{il\varphi}+\sum_{l,n}\mathscr{C}_{ln}(z,t)\frac{\partial}{\partial r}J_l(\alpha_{ln}r)e^{il\varphi}\right)\\&+\frac{\partial}{r\partial\varphi}\left(\sum_{l,n}\mathscr{D}_{ln}(z,t)\frac{il}{r}J_l(\alpha_{ln}r)e^{il\varphi}+\sum_{l,n}\mathscr{F}_{ln}(z,t)\frac{\partial}{\partial r}J_l(\alpha_{ln}r)e^{il\varphi}\right)\\&+\frac{\partial}{\partial z}\left(\sum_{l,n}\mathscr{A}_{ln}(z,t)J_l(\alpha_{nl}r)e^{il\varphi}\right)=0,\end{aligned} \tag{9.67}$$

$$\frac{\partial}{r\partial r}\left(r(\sum_{l,n}\mathscr{L}_{ln}(z,t)\frac{il}{r}J_l(\alpha_{ln}r)e^{il\varphi}+\sum_{l,n}\mathscr{M}_{ln}(z,t)\frac{\partial}{\partial r}J_l(\alpha_{ln}r)e^{il\varphi})\right)$$
$$+\frac{\partial}{r\partial\varphi}\left(\sum_{l,n}\mathscr{R}_{ln}(z,t)\frac{il}{r}J_l(\alpha_{ln}r)e^{il\varphi}+\sum_{l,n}\mathscr{S}_{ln}(z,t)\frac{\partial}{\partial r}J_l(\alpha_{ln}r)e^{il\varphi}\right)$$
$$+\frac{\partial}{\partial z}\sum_{l,n}\mathscr{K}_{ln}(z,t)J_l(\alpha_{nl}r)e^{il\varphi}=0. \tag{9.68}$$

For further calculations, the Bessel function properties will be useful:

$$\frac{\partial}{\partial r}J_l(r)=\frac{1}{2}\left(J_{l-1}(r)-J_{l+1}(r)\right), \tag{9.69}$$
$$\frac{l}{r}J_l(r)=\frac{1}{2}\left(J_{l-1}(r)+J_{l+1}(r)\right).$$

Both Eqs (9.67) and (9.68) have similar forms. After solving the first equation (9.67), the acquired solution can be applied to the second one (9.68).

$$\sum_{l,n}\Big(il\mathscr{B}_{ln}(z,t)\frac{\partial}{r\partial r}J_l(\alpha_{ln}r)+\mathscr{C}_{ln}(z,t)\frac{\partial}{r\partial r}r\frac{\partial}{\partial r}J_l(\alpha_{ln}r)+ \tag{9.70}$$
$$il\mathscr{D}_{ln}(z,t)\frac{\partial}{r\partial\varphi}\frac{1}{r}J_l(\alpha_{ln}r)+\mathscr{F}_{ln}(z,t)\frac{\partial}{r\partial\varphi}\frac{\partial}{\partial r}J_l(\alpha_{ln}r)+$$
$$\frac{\partial}{\partial z}\mathscr{A}_{ln}(z,t)J_l(\alpha_{nl}r)\Big)e^{il\varphi}=0.$$

The order of derivatives of the corresponding terms can be diminished by the application of the Bessel equation

$$\frac{\partial^2}{\partial r^2}J_l(r)=-\frac{1}{r}\frac{\partial}{\partial r}J_l(r)-\left(1-\frac{l^2}{r^2}\right)J_l(r). \tag{9.71}$$

So, by means of the (9.71), we do transformations, rescaling $r\to\alpha_{ln}r$, for any value of n, l. For a reader convenience we give more details, connected with the rescaling $\rho=\alpha r$, repeating the Bessel equation as:

$$\frac{\partial^2}{\partial\rho^2}J_l(\rho)+\frac{1}{\rho}\frac{\partial}{\partial\rho}J_l(\rho)=\frac{1}{\rho}\frac{\partial}{\partial\rho}\rho\frac{\partial}{\partial\rho}J_l(\rho)=-\left(1-\frac{l^2}{\rho^2}\right)J_l(\rho). \tag{9.72}$$

The change of derivatives are written via

$$\frac{\partial}{\partial r}J_l(\rho)=\frac{\partial}{\partial\rho}J_l(\rho)\frac{\partial\rho}{\partial r},$$

where $\frac{\partial\rho}{\partial r}=\alpha$.

Then

$$\begin{aligned}&\tfrac{1}{r}\tfrac{\partial}{\partial r}r\tfrac{\partial}{\partial r}J_l(\rho) = \alpha^2\tfrac{1}{\rho}\tfrac{\partial}{\partial \rho}\rho\tfrac{\partial}{\partial \rho}J_l(\rho)\\ &= -\alpha^2(1-\tfrac{l^2}{\rho^2})J_l(\rho) = -\alpha^2(1-\tfrac{l^2}{\alpha^2 r^2})J_l(\rho) = -\alpha^2 J_l(\rho) + \tfrac{l^2}{r^2}J_l(\rho),\end{aligned} \tag{9.73}$$

the indexes by α are omitted for shorthand. By this result application, the terms of the sum are transformed as

$$\begin{aligned}&il\mathscr{B}_{ln}(z,t)\tfrac{\partial}{r\partial r}J_l(\alpha_{ln}r) + \mathscr{C}_{ln}(z,t)\tfrac{\partial}{r\partial r}J_l(\alpha_{ln}r)+\\ &\mathscr{C}_{ln}(z,t)(-\tfrac{1}{r}\tfrac{\partial}{\partial r}J_l(\alpha_{ln}r) - \alpha_{ln}^2 J_l(\alpha_{ln}r) + \tfrac{l^2}{r^2}J_l(\alpha_{ln}r))+\\ &il\mathscr{D}_{ln}(z,t)\tfrac{il}{r^2}J_l(\alpha_{ln}r) + \mathscr{F}_{ln}(z,t)\tfrac{il}{r}\tfrac{\partial}{\partial r}J_l(\alpha_{ln}r) + \tfrac{\partial}{\partial z}\mathscr{A}_{ln}(z,t)J_l(\alpha_{nl}r).\end{aligned}$$

Now, rearranging the result applied to Eq. (9.70), we require each term of the sum to be zero identically

$$\begin{aligned}&il\left(\mathscr{B}_{ln}(z,t)+\mathscr{F}_{ln}(z,t)\right)\frac{\partial}{r\partial r}J_l(\alpha_{ln}r) + \left(\mathscr{C}_{ln}(z,t)l^2 + (il)^2\mathscr{D}_{ln}(z,t)\right)\\ &\frac{1}{r^2}J_l(\alpha_{ln}r) + \left(\frac{\partial}{\partial z}\mathscr{A}_{ln}(z,t) - \mathscr{C}_{ln}(z,t)\alpha_{ln}^2\right)J_l(\alpha_{nl}r) = 0.\end{aligned}$$

Hence, with use of independence of $\frac{\partial}{r\partial r}J_l(\alpha_{ln}r)$, $\frac{1}{r^2}J_l(\alpha_{ln}r)$ and $J_l(\alpha_{nl}r)$ we can write the equations for coefficients, which guarantee the validity of (9.70)

$$\mathscr{B}_{ln}(z,t) + \mathscr{F}_{ln}(z,t) = 0, \tag{9.74}$$

$$\mathscr{C}_{ln}(z,t) - \mathscr{D}_{ln}(z,t) = 0, \tag{9.75}$$

$$\frac{\partial}{\partial z}\mathscr{A}_{ln}(z,t) - \mathscr{C}_{ln}(z,t)\alpha_{ln}^2 = 0. \tag{9.76}$$

Lets take a closer look at the second Maxwell equation (9.68), having the sum of expressions:

$$\begin{aligned}&il\mathscr{L}_{ln}(z,t)\frac{\partial}{r\partial r}J_l(\alpha_{ln}r)e^{il\varphi} + \mathscr{M}_{ln}(z,t)\frac{\partial}{r\partial r}r\frac{\partial}{\partial r}J_l(\alpha_{ln}r)e^{il\varphi}\\ &+il\mathscr{R}_{ln}(z,t)\frac{1}{r^2}\frac{\partial}{\partial\varphi}J_l(\alpha_{ln}r)e^{il\varphi} + \frac{\partial}{\partial\varphi}\mathscr{S}_{ln}(z,t)\frac{\partial}{r\partial r}J_l(\alpha_{ln}r)e^{il\varphi}\\ &+\frac{\partial}{\partial z}\mathscr{K}_{ln}(z,t)J_l(\alpha_{nl}r)e^{il\varphi}.\end{aligned} \tag{9.77}$$

Now, based on analogy between Eqs (9.67) and (9.68), we differentiate with respect to φ and require that

$$\begin{aligned}&il\mathscr{L}_{ln}(z,t)\frac{\partial}{r\partial r}J_l(\alpha_{ln}r) + \mathscr{M}_{ln}(z,t)\frac{\partial}{r\partial r}J_l(\alpha_{ln}r) + \mathscr{M}_{ln}(z,t)\frac{\partial^2}{\partial r^2}J_l(\alpha_{ln}r)+\\ &il\mathscr{R}_{ln}(z,t)\frac{il}{r^2}J_l(\alpha_{ln}r) + \mathscr{S}_{ln}(z,t)\frac{il}{r}\frac{\partial}{\partial r}J_l(\alpha_{ln}r) + \frac{\partial}{\partial z}\mathscr{K}_{ln}(z,t)J_l(\alpha_{nl}r) = 0.\end{aligned} \tag{9.78}$$

Using again the Bessel equation (9.71) and performing similar steps, we obtain the relation between coefficients that can be presented by the system of equations

$$\mathscr{L}_{ln}(z,t) + \mathscr{S}_{ln}(z,t) = 0, \tag{9.79}$$

$$\mathscr{M}_{ln}(z,t) - \mathscr{R}_{ln}(z,t) = 0, \tag{9.80}$$

$$\frac{\partial}{\partial z}\mathscr{K}_{ln}(z,t) - \mathscr{M}_{ln}(z,t)\alpha_{ln}^2 = 0. \tag{9.81}$$

The next three equations of the Maxwell system for the fields in the Bessel basis have the form

$$\frac{il}{r}\mathscr{A}_{ln}(z,t)J_l(\alpha_{nl}r) - \frac{\partial}{\partial z}\left(\mathscr{D}_{ln}(z,t)\frac{il}{r}J_l(\alpha_{ln}r) + \mathscr{F}_{ln}(z,t)\frac{\partial}{\partial r}J_l(\alpha_{ln}r)\right)$$

$$= -\frac{1}{c}\frac{\partial}{\partial t}\left(\mathscr{L}_{ln}(z,t)\frac{il}{r}J_l(\alpha_{ln}r) + \mathscr{M}_{ln}(z,t)\frac{\partial}{\partial r}J_l(\alpha_{ln}r)\right), \tag{9.82}$$

$$\frac{\partial}{\partial z}\left(\mathscr{B}_{ln}(z,t)\frac{il}{r}J_l(\alpha_{ln}r) + \mathscr{C}_{ln}(z,t)\frac{\partial}{\partial r}J_l(\alpha_{ln}r)\right) - \frac{\partial}{\partial r}\mathscr{A}_{ln}(z,t)J_l(\alpha_{nl}r)$$

$$= -\frac{1}{c}\frac{\partial}{\partial t}\left(\mathscr{R}_{ln}(z,t)\frac{il}{r}J_l(\alpha_{ln}r) + \mathscr{S}_{ln}(z,t)\frac{\partial}{\partial r}J_l(\alpha_{ln}r)\right), \tag{9.83}$$

$$\frac{1}{r}\left(\frac{\partial}{\partial r}r\left(\mathscr{D}_{ln}(z,t)\frac{il}{r}J_l(\alpha_{ln}r) + \mathscr{F}_{ln}(z,t)\frac{\partial}{\partial r}J_l(\alpha_{ln}r)\right)\right.$$

$$\left. - \mathscr{B}_{ln}(z,t)\frac{-l^2}{r}J_l(\alpha_{ln}r) - \mathscr{C}_{ln}(z,t)\frac{il\partial}{\partial r}J_l(\alpha_{ln}r)\right) = -\frac{1}{c}\frac{\partial}{\partial t}\mathscr{K}_{ln}(z,t)J_l(\alpha_{nl}r), \tag{9.84}$$

we don't repeat the relations obtained before, having

$$\mathscr{A}_{ln}(z,t) - \frac{\partial}{\partial z}\mathscr{D}_{ln}(z,t) = -\frac{1}{c}\frac{\partial}{\partial t}\mathscr{L}_{ln}(z,t), \tag{9.85}$$

$$-\frac{\partial}{\partial z}\mathscr{F}_{ln}(z,t) = -\frac{1}{c}\frac{\partial}{\partial t}\mathscr{M}_{ln}(z,t), \tag{9.86}$$

$$\frac{\partial}{\partial z}\mathscr{B}_{ln}(z,t) = -\frac{1}{c}\frac{\partial}{\partial t}\mathscr{R}_{ln}(z,t), \tag{9.87}$$

$$\frac{\partial}{\partial z}\mathscr{C}_{ln}(z,t) - \mathscr{A}_{ln}(z,t) = -\frac{1}{c}\frac{\partial}{\partial t}\mathscr{S}_{ln}(z,t), \tag{9.88}$$

$$-\mathscr{F}_{ln}(z,t)\alpha_{ln}^2 = -\frac{1}{c}\frac{\partial}{\partial t}\mathscr{K}_{ln}(z,t). \tag{9.89}$$

$$\frac{il}{r}\mathcal{K}_{ln}(z,t)J_l(\alpha_{nl}r) - \frac{\partial}{\partial z}\left(\mathcal{R}_{ln}(z,t)\frac{il}{r}J_l(\alpha_{ln}r) + \mathcal{S}_{ln}(z,t)\frac{\partial}{\partial r}J_l(\alpha_{ln}r)\right)$$

$$= \frac{\varepsilon\mu}{c}\frac{\partial}{\partial t}\left(\mathcal{B}_{ln}(z,t)\frac{il}{r}J_l(\alpha_{ln}r) + \mathcal{C}_{ln}(z,t)\frac{\partial}{\partial r}J_l(\alpha_{ln}r)\right), \tag{9.90}$$

$$\frac{\partial}{\partial z}\left(\mathcal{L}_{ln}(z,t)\frac{il}{r}J_l(\alpha_{ln}r) + \mathcal{M}_{ln}(z,t)\frac{\partial}{\partial r}J_l(\alpha_{ln}r)\right) - \frac{\partial}{\partial r}\mathcal{K}_{ln}(z,t)J_l(\alpha_{nl}r)$$

$$= \frac{\varepsilon\mu}{c}\frac{\partial}{\partial t}\left(\mathcal{D}_{ln}(z,t)\frac{il}{r}J_l(\alpha_{ln}r) + \mathcal{F}_{ln}(z,t)\frac{\partial}{\partial r}J_l(\alpha_{ln}r)\right), \tag{9.91}$$

$$\frac{1}{r}\left(\frac{\partial}{\partial r}r\left(\mathcal{R}_{ln}(z,t)\frac{il}{r}J_l(\alpha_{ln}r) + \mathcal{S}_{ln}(z,t)\frac{\partial}{\partial r}J_l(\alpha_{ln}r)\right)\right.$$

$$\left. - il(\mathcal{L}_{ln}(z,t)\frac{il}{r}J_l(\alpha_{ln}r) + \mathcal{M}_{ln}(z,t)\frac{\partial}{\partial r}J_l(\alpha_{ln}r))\right) = \frac{\varepsilon\mu}{c}\frac{\partial}{\partial t}\mathcal{A}_{ln}(z,t)J_l(\alpha_{nl}r). \tag{9.92}$$

Directly from the above equations the relations between coefficients can be written as follows

$$\mathcal{K}_{ln}(z,t) - \frac{\partial}{\partial z}\mathcal{R}_{ln}(z,t) = \frac{\varepsilon\mu}{c}\frac{\partial}{\partial t}\mathcal{B}_{ln}(z,t), \tag{9.93}$$

$$-\frac{\partial}{\partial z}\mathcal{S}_{ln}(z,t) = \frac{\varepsilon\mu}{c}\frac{\partial}{\partial t}\mathcal{C}_{ln}(z,t), \tag{9.94}$$

$$\frac{\partial}{\partial z}\mathcal{L}_{ln}(z,t) = \frac{\varepsilon\mu}{c}\frac{\partial}{\partial t}\mathcal{D}_{ln}(z,t), \tag{9.95}$$

$$\frac{\partial}{\partial z}\mathcal{M}_{ln}(z,t) - \mathcal{K}_{ln}(z,t) = \frac{\varepsilon\mu}{c}\frac{\partial}{\partial t}\mathcal{F}_{ln}(z,t), \tag{9.96}$$

$$-\mathcal{S}_{ln}(z,t)\alpha_{ln}^2 = \frac{\varepsilon\mu}{c}\frac{\partial}{\partial t}\mathcal{A}_{ln}(z,t). \tag{9.97}$$

Excluding the algebraic relations (9.74) and (9.79), we arrive at the the complete set of six differential equations for the coefficient functions:

$$\frac{\varepsilon\mu}{c}\frac{\partial}{\partial t}\mathcal{A}_{ln}(z,t) = \alpha_{ln}^2\mathcal{L}_{ln}(z,t). \tag{9.98a}$$

$$\frac{\varepsilon\mu}{c}\frac{\partial}{\partial t}\mathcal{B}_{ln}(z,t) = \mathcal{K}_{ln}(z,t) - \frac{\partial}{\partial z}\mathcal{M}_{ln}(z,t), \tag{9.98b}$$

$$\frac{\varepsilon\mu}{c}\frac{\partial}{\partial t}\mathcal{C}_{ln}(z,t) = \frac{\partial}{\partial z}\mathcal{L}_{ln}(z,t), \tag{9.98c}$$

$$\frac{1}{c}\frac{\partial}{\partial t}\mathcal{K}_{ln}(z,t) = -\alpha_{ln}^2\mathcal{B}_{ln}(z,t), \tag{9.98d}$$

$$\frac{1}{c}\frac{\partial}{\partial t}\mathcal{L}_{ln}(z,t) = -\mathcal{A}_{ln}(z,t) + \frac{\partial}{\partial z}\mathcal{C}_{ln}(z,t), \tag{9.98e}$$

$$\frac{1}{c}\frac{\partial}{\partial t}\mathcal{M}_{ln}(z,t) = -\frac{\partial}{\partial z}\mathcal{B}_{ln}(z,t). \tag{9.98f}$$

Combining the first and fifth equations from (9.98a), taking the differential relation from (9.74), we obtain the wave equation for $\mathcal{A}_{ln}(z,t)$:

$$\frac{\varepsilon\mu}{c^2}\frac{\partial^2}{\partial t^2}\mathcal{A}_{ln}(z,t) = \frac{\partial^2}{\partial z^2}\mathcal{A}_{ln}(z,t) - \alpha_{ln}^2\mathcal{A}_{ln}(z,t). \tag{9.99}$$

This is a direct test of the rather tedious calculations we made. The equation of such form for $\mathcal{K}_{ln}(z,t)$ is derived identically. Similarly - for others.

9.5 DYNAMICAL PROJECTING OPERATORS

9.5.1 Z-EVOLUTION SYSTEM AND TRANSITION TO ω-DOMAIN

Considering the method of construction of dynamic projection operators for a Cauchy problem, there will be six projection operators for such a system (9.98a). In the situation of a boundary regime problem that fits an excitation of a fiber at an end (e.g. $z = 0$), it would be convenient to reduce the system to four equations with z- derivatives on the L.H.S. Then it will be possible to associate two new variables to the electric component with the left and right direction of wave propagation and two variables with magnetic component with left and right directions of wave propagation. Hence, taking from the system two relations

$$\mathcal{L}_{ln}(z,t) = \frac{\partial}{\partial t}\frac{\varepsilon\mu}{c\alpha_{ln}^2}\mathcal{A}_{ln}(z,t), \tag{9.100a}$$

$$\mathcal{B}_{ln}(z,t) = -\frac{1}{c\alpha_{ln}^2}\frac{\partial}{\partial t}\mathcal{K}_{ln}(z,t), \tag{9.100b}$$

we exclude $\mathcal{L}_{ln}$ and $\mathcal{B}_{ln}(z,t)$ and, next $\mathcal{F}_{ln}(z,t)$ and $\mathcal{S}_{ln}(z,t)$ going to the system of z-evolution system of four equations.

$$\frac{\partial}{\partial z}\mathcal{A}_{ln}(z,t) = \alpha_{ln}^2\mathcal{C}_{ln}(z,t), \tag{9.101a}$$

$$\frac{\partial}{\partial z}\mathcal{C}_{ln}(z,t) = \left(\frac{\varepsilon\mu}{\alpha_{ln}^2 c^2}\frac{\partial^2}{\partial t^2} + 1\right)\mathcal{A}_{ln}(z,t), \tag{9.101b}$$

$$\frac{\partial}{\partial z}\mathcal{K}_{ln}(z,t) = \alpha_{ln}^2\mathcal{M}_{ln}(z,t), \tag{9.101c}$$

$$\frac{\partial}{\partial z}\mathcal{M}_{ln}(z,t) = \left(\frac{\varepsilon\mu}{\alpha_{ln}^2 c^2}\frac{\partial^2}{\partial t^2} + 1\right)\mathcal{K}_{ln}(z,t). \tag{9.101d}$$

Note, once more, that the system has the obvious corollary in a form of 1D wave equations for $\mathscr{A}_{ln}(z,t)$ and $\mathscr{K}_{ln}(z,t)$.

With the complete set of four equations, it is possible to present the system in a form of matrix operator z-evolution equation:

$$\partial_z \Psi(z,t) - \hat{L}\Psi(z,t) = 0. \tag{9.102}$$

Following the procedure presented in the introduction of this book, see also Section 7.3, a form of evolution equation is the opening to the acquiring of the projection operators. The evolution operator $\hat{L}$ has been constructed and applied to the evolution equation

$$\partial_z \begin{pmatrix} \mathscr{A}_{ln}(z,t) \\ \mathscr{C}_{ln}(z,t) \\ \mathscr{K}_{ln}(z,t) \\ \mathscr{M}_{ln}(z,t) \end{pmatrix} = \begin{pmatrix} 0 & \alpha_{ln}^2 & 0 & 0 \\ \frac{\partial^2}{\partial t^2}\frac{\varepsilon\mu}{\alpha_{ln}^2 c^2}+1 & 0 & 0 & 0 \\ 0 & 0 & 0 & \alpha_{ln}^2 \\ 0 & 0 & \frac{\partial^2}{\partial t^2}\frac{\varepsilon\mu}{\tilde{\alpha}_{ln}^2 c^2}+1 & 0 \end{pmatrix} \begin{pmatrix} \mathscr{A}_{ln}(z,t) \\ \mathscr{C}_{ln}(z,t) \\ \mathscr{K}_{ln}(z,t) \\ \mathscr{M}_{ln}(z,t) \end{pmatrix}. \tag{9.103}$$

The Fourier transformation in the time domain provides the solution for the system and leads it to the current form:

$$\frac{\partial}{\partial z}\tilde{\mathscr{A}}_{ln}(z,\omega) = \alpha_{ln}^2 \tilde{\mathscr{C}}_{ln}(z,\omega), \tag{9.104a}$$

$$\frac{\partial}{\partial z}\tilde{\mathscr{C}}_{ln}(z,\omega) = \left(-\omega^2 \frac{\varepsilon\mu}{\alpha_{ln}^2 c^2}+1\right)\tilde{\mathscr{A}}_{ln}(z,\omega), \tag{9.104b}$$

$$\frac{\partial}{\partial z}\tilde{\mathscr{K}}_{ln}(z,\omega) = \alpha_{ln}^2 \tilde{\mathscr{M}}_{ln}(z,\omega), \tag{9.104c}$$

$$\frac{\partial}{\partial z}\tilde{\mathscr{M}}_{ln}(z,\omega) = \left(-\omega^2 \frac{\varepsilon\mu}{\alpha_{ln}^2 c^2}+1\right)\tilde{\mathscr{K}}_{ln}(z,\omega). \tag{9.104d}$$

This system of ODE is solved via a linear component, of the exponential solution; for the first component, we write $\tilde{\mathscr{A}}_{ln}(z,\omega) = \breve{\mathscr{A}}_{ln}(k,\omega)e^{ikz}$, which will be adopted to other variables. Applying this solution to our system, the result looks like

$$ik\breve{\mathscr{A}}_{ln}(k,\omega) = \alpha_{ln}^2 \breve{\mathscr{C}}_{ln}(k,\omega), \tag{9.105a}$$

$$ik\breve{\mathscr{C}}_{ln}(k,\omega) = \left(-\omega^2 \frac{\varepsilon\mu}{\alpha_{ln}^2 c^2}+1\right)\breve{\mathscr{A}}_{ln}(k,\omega), \tag{9.105b}$$

$$ik\breve{\mathscr{K}}_{ln}(k,\omega) = \alpha_{ln}^2 \breve{\mathscr{M}}_{ln}(k,\omega), \tag{9.105c}$$

$$ik\breve{\mathscr{M}}_{ln}(k,\omega) = \left(-\omega^2 \frac{\varepsilon\mu}{\alpha_{ln}^2 c^2}+1\right)\breve{\mathscr{K}}_{ln}(k,\omega). \tag{9.105d}$$

This is, as in previous chapters, an eigenvalue problem. The matrix (9.103) and one that stands in (9.105) is the direct sum of equal matrices. It means, that the eigenvalues

$$k_{nl} = \pm\sqrt{\frac{\varepsilon\mu}{c^2}\omega^2 - \alpha_{ln}^2} \tag{9.106}$$

are degenerate. Each subspace with fixed $k_{nl}(\omega)$ contain two eigenvectors of the matrix L

$$\begin{pmatrix} 0 & \alpha_{ln}^2 & 0 & 0 \\ (1-\omega^2\frac{\varepsilon\mu}{\alpha_{ln}^2c^2}) & 0 & 0 & 0 \\ 0 & 0 & 0 & \alpha_{ln}^2 \\ 0 & 0 & (1-\omega^2\frac{\varepsilon\mu}{\alpha_{ln}^2c^2}) & 0 \end{pmatrix}, \tag{9.107}$$

that correspond to two polarizations. The preceding calculations lead also to the dispersion relation. In this place, we should choose the positive value of k_{nl}^2 by a frequency choice above the cutoff one $\omega > \omega_{cutoff} = \frac{\alpha_{ln}c}{\sqrt{\varepsilon\mu}}$ because the wave vector k_{nl} itself should have the real value. In eq (9.106) we have positive and negative values of the wave vector $k_{nl}(\omega)$. The positive is connected to the wave propagating to the right direction and negative is connected to the wave propagating to the left. Solving the eigenvalue problem for the matrix (9.107) the components of state the vector can be connected by

$$\breve{\mathscr{A}}_{ln}(\omega) = \frac{\pm i\alpha_{ln}^2}{\sqrt{\frac{\varepsilon\mu}{c^2}\omega^2 - \alpha_{ln}^2}}\breve{\mathscr{C}}_{ln}(\omega), \tag{9.108a}$$

$$\breve{\mathscr{M}}_{ln}(\omega) = \frac{\pm i}{\alpha_{ln}^2}\sqrt{\frac{\varepsilon\mu}{c^2}\omega^2 - \alpha_{ln}^2}\,\breve{\mathscr{K}}_{ln}(\omega). \tag{9.108b}$$

The signs $\pm$ mark the direction of wave propagation; k is fixed and omitted. The possible choice of orthogonal vectors for both polarizations within directed waves subspace is shown here:

$$\Psi_{11} = \frac{1}{2}\begin{pmatrix} -i\alpha_{ln}^2\tilde{k}_{nl}^{-1} \\ 1 \\ 0 \\ 0 \end{pmatrix}\breve{\mathscr{C}}_{ln}(\omega), \tag{9.109}$$

$$\Psi_{21} = \frac{1}{2}\begin{pmatrix} i\alpha_{ln}^2\tilde{k}_{nl}^{-1} \\ 1 \\ 0 \\ 0 \end{pmatrix}\breve{\mathscr{C}}_{ln}(\omega), \tag{9.110}$$

$$\Psi_{12} = \frac{1}{2}\begin{pmatrix} 0 \\ 0 \\ -i\alpha_{ln}^2\tilde{k}_{nl}^{-1} \\ 1 \end{pmatrix}\breve{\mathscr{M}}_{ln}(\omega), \tag{9.111}$$

$$\Psi_{22} = \frac{1}{2} \begin{pmatrix} 0 \\ 0 \\ i\alpha_{ln}^2 \tilde{k}_{nl}^{-1} \\ 1 \end{pmatrix} \breve{\mathscr{M}}_{ln}(\omega), \tag{9.112}$$

To simplify calculations, a variable $\tilde{k}_{nl} = \sqrt{\omega^2 \frac{\varepsilon\mu}{c^2} - \alpha_{ln}^2}$ is used in the projection operator derivation. By the standard algorithm, the first direction and polarization corresponds to

$$P_{11} = \frac{1}{2} \begin{pmatrix} 1 & -\frac{i}{\tilde{k}_{ln}}\alpha_{ln}^2 & 0 & 0 \\ i\frac{\tilde{k}_{ln}}{\alpha_{ln}^2} & 1 & 0 & 0 \\ 0 & 0 & 0 & 0 \\ 0 & 0 & 0 & 0 \end{pmatrix}. \tag{9.113}$$

The second polarization fo same direcion is given by

$$P_{21} = \frac{1}{2} \begin{pmatrix} 1 & \frac{i}{\tilde{k}_{ln}}\alpha_{ln}^2 & 0 & 0 \\ -i\frac{\tilde{k}_{ln}}{\alpha_{ln}^2} & 1 & 0 & 0 \\ 0 & 0 & 0 & 0 \\ 0 & 0 & 0 & 0 \end{pmatrix}. \tag{9.114}$$

For the opposite direction, the first polarization operator is

$$P_{12} = \frac{1}{2} \begin{pmatrix} 0 & 0 & 0 & 0 \\ 0 & 0 & 0 & 0 \\ 0 & 0 & 1 & \frac{i}{\tilde{k}_{ln}}\alpha_{ln}^2 \\ 0 & 0 & -i\frac{\tilde{k}_{ln}}{\alpha_{ln}^2} & 1 \end{pmatrix}, \tag{9.115}$$

and the second one is

$$P_{22} = \frac{1}{2} \begin{pmatrix} 0 & 0 & 0 & 0 \\ 0 & 0 & 0 & 0 \\ 0 & 0 & 1 & -\frac{i}{\tilde{k}_{ln}}\alpha_{ln}^2 \\ 0 & 0 & i\frac{\tilde{k}_{ln}}{\alpha_{ln}^2} & 1 \end{pmatrix} \tag{9.116}$$

Action of the first projector to the z-evolution equation, after commutation with L and ∂_z operators, looks like

$$\partial_z P_{11}\Psi(z,\omega) - \hat{L}P_{11}\Psi(z,\omega) = 0. \tag{9.117}$$

The action of the projectors on the vectors of state defines the mode variables consequently by

$$\hat{P}_{11}\Psi(z,\omega) = \begin{pmatrix} \mathscr{A}_{ln}(z,\omega) + i\alpha_{ln}^2 \tilde{k}_{nl}^{-1}\mathscr{C}_{ln}(z,\omega) \\ \frac{i}{-\alpha_{ln}^2}\tilde{k}_{nl}\mathscr{A}_{ln}(z,\omega) + \mathscr{C}_{ln}(z,\omega) \\ 0 \\ 0 \end{pmatrix} = \begin{pmatrix} i\alpha_{ln}^2 \tilde{k}_{nl}^{-1}\Lambda_1(z,\omega) \\ \Lambda_1(z,\omega) \\ 0 \\ 0 \end{pmatrix} \tag{9.118}$$

$$\hat{P}_{12}\Psi(z,\omega) = \begin{pmatrix} 0 \\ 0 \\ \mathscr{K}_{ln}(z,\omega) + i\alpha_{ln}^2\tilde{k}_{nl}^{-1}\mathscr{M}_{ln}(z,\omega) \\ \frac{-i}{\alpha_{ln}^2}\tilde{k}_{nl}\mathscr{K}_{ln}(z,\omega) + \mathscr{M}_{ln}(z,\omega) \end{pmatrix} = \begin{pmatrix} 0 \\ 0 \\ -i\tilde{\alpha}_{ln}^2\hat{k}_{nl}^{-1}\Lambda_2(z,\omega) \\ \Lambda_2(z,\omega) \end{pmatrix}, \tag{9.119}$$

$$\hat{P}_{21}\Psi(z,\omega) = \begin{pmatrix} \mathscr{A}_{ln}(z,\omega) - i\alpha_{ln}^2\tilde{k}_{nl}^{-1}\mathscr{C}_{ln}(z,\omega) \\ \frac{i}{\alpha_{ln}^2}\tilde{k}_{nl}\mathscr{A}_{ln}(z,\omega) + \mathscr{C}_{ln}(z,\omega) \\ 0 \\ 0 \end{pmatrix} = \begin{pmatrix} i\alpha_{ln}^2\tilde{k}_{nl}^{-1}\Pi_1(z,\omega) \\ \Pi_1(z,\omega) \\ 0 \\ 0 \end{pmatrix}, \tag{9.120}$$

$$\hat{P}_{22}\Psi(z,\omega) = \begin{pmatrix} 0 \\ 0 \\ \mathscr{K}_{ln}(z,\omega) - i\alpha_{ln}^2\tilde{k}_{nl}^{-1}\mathscr{M}_{ln}(z,\omega) \\ \frac{i}{\alpha_{ln}^2}\tilde{k}_{nl}\mathscr{K}_{ln}(z,\omega) + \mathscr{M}_{ln}(z,\omega) \end{pmatrix} = \begin{pmatrix} 0 \\ 0 \\ i\alpha_{ln}^2\tilde{k}_{nl}^{-1}\Pi_2(z,\omega) \\ \Pi_2(z,\omega) \end{pmatrix}. \tag{9.121}$$

At this point, Λ_1 can be associated with the polarized electric wave (TE), propagating to the left. Λ_2 is associated with the polarized magnetic wave (TM), propagating in the same direction. Waves propagating in the opposite direction are described by Π_1 associated to electric wave and Π_2 associated to magnetic one.

Let us list the mode fields expressions:

$$\begin{aligned} \Lambda_1(z,\omega) &= -\frac{i}{\alpha_{ln}^2}\tilde{k}_{nl}\mathscr{A}_{ln}(z,\omega) + \mathscr{C}_{ln}(z,\omega), \\ \Lambda_2(z,\omega) &= -\frac{i}{\alpha_{ln}^2}\tilde{k}_{nl}\mathscr{K}_{ln}(z,\omega) + \mathscr{M}_{ln}(z,\omega), \\ \Pi_1(z,\omega) &= \frac{i}{\alpha_{ln}^2}\tilde{k}_{nl}\mathscr{A}_{ln}(z,\omega) + \mathscr{C}_{ln}(z,\omega), \\ \Pi_2(z,\omega) &= \frac{i}{\alpha_{ln}^2}\tilde{k}_{nl}\mathscr{K}_{ln}(z,\omega) + \mathscr{M}_{ln}(z,\omega). \end{aligned} \tag{9.122}$$

The inverse transformation is

$$\begin{aligned} \mathscr{A}_{ln}(z,\omega) &= \frac{i\alpha_{ln}^2}{2\tilde{k}_{nl}}(\Lambda_1(z,\omega) - \Pi_1(z,\omega)), \\ \mathscr{C}_{ln}(z,\omega) &= \frac{1}{2}(\Lambda_1(z,\omega) + \Pi_1(z,\omega)), \\ \mathscr{K}_{ln}(z,\omega) &= \frac{i\alpha_{ln}^2}{2\tilde{k}_{nl}}(\Lambda_2(z,\omega) - \Pi_2(z,\omega)), \\ \mathscr{M}_{ln}(z,\omega) &= \frac{1}{2}(\Lambda_2(z,\omega) + \Pi_2(z,\omega)). \end{aligned} \tag{9.123}$$

In the further analysis, only the first projection operator P_{11} will be used for illustration. Acting on the state vector $\Psi(\mathscr{A},\mathscr{C},\mathscr{K},\mathscr{M})$ with P_{11},

$$\partial_z P_{11}\Psi(\mathscr{A},\mathscr{C},\mathscr{K},\mathscr{M}) - \hat{L}P_{11}\Psi(\mathscr{A},\mathscr{C},\mathscr{K},\mathscr{M}) = 0, \tag{9.124}$$

the equation describing electric wave propagating in the left direction is derived, reading the second line of projected (9.103), and taking (10.8) into account. In ω-domain it yields

$$\partial_z \Lambda_1 - i\alpha_{ln}^2 \left(1 - \omega^2 \frac{\varepsilon\mu}{\alpha_{ln}^2 c^2}\right) \tilde{k}_{nl}^{-1} \Lambda_1 = 0. \tag{9.125}$$

The identity

$$\alpha_{ln}^2 \left(1 - \omega^2 \frac{\varepsilon\mu}{\alpha_{ln}^2 c^2}\right) \tilde{k}_{nl}^{-1} = \left(\alpha_{ln}^2 - \omega^2 \frac{\varepsilon\mu}{c^2}\right) \tilde{k}_{nl}^{-1} = -\tilde{k}_{nl}^2 \tilde{k}_{nl}^{-1} = \tilde{k}_{nl}$$

allows us to finalize the derivation of evolution for the polarized TE left mode

$$\partial_z \Lambda_1 + i\tilde{k}_{nl} \Lambda_1 = 0. \tag{9.126}$$

with obvious physical sense. A transition to t-domain put at the place of the guide wavevector $\tilde{k}_{nl}$, look (9.106) the correspondent integral operator $\hat{k}_{nl}$. We sould specify it with the material dispersion account in the next section.

9.5.2 PROJECTION OPERATORS IN TIME DOMAIN. DISPERSION ACCOUNT

As it follows from previous chapter, a material dispersion account may be made by the introduction of the susceptibility and permeability operators $\hat{\mu}, \hat{\varepsilon}$ into the material equations (9.66) like the ones in Chapter 8, Eqs. (8.18). After a Fourier transformation by time, we obtain the functions of frequency $\varepsilon(\omega), \mu(\omega)$ that enter the formalism exactly at the same places as constant parameters ε, μ of the previous section. Taking these silica optic fibers as important example, we put $\mu = 1$ and left The dielectric susceptibility coefficient $\varepsilon(\omega)$ either originated from quantum version of the Lorentz formula (see e.g. [7,10]) or directly from phenomenology [8,9]; for example, it may be approximated as a Taylor expansion at $\omega = \infty$ point at [11], which coincides with Drude formula (in Chapter 8),

$$\varepsilon(\omega) \approx 1 + 4\pi\chi(\omega) \approx 1 - 4\pi\chi_0\omega^{-2}. \tag{9.127}$$

Taking into consideration that ε is dependent on ω, from that moment $\hat{\varepsilon}$ will be treated as an operator. Hence, the wave vector $\hat{k}_{nl}$ will also be an operator as it depends on $\hat{\varepsilon}$. Including the ε dependency over ω (9.127) it is necessary to rewrite the dispersion relation Eq. (9.106) to account for both material and waveguide dispersion now:

$$k_{nl} = \omega \sqrt{\frac{(1 - 4\pi\chi_0\omega^{-2})\mu}{c^2} - \frac{\alpha_{nl}^2}{\omega^2}}. \tag{9.128}$$

or

$$k_{nl} = \frac{\sqrt{\mu}\omega}{c} \sqrt{1 - \frac{\alpha_{ln}^2 c^2 + 4\pi\mu\chi_0}{\mu\omega^2}}, \tag{9.129}$$

In some frequency range and chosen mode with given l, n such that $\frac{\mu\omega^2}{\alpha_{ln}^2 c^2 - 4\pi\mu\chi_0} \gg 1$, the right-hand side of the expression (9.129) may be expanded in a Taylor series over $\omega^{-2} \to 0$, or, for high frequency,

$$k_{nl} = \frac{\sqrt{\mu}\omega}{c}\left(1 - \frac{\alpha_{ln}^2 c^2 + 4\pi\mu\chi_0}{2\mu\omega^2}\right) = \frac{\sqrt{\mu}}{c}\left(\omega - \frac{\alpha_{ln}^2 c^2 + 4\pi\mu\chi_0}{2\mu\omega}\right), \tag{9.130}$$

The approximate equation (9.126) in ω-domain looks like

$$\partial_z \Lambda(z,\omega) + i\frac{\sqrt{\mu}}{c}\left(\omega - \frac{\alpha_{ln}^2 c^2 + 4\pi\mu\chi_0}{2\mu\omega}\right)\Lambda(z,\omega) = 0. \tag{9.131}$$

After the inverse Fourier transformation, the dispersion relation (9.106) in z,t representation (t-domain) with dielectric susceptibility (material) coefficient (9.127) included, has the form

$$\partial_z \Lambda(z,t) + \frac{\sqrt{\mu}}{c}\left(\partial_t + \frac{\alpha_{ln}^2 c^2 + 4\pi\mu\chi_0}{2\mu}\partial_t^{-1}\right)\Lambda(z,t) = 0. \tag{9.132}$$

It is possible to work with the projection operators (9.113) directly in the t-representation, using the resulting form via the integral operator $\hat{k}_{nl}$ or its approximate version.

$$\hat{P}_{11} = \frac{1}{2}\begin{pmatrix} 1 & -i\alpha_{ln}^2 \hat{k}_{nl}^{-1} & 0 & 0 \\ i\alpha_{ln}^{-2}\hat{k}_{nl} & 1 & 0 & 0 \\ 0 & 0 & 0 & 0 \\ 0 & 0 & 0 & 0 \end{pmatrix}, \hat{P}_{21} = \frac{1}{2}\begin{pmatrix} 1 & i\alpha_{ln}^2 \hat{k}_{nl}^{-1} & 0 & 0 \\ -i\alpha_{ln}^{-2}\hat{k}_{nl} & 1 & 0 & 0 \\ 0 & 0 & 0 & 0 \\ 0 & 0 & 0 & 0 \end{pmatrix}$$

$$\hat{P}_{12} = \frac{1}{2}\begin{pmatrix} 0 & 0 & 0 & 0 \\ 0 & 0 & 0 & 0 \\ 0 & 0 & 1 & i\alpha_{ln}^2 \hat{k}_{nl}^{-1} \\ 0 & 0 & -i\alpha_{ln}^{-2}\hat{k}_{nl}^{-1} & 1 \end{pmatrix}, \hat{P}_{22} = \frac{1}{2}\begin{pmatrix} 0 & 0 & 0 & 0 \\ 0 & 0 & 0 & 0 \\ 0 & 0 & 1 & -i\alpha_{ln}^2 \hat{k}_{nl}^{-1} \\ 0 & 0 & i\alpha_{ln}^{-2}\hat{k}_{nl}^{-1} & 1 \end{pmatrix}. \tag{9.133}$$

9.6 INCLUDING NONLINEARITY

The importance of the nonlinear part of polarization is unquestionable [14]; the nonlinearity may drastically change the character of wave propagation, leading to higher harmonics generation or results in solitons formation [15]. As it is conventionally performed, in the case of weak nonlinearity considered here, we use linear links for the variables in the nonlinear terms. It leads to a complex calculation, which we would like to simplify in some way. Going back to the definition of **D** (7.5) and the definition of polarization vector (7.6), the nonlinear Maxwell equations in cylindrical coordinates have the form

$$\frac{\partial}{r\partial r}(r(D_L)_r)+\frac{\partial(D_L)_\varphi}{r\partial\varphi}+\frac{\partial(D_L)_z}{\partial z}$$

$$+\frac{\partial}{r\partial r}(r(P_{NL})_r)+\frac{\partial(P_{NL})_\varphi}{r\partial\varphi}+\frac{\partial(P_{NL})_z}{\partial z}=0, \tag{9.134}$$

$$\frac{\partial}{r\partial r}(rB_r)+\frac{\partial B_\varphi}{r\partial\varphi}+\frac{\partial B_z}{\partial z}=0, \tag{9.135}$$

$$\left(\frac{\partial E_z}{r\partial\varphi}-\frac{\partial E_\varphi}{\partial z}\right)=-\frac{1}{c}\frac{\partial B_r}{\partial t}, \tag{9.136}$$

$$\left(\frac{\partial E_r}{\partial z}-\frac{\partial E_z}{\partial r}\right)=-\frac{1}{c}\frac{\partial B_\varphi}{\partial t}, \tag{9.137}$$

$$\frac{1}{r}\left(\frac{\partial(rE_\varphi)}{\partial r}-\frac{\partial E_r}{\partial\varphi}\right)=-\frac{1}{c}\frac{\partial B_z}{\partial t}, \tag{9.138}$$

$$\left(\frac{\partial H_z}{r\partial\varphi}-\frac{\partial H_\varphi}{\partial z}\right)=\frac{1}{c}\frac{\partial(D_L)_r}{\partial t}+\frac{1}{c}\frac{\partial(P_{NL})_r}{\partial t}, \tag{9.139}$$

$$\left(\frac{\partial H_r}{\partial z}-\frac{\partial H_z}{\partial r}\right)=\frac{1}{c}\frac{\partial(D_L)_\varphi}{\partial t}+\frac{1}{c}\frac{\partial(P_{NL})_\varphi}{\partial t}, \tag{9.140}$$

$$\frac{1}{r}\left(\frac{\partial(rH_\varphi)}{\partial r}-\frac{\partial H_r}{\partial\varphi}\right)=\frac{1}{c}\frac{\partial(D_L)_z}{\partial t}+\frac{1}{c}\frac{\partial(P_{NL})_z}{\partial t}, \tag{9.141}$$

where the linear part of the electric induction still represents by integral operator

$$\mathbf{D}_L=\int_{-\infty}^{\infty}\int_{-\infty}^{\infty}\varepsilon(\omega)e^{-i\omega(t-t')}d\omega\mathbf{E}(t')dt'=\hat{\varepsilon}\mathbf{E}. \tag{9.142}$$

Further, the electromagnetic field components (9.57) will be presented with the substitution of derived linear relations for the introduced coefficients (9.101), using only four two-dimensional fields

$$E_z(r,\varphi,z,t) = \sum_{l,n} \mathscr{A}_{ln}(z,t) J_l(\alpha_{nl} r) e^{il\varphi} + c.c., \tag{9.143a}$$

$$E_r(r,\varphi,z,t) = -\sum_{l,n} \frac{\partial}{\partial t} \frac{1}{c\alpha_{ln}^2} \mathscr{K}_{ln}(z,t) \frac{il}{r} J_l(\alpha_{ln} r) e^{il\varphi}$$

$$+ \sum_{l,n} \mathscr{C}_{ln}(z,t) \frac{\partial}{\partial r} J_l(\alpha_{ln} r) e^{il\varphi} + c.c., \tag{9.143b}$$

$$E_\varphi(r,\varphi,z,t) = \sum_{l,n} \mathscr{C}_{ln}(z,t) \frac{il}{r} J_l(\alpha_{ln} r) e^{il\varphi} + \sum_{l,n} \frac{\partial}{\partial t} \frac{1}{c\alpha_{ln}^2} \mathscr{K}_{ln}(z,t) \frac{\partial}{\partial r} J_l(\alpha_{ln} r) e^{il\varphi} + c.c., \tag{9.143c}$$

$$B_z(r,\varphi,z,t) = \sum_{l,n} \mathscr{K}_{ln}(z,t) J_l(\alpha_{nl} r) e^{il\varphi} + c.c., \tag{9.143d}$$

$$B_r(r,\varphi,z,t) = \sum_{l,n} \mathscr{M}_{ln}(z,t) \frac{\partial}{\partial r} J_l(\alpha_{ln} r) e^{il\varphi}$$

$$+ \sum_{l,n} \frac{\partial}{\partial t} \frac{\hat{\varepsilon}\mu}{\alpha_{ln}^2 c} \mathscr{A}_{ln}(z,t) \frac{il}{r} J_l(\alpha_{ln} r) e^{il\varphi} + c.c., \tag{9.143e}$$

$$B_\varphi(r,\varphi,z,t) = \sum_{l,n} \mathscr{M}_{ln}(z,t) \frac{il}{r} J_l(\alpha_{ln} r) e^{il\varphi}$$

$$- \sum_{l,n} \frac{\partial}{\partial t} \frac{\hat{\varepsilon}\mu}{\alpha_{ln}^2 c} \mathscr{A}_{ln}(z,t) \frac{\partial}{\partial r} J_l(\alpha_{ln} r) e^{il\varphi} + c.c.. \tag{9.143f}$$

The nonlinear part $(\mathbf{P}_{NL})_i$ in previous chapters was conventionally expressed in Cartesian coordinates. Each component $(\mathbf{P}_{NL})_i$ should be transformed to cylindrical coordinates. To simplify the calculations, one can use the third-order susceptibility tensor expression [12,13]. The components $(\mathbf{P}_{NL})_i$ for isotropic materials with a center of symmetry may be reduced as $\chi_{ijkl} = \chi_{iikk}\delta_{ij}\delta_{kl} + \chi_{ijij}\delta_{ik}\delta_{jl} + \chi_{ijji}\delta_{il}\delta_{jk}$, or in full view,

$$\begin{aligned}
(\mathbf{P}_{NL})_x =& \chi_{xxxx}E_x^3 + \chi_{xxyy}E_xE_y^2 + \chi_{xxzz}E_xE_zE_z + \chi_{xxxx}E_xE_xE_x \\
&+ \chi_{xyxy}E_yE_xE_y + \chi_{xzxz}E_zE_xE_z + \chi_{xxxx}E_xE_xE_x + \chi_{xyyx}E_yE_yE_x + \chi_{xzzx}E_zE_zE_x \\
(\mathbf{P}_{NL})_y =& \chi_{yyyy}E_yE_yE_y + \chi_{yyxx}E_yE_xE_x + \chi_{yyzz}E_yE_zE_z + \chi_{yyyy}E_yE_yE_y \\
&+ \chi_{yxyx}E_xE_yE_x + \chi_{yzyz}E_zE_yE_z + \chi_{yyyy}E_yE_yE_y + \chi_{yxxy}E_xE_xE_y + \chi_{yzzy}E_zE_zE_y \\
(\mathbf{P}_{NL})_z =& \chi_{zzzz}E_zE_zE_z + \chi_{zzyy}E_zE_yE_y + \chi_{zzxx}E_zE_xE_x + \chi_{zzzz}E_zE_zE_z \\
&+ \chi_{zyzy}E_yE_zE_y + \chi_{zxzx}E_xE_zE_x + \chi_{zzzz}E_zE_zE_z + \chi_{zyyz}E_yE_yE_z + \chi_{zxxz}E_xE_xE_z
\end{aligned} \tag{9.144}$$

With use of properties given in the mentioned books [12,13],

$$\chi_{xxxx} = \chi_{yyyy}, \quad \chi_{xxyy} = \chi_{yyxx}, \quad \chi_{xyxy} = \chi_{yxyx}, \quad \chi_{xyyx} = \chi_{yxxy}, \tag{9.145}$$

the expressions for the nonlinear polarization vector simplifies as

$$\begin{aligned}
(\mathbf{P}_{NL})_x &= 3\chi_{xxxx}E_x^3 + (\chi_{xxyy} + \chi_{xyxy} + \chi_{xyyx})E_y^2 E_x + (\chi_{xxzz} + \chi_{xzxz} + \chi_{xzzx})E_z^2 E_x, \\
(\mathbf{P}_{NL})_y &= 3\chi_{yyyy}E_y^3 + (\chi_{yyxx} + \chi_{yxyx} + \chi_{yxxy})E_x^2 E_y + (\chi_{yyzz} + \chi_{yzyz} + \chi_{yzzy})E_z^2 E_y, \\
(\mathbf{P}_{NL})_z &= 3\chi_{zzzz}E_z^3 + (\chi_{zzyy} + \chi_{zyzy} + \chi_{zyyz})E_y^2 E_z + (\chi_{zzxx} + \chi_{zxzx} + \chi_{zxxz})E_x^2 E_z.
\end{aligned} \tag{9.146}$$

One more property from the mentioned monograph $\chi_{xxxx} = \chi_{yyyy} = \chi_{zzzz} = \chi_{xxyy} + \chi_{xyyx} + \chi_{xyxy}$ finally simplifies the above expressions (9.146) for the components

$$\begin{aligned}
(\mathbf{P}_{NL})_x &= \chi_{zzzz}\left(3E_x^3 + E_y^2 E_x + E_z^2 E_x\right), \\
(\mathbf{P}_{NL})_y &= \chi_{zzzz}\left(3E_y^3 + E_x^2 E_y + E_z^2 E_y\right), \\
(\mathbf{P}_{NL})_z &= \chi_{zzzz}\left(3E_z^3 + E_y^2 E_z + E_x^2 E_z\right).
\end{aligned} \tag{9.147}$$

Expressions (9.146) are generalized as compared to Eq. (7.182) obtained in Section 7.4. In an optical waveguide, the polarization vector has the only transversal to the propagation directions variables, which the propagation direction coordinates neglect. However, in this model this assumption will not be used at this time. The polarization in cylindrical coordinates has the form $E_x = E_r \cos\varphi - E_\varphi \sin\varphi$, $E_y = E_r \sin\varphi + E_\varphi \cos\varphi$, $E_z = E_z$.

$$\begin{aligned}
(\mathbf{P}_{NL})_r =& \chi_{zzzz}\left(3(E_r\cos\varphi - E_\varphi\sin\varphi)^3 + (E_r\sin\varphi + E_\varphi\cos\varphi)^2\right. \\
&\left.\times(E_r\cos\varphi - E_\varphi\sin\varphi) + E_z^2(E_r\cos\varphi - E_\varphi\sin\varphi)\right), \\
(\mathbf{P}_{NL})_\varphi =& \chi_{zzzz}\left(3(E_r\sin\varphi + E_\varphi\cos\varphi)^3 + (E_r\cos\varphi - E_\varphi\sin\varphi)^2\right. \\
&\left.(E_r\sin\varphi + E_\varphi\cos\varphi) + E_z^2(E_r\sin\varphi + E_\varphi\cos\varphi)\right), \\
(\mathbf{P}_{NL})_z =& \chi_{zzzz}\left(3E_z^3 + (E_r\sin\varphi + E_\varphi\cos\varphi)^2 E_z + (E_r\cos\varphi - E_\varphi\sin\varphi)^2 E_z\right).
\end{aligned} \tag{9.148}$$

After few steps with use of trigonometric function transformations the formulas for the polarization coordinates (9.148) is written as

$$\begin{aligned}
(\mathbf{P}_{NL})_r =& \chi_{zzzz}\Big(3(E_r^3\cos^3\varphi - 3E_r^2E_\varphi\cos^2\varphi\sin\varphi + 3E_rE_\varphi^2\cos\varphi\sin^2\varphi - E_\varphi^3\sin^3\varphi) \\
&+ (E_r^2\sin^2\varphi + 2E_rE_\varphi\sin\varphi\cos\varphi + E_\varphi^2\cos^2\varphi)(E_r\cos\varphi - E_\varphi\sin\varphi) \\
&+ E_z^2(E_r\cos\varphi - E_\varphi\sin\varphi)\Big),
\end{aligned}$$

$$(\mathbf{P}_{NL})_\varphi = \chi_{zzzz}\Big(3(E_r^3\sin^3\varphi + 3E_r^2E_\varphi\sin^2\varphi\cos\varphi + 3E_\varphi^2E_r\cos^2\varphi\sin\varphi + E_\varphi^3\cos^3\varphi)$$
$$+ (E_r^2\cos^2\varphi - 2E_r\cos\varphi E_\varphi\sin\varphi + E_\varphi^2\sin^2\varphi)(E_r\sin\varphi + E_\varphi\cos\varphi)$$
$$+ E_z^2(E_r\sin\varphi + E_\varphi\cos\varphi)\Big),$$
$$(\mathbf{P}_{NL})_z = \chi_{zzzz}\Big(3E_z^3 + (E_r^2\sin^2\varphi + 2E_rE_\varphi\sin\varphi\cos\varphi + E_\varphi^2\cos^2\varphi)E_z$$
$$+ (E_r^2\cos^2\varphi - 2E_rE_\varphi\cos\varphi\sin\varphi + E_\varphi^2\sin^2\varphi)E_z\Big). \qquad (9.149)$$

After such tedious work, we arrive at some rather concise expressions:

$$(\mathbf{P}_{NL})_r = \chi_{zzzz}\Big(E_r^3(3\cos^3\varphi + \sin^2\varphi\cos\varphi) - E_r^2E_\varphi(7\sin\varphi\cos^2\varphi + \sin^3\varphi)$$
$$+ E_rE_\varphi^2(7\sin^2\varphi\cos\varphi + \cos^3\varphi) - E_\varphi^3(3\sin^3\varphi + \sin\varphi\cos^2\varphi)$$
$$+ E_z^2(E_r\cos\varphi - E_\varphi\sin\varphi)\Big),$$
$$(\mathbf{P}_{NL})_\varphi = \chi_{zzzz}\Big(E_r^3(3\sin^3\varphi + \sin\varphi\cos^2\varphi) + E_r^2E_\varphi(7\sin^2\varphi\cos\varphi + \cos^3\varphi)$$
$$+ E_\varphi^2E_r(7\cos^2\varphi\sin\varphi + \sin^3\varphi) + E_\varphi^3(3\cos^3\varphi + \sin^2\varphi\cos\varphi)$$
$$+ E_z^2(E_r\sin\varphi + E_\varphi\cos\varphi)\Big),$$
$$(\mathbf{P}_{NL})_z = \chi_{zzzz}\Big(3E_z^3 + (E_r^2 + E_\varphi^2)E_z\Big). \qquad (9.150)$$

At this point, the nonlinear part can be inserted into the Maxwell system (9.134). Presented in the basis of (9.143) Maxwell system will be reduced to a form which, after simplification, can be described as the system of the following equations:

$$\sum_{l,n}\Big(\frac{\partial}{\partial z}\mathscr{K}_{ln}(z,t)\Big)J_l(\alpha_{nl}r)e^{il\varphi} - \mathscr{M}_{ln}(z,t)\alpha_{ln}^2 + +c.c. = 0, \qquad (9.151a)$$

$$\sum_{l,n}\Big(\mathscr{A}_{ln}(z,t) - \frac{\partial}{\partial z}\mathscr{C}_{ln}(z,t) + \frac{\partial^2}{\partial t^2}\frac{\hat{\varepsilon}\mu}{\alpha_{ln}^2c^2}\mathscr{A}_{ln}(z,t)\Big)\frac{il}{r}J_l(\alpha_{ln}r)e^{il\varphi} + c.c. = 0$$
$$(9.151b)$$

$$\sum_{l,n}\Big(\mathscr{K}_{ln}(z,t) - \frac{\partial}{\partial z}\mathscr{M}_{ln}(z,t) + \frac{\hat{\varepsilon}\mu}{c}\frac{\partial^2}{\partial t^2}\frac{1}{c\alpha_{ln}^2}\mathscr{K}_{ln}(z,t)\Big)\frac{il}{r}J_l(\alpha_{ln}r)e^{il\varphi} + c.c.$$
$$(9.151c)$$
$$= \sum_{l,n}\frac{\partial}{\partial t}\Big(-\frac{\partial}{\partial z}\frac{\hat{\varepsilon}\mu}{\alpha_{ln}^2c}\mathscr{A}_{ln}(z,t) + \frac{\hat{\varepsilon}\mu}{c}\mathscr{C}_{ln}(z,t)\Big)\frac{\partial}{\partial r}J_l(\alpha_{ln}r)e^{il\varphi} + \frac{\mu}{c}\frac{\partial}{\partial t}(P_{NL})_r + c.c.$$
$$(9.151d)$$

$$\sum_{l,n}\left(-\hat{\varepsilon}\mu\frac{\partial^2}{\partial t^2}\frac{1}{c^2\alpha_{ln}^2}\mathscr{K}_{ln}(z,t)+\frac{\partial}{\partial z}\mathscr{M}_{ln}(z,t)-\mathscr{K}_{ln}(z,t)\right)\frac{\partial}{\partial r}J_l(\alpha_{nl}r)e^{il\varphi}+c.c. \tag{9.151e}$$

$$=\sum_{l,n}\left(\frac{\partial}{\partial z}\frac{\partial}{\partial t}\frac{-\hat{\varepsilon}\mu}{\alpha_{ln}^2 c}\mathscr{A}_{ln}(z,t)+\frac{1}{c}\frac{\partial}{\partial t}\hat{\varepsilon}\mu\mathscr{C}_{ln}(z,t)\right)\frac{il}{r}J_l(\alpha_{ln}r)e^{il\varphi}$$
$$+\frac{\mu}{c}\frac{\partial}{\partial t}(P_{NL})_{\varphi}+c.c. \tag{9.151f}$$

This leads to the evolution equation in standard form $\partial_z\Psi(z,t)-\hat{L}\Psi(z,t)=\mathscr{N}(\Psi)$, which allows us to launch the procedure of principal modes evolution extraction. However to realize the program, the third and fourth equation have to be rebuilt with use of Bessel functions properties (9.69). To form a nonlinear term $\mathscr{N}(\Psi)$ it is crucial to find the proper relation between variables $\mathscr{A}_{ln}$, $\mathscr{C}_{ln}$ and variables $\mathscr{K}_{ln}$, $\mathscr{M}_{ln}$ in linear approximation, so that the nonlinear part may be expressed in the same variables. By simple algebraic operations, Bessel functions $J_{l-1}(\alpha_{nl}r)$ and $J_{l+1}(\alpha_{nl}r)$ will be separated,

$$\sum_{l,n}i\left(\frac{\hat{\varepsilon}\mu}{c^2\alpha_{ln}^2}\frac{\partial^2}{\partial t^2}\mathscr{K}_{ln}(z,t)+\mathscr{K}_{ln}(z,t)-\frac{\partial}{\partial z}\mathscr{M}_{ln}(z,t)\right)J_{l-1}(\alpha_{ln}r)e^{il\varphi}+c.c.$$
$$=\sum_{l,n}\frac{\hat{\varepsilon}\mu}{c}\frac{\partial}{\partial t}\left(\mathscr{C}_{ln}(z,t)-\frac{\partial}{\partial z}\frac{1}{\alpha_{ln}^2}\mathscr{A}_{ln}(z,t)+\right)J_{l-1}(\alpha_{nl}r)e^{il\varphi}$$
$$+\frac{\mu}{c}\frac{\partial(P_{NL})_r}{\partial t}-\frac{i\mu}{c}\frac{\partial}{\partial t}(P_{NL})_{\varphi}, \tag{9.152}$$

almost the same way we have with $J_{l+1}(\alpha_{ln}r)$.

Now the results can be multiplied by $e^{-il'\varphi}$ and integrated over φ to rely upon the exponents orthogonality:

$$\int_0^{2\pi}\sum_{l,n}i\left(\frac{\hat{\varepsilon}\mu}{c^2\alpha_{ln}^2}\frac{\partial^2}{\partial t^2}\mathscr{K}_{ln}(z,t)+\mathscr{K}_{ln}(z,t)-\frac{\partial}{\partial z}\mathscr{M}_{ln}(z,t)\right)J_{l-1}(\alpha_{nl}r)e^{il\varphi}e^{-il'\varphi}d\varphi$$
$$+c.c.=\int_0^{2\pi}\sum_{l,n}\frac{\hat{\varepsilon}\mu}{c}\frac{\partial}{\partial t}\left(\mathscr{C}_{ln}(z,t)-\frac{\partial}{\partial z}\frac{1}{\alpha_{ln}^2}\mathscr{A}_{ln}(z,t)+\right)J_{l-1}(\alpha_{ln}r)e^{il\varphi}e^{-il'\varphi}d\varphi$$
$$+\int_0^{2\pi}e^{-il'\varphi}\left(\frac{\mu}{c}\frac{\partial(P_{NL})_r}{\partial t}-\frac{i\mu}{c}\frac{\partial}{\partial t}(P_{NL})_{\varphi}\right)d\varphi, \tag{9.153}$$

acting similarly with $J_{l+1}(\alpha_{ln}r)$-terms.

The orthogonality of Bessel functions also should be used (both equations will be multiplied by Bessel function $rJ_{l\pm1}(\alpha_{ln'}r)$ and integrated over $r \in [0, r_0]$), which can be applied to the above equation, and here we show the first of them:

$$\int_0^{r_0} \sum_n i\left(\frac{\hat{\varepsilon}\mu}{c^2\alpha_{ln}^2}\frac{\partial^2}{\partial t^2}\mathscr{K}_{ln}(z,t) + \mathscr{K}_{ln}(z,t) - \frac{\partial}{\partial z}\mathscr{M}_{ln}(z,t)\right) rJ_{l-1}(\alpha_{ln}r)J_{l-1}(\alpha_{ln'}r)dr$$

$$+c.c. - \int_0^{r_0} \sum_n \frac{\hat{\varepsilon}\mu}{c}\frac{\partial}{\partial t}\left(\mathscr{C}_{ln}(z,t) - \frac{\partial}{\partial z}\frac{1}{\alpha_{ln}^2}\mathscr{A}_{ln}(z,t) + \right) J_{l-1}(\alpha_{ln}r)rJ_{l-1}(\alpha_{ln'}r)dr$$

$$= \int_0^{r_0}\int_0^{2\pi} e^{-il\varphi}\left(\frac{\mu}{c}\frac{\partial(P_{NL})_r}{\partial t} - \frac{i\mu}{c}\frac{\partial}{\partial t}(P_{NL})_\varphi\right) rJ_{l-1}(\alpha_{ln'}r)drd\varphi. \tag{9.154}$$

Based on the relation of Bessel functions orthogonality

$$\int_0^{r_0} J_{l\pm1}(\alpha_{ln}r)rJ_{l\pm1}(\alpha_{ln'}r)dr = N_{l\pm1}\delta_{n,n'}, \tag{9.155}$$

and the celebrated

$$\int_0^{2\pi} e^{-il'\varphi}e^{il\varphi}d\phi = 2\pi\delta_{ll'},$$

the sums from our equations go away, which gives

$$i\left(\frac{\hat{\varepsilon}\mu}{c^2\alpha_{ln}^2}\frac{\partial^2}{\partial t^2}\mathscr{K}_{ln}(z,t) + \mathscr{K}_{ln}(z,t) - \frac{\partial}{\partial z}\mathscr{M}_{ln}(z,t)\right) N_{l-1}$$

$$+c.c. - \frac{\hat{\varepsilon}\mu}{c}\frac{\partial}{\partial t}\left(\mathscr{C}_{ln}(z,t) - \frac{\partial}{\partial z}\frac{1}{\alpha_{ln}^2}\mathscr{A}_{ln}(z,t) + \right) N_{l-1}$$

$$= \frac{1}{2\pi}\int_0^{r_0}\int_0^{2\pi} e^{-il\varphi}\left(\frac{\mu}{c}\frac{\partial(P_{NL})_r}{\partial t} - \frac{i\mu}{c}\frac{\partial}{\partial t}(P_{NL})_\varphi\right) rJ_{l-1}(\alpha_{ln}r)drd\varphi, \tag{9.156}$$

and

$$i\left(\frac{\hat{\varepsilon}\mu}{c^2\alpha_{ln}^2}\frac{\partial^2}{\partial t^2}\mathscr{K}_{ln}(z,t) + \mathscr{K}_{ln}(z,t) - \frac{\partial}{\partial z}\mathscr{M}_{ln}(z,t)\right) N_{l+1}$$

$$+c.c. + \frac{\hat{\varepsilon}\mu}{c}\frac{\partial}{\partial t}\left(\mathscr{C}_{ln}(z,t) - \frac{\partial}{\partial z}\frac{1}{\alpha_{ln}^2}\mathscr{A}_{ln}(z,t) + \right) N_{l+1}$$

$$= \frac{1}{2\pi}\int_0^{r_0}\int_0^{2\pi} e^{-il\varphi}\left(\frac{\mu}{c}\frac{\partial(P_{NL})_r}{\partial t} + \frac{i\mu}{c}\frac{\partial}{\partial t}(P_{NL})_\varphi\right) rJ_{l+1}(\alpha_{ln}r)drd\varphi. \tag{9.157}$$

To separate the expressions $\frac{\hat{\varepsilon}\mu}{c^2\alpha_{ln}^2}\frac{\partial^2}{\partial t^2}\mathscr{K}_{ln}(z,t)-\mathscr{K}_{ln}(z,t)+\frac{\partial}{\partial z}\mathscr{M}_{ln}(z,t)$ and $\mathscr{C}_{ln}(z,t)-\frac{\partial}{\partial z}\frac{1}{\alpha_{ln}^2}\mathscr{A}_{ln}(z,t)+$ from the above system it is necessary to divide each equation by the normalization parameter $N_{l\pm1}$ and to sum by side both equations. The result

$$i\left(\frac{\hat{\varepsilon}\mu}{c^2\alpha_{ln}^2}\frac{\partial^2}{\partial t^2}\mathscr{K}_{ln}(z,t)-\mathscr{K}_{ln}(z,t)+\frac{\partial}{\partial z}\mathscr{M}_{ln}(z,t)\right)$$
$$+c.c.=\frac{1}{4\pi N_{l-1}}\int_0^{r_0}\int_0^{2\pi}e^{-il\varphi}\left(\frac{\mu}{c}\frac{\partial(P_{NL})_r}{\partial t}-\frac{i\mu}{c}\frac{\partial}{\partial t}(P_{NL})_\varphi\right)rJ_{l-1}(\alpha_{ln'}r)drd\varphi,$$
$$+\frac{1}{4\pi N_{l+1}}\int_0^{r_0}\int_0^{2\pi}e^{-il\varphi}\left(\frac{\mu}{c}\frac{\partial(P_{NL})_r}{\partial t}+\frac{i\mu}{c}\frac{\partial}{\partial t}(P_{NL})_\varphi\right)rJ_{l+1}(\alpha_{ln}r)drd\varphi, \tag{9.158}$$

and one obtained by subtraction of the equations.

$$\mathscr{C}_{ln}(z,t)-\frac{\partial}{\partial z}\frac{1}{\alpha_{ln}^2}\mathscr{A}_{ln}(z,t)+c.c.$$
$$=-\frac{1}{4\pi N_{l-1}}\int_0^{r_0}\int_0^{2\pi}e^{-il\varphi}\left(\frac{\mu}{c}\frac{\partial(P_{NL})_r}{\partial t}-\frac{i\mu}{c}\frac{\partial}{\partial t}(P_{NL})_\varphi\right)rJ_{l-1}(\alpha_{ln'}r)drd\varphi,$$
$$+\frac{1}{4\pi N_{l+1}}\int_0^{r_0}\int_0^{2\pi}e^{-il\varphi}\left(\frac{\mu}{c}\frac{\partial(P_{NL})_r}{\partial t}+\frac{i\mu}{c}\frac{\partial}{\partial t}(P_{NL})_\varphi\right)rJ_{l+1}(\alpha_{ln}r)drd\varphi, \tag{9.159}$$

close the basic system.

Finally, for the choice of real amplitudes, the system of four equations

$$\frac{\partial}{\partial z}\mathscr{A}_{ln}(z,t)-\alpha_{ln}^2\mathscr{C}_{ln}(z,t)=\mathscr{N}_1 \tag{9.160}$$

$$\frac{\partial}{\partial z}\mathscr{C}_{ln}(z,t)-\frac{\partial^2}{\partial t^2}\frac{\hat{\varepsilon}\mu}{\alpha_{ln}^2c^2}\mathscr{A}_{ln}(z,t)-\mathscr{A}_{ln}(z,t))=0 \tag{9.161}$$

$$\frac{\partial}{\partial z}\mathscr{K}_{ln}(z,t)-\alpha_{ln}^2\mathscr{M}_{ln}(z,t)=0, \tag{9.162}$$

$$\frac{\partial}{\partial z}\mathscr{M}_{ln}(z,t)-\frac{\hat{\varepsilon}\mu}{c^2\alpha_{ln}^2}\frac{\partial^2}{\partial t^2}\mathscr{K}_{ln}(z,t)-\mathscr{K}_{ln}(z,t)=\mathscr{N}_2 \tag{9.163}$$

defines no new, “nonlinear” amplitudes, where

$$\mathscr{N}_1=\frac{\alpha_{ln}^2}{4\pi N_{l-1}}\int_0^{r_0}\int_0^{2\pi}e^{-il\varphi}\left(\frac{\mu}{c}\frac{\partial(P_{NL})_r}{\partial t}-\frac{i\mu}{c}\frac{\partial}{\partial t}(P_{NL})_\varphi\right)rJ_{l-1}(\alpha_{ln'}r)drd\varphi,$$
$$-\frac{\alpha_{ln}^2}{4\pi N_{l+1}}\int_0^{r_0}\int_0^{2\pi}e^{-il\varphi}\left(\frac{\mu}{c}\frac{\partial(P_{NL})_r}{\partial t}+\frac{i\mu}{c}\frac{\partial}{\partial t}(P_{NL})_\varphi\right)rJ_{l+1}(\alpha_{ln}r)drd\varphi, \tag{9.164}$$

$$\mathcal{N}_2 = \frac{-i\alpha_{ln}^2}{4\pi N_{l-1}} \int_0^{r_0} \int_0^{2\pi} e^{-il\varphi} \left(\frac{\mu}{c} \frac{\partial (P_{NL})_r}{\partial t} - \frac{i\mu}{c} \frac{\partial}{\partial t} (P_{NL})_\varphi \right) r J_{l-1}(\alpha_{ln'} r) dr d\varphi,$$

$$+ \frac{-i\alpha_{ln}^2}{4\pi N_{l+1}} \int_0^{r_0} \int_0^{2\pi} e^{-il\varphi} \left(\frac{\mu}{c} \frac{\partial (P_{NL})_r}{\partial t} + \frac{i\mu}{c} \frac{\partial}{\partial t} (P_{NL})_\varphi \right) r J_{l+1}(\alpha_{ln} r) dr d\varphi \tag{9.165}$$

can be presented in a form $\partial_z \Psi(z,t) - \hat{L}\Psi(z,t) = \mathcal{N}(\Psi)$ where

$$\Psi(z,t) = \begin{pmatrix} \mathcal{A}_{ln}(z,t) \\ \mathcal{C}_{ln}(z,t) \\ \mathcal{K}_{ln}(z,t) \\ \mathcal{M}_{ln}(z,t) \end{pmatrix}, \quad \hat{L} = \begin{pmatrix} 0 & \alpha_{ln}^2 & 0 & 0 \\ \frac{\hat{\varepsilon}\mu}{\alpha_{ln}^2 c^2} \frac{\partial^2}{\partial t^2} + 1 & 0 & 0 & 0 \\ 0 & 0 & 0 & \alpha_{ln}^2 \\ 0 & 0 & \frac{\hat{\varepsilon}\mu}{\alpha_{ln}^2 c^2} \frac{\partial^2}{\partial t^2} + 1 & 0 \end{pmatrix},$$

$$\mathcal{N}(\Psi) = \begin{pmatrix} \mathcal{N}_1 \\ 0 \\ 0 \\ \mathcal{N}_2 \end{pmatrix}. \tag{9.166}$$

9.6.1 APPLICATION OF PROJECTION OPERATORS

As all components to the equation are given, it is possible to proceed with the procedure of the projection operator's application. The P_{11} operator (look the expressions (9.133)) is chosen as the exemplary one applied to the presented equation (9.166)

$$\frac{\partial}{\partial z} P_{11} \begin{pmatrix} \mathcal{A}_{ln}(z,t) \\ \mathcal{C}_{ln}(z,t) \\ \mathcal{K}_{ln}(z,t) \\ \mathcal{M}_{ln}(z,t) \end{pmatrix} - P_{11} \begin{pmatrix} 0 & \alpha_{ln}^2 & 0 & 0 \\ +\frac{\partial^2}{\partial t^2} \frac{\hat{\varepsilon}\mu}{\alpha_{ln}^2 c^2} + 1 & 0 & 0 & 0 \\ 0 & 0 & 0 & \alpha_{ln}^2 \\ 0 & 0 & \frac{\partial^2}{\partial t^2} \frac{\hat{\varepsilon}\mu}{\alpha_{ln}^2 c^2} + 1 & 0 \end{pmatrix}$$

$$\times \begin{pmatrix} \mathcal{A}_{ln}(z,t) \\ \mathcal{C}_{ln}(z,t) \\ \mathcal{K}_{ln}(z,t) \\ \mathcal{M}_{ln}(z,t) \end{pmatrix} = P_{11} \begin{pmatrix} \mathcal{N}_1 \\ 0 \\ 0 \\ \mathcal{N}_2 \end{pmatrix}. \tag{9.167}$$

As the result of projection, the equation is acquired:

$$\frac{\partial}{\partial z}\begin{pmatrix} i\alpha_{ln}^2\hat{k}_{nl}^{-1}\Lambda_1(z,t) \\ \Lambda_1(z,t) \\ 0 \\ 0 \end{pmatrix} - \begin{pmatrix} 0 & \alpha_{ln}^2 & 0 & 0 \\ \frac{\partial^2}{\partial t^2}\frac{\hat{o}ver\hat{\varepsilon}\mu}{\alpha_{ln}^2 c^2}+1 & 0 & 0 & 0 \\ 0 & 0 & 0 & \alpha_{ln}^2 \\ 0 & 0 & \frac{\partial^2}{\partial t^2}\frac{\hat{o}ver\hat{\varepsilon}\mu}{\alpha_{ln}^2 c^2}+1 & 0 \end{pmatrix}$$

$$\times \begin{pmatrix} i\alpha_{ln}^2\hat{k}_{nl}^{-1}\Lambda_1(z,t) \\ \Lambda_1(z,t) \\ 0 \\ 0 \end{pmatrix} = \frac{1}{2}\begin{pmatrix} \mathscr{N}_1 \\ \frac{i}{\alpha_{ln}^2}\hat{k}_{nl}\mathscr{N}_1 \\ 0 \\ 0 \end{pmatrix}.$$

Comparing the dispersion relation (9.106) with the $\hat{L}$ operator, it is possible to replace $\frac{\partial^2}{\partial t^2}\frac{\hat{\varepsilon}\mu}{\alpha_{ln}^2 c^2}+1$ by $\alpha_{ln}^{-2}\hat{k}_{nl}^2$. The final general equation describing the propagation of the polarized, nonlinear electromagnetic wave to the left side has the form

$$\frac{\partial}{\partial z}\Lambda_1(z,t) + i\hat{k}_{nl}\Lambda_1(z,t) = \frac{1}{2}(i\alpha_{ln}^{-2}\hat{k}_{nl})\mathscr{N}_1, \tag{9.168}$$

which can be rewritten in an explicit operator form (recall the integral operator $\hat{k}_{nl}$ definition):

$$\frac{\partial}{\partial z}\Lambda_1(z,t) + i\hat{k}_{nl}\Lambda_1(z,t) = \frac{i\mu\hat{k}_{nl}}{8\pi c N_{l-1}}\int_0^{r_0}\int_0^{2\pi} e^{-il\varphi}\left(\frac{\partial (P_{NL})_r}{\partial t} - i\frac{\partial}{\partial t}(P_{NL})_\varphi\right)$$
$$\times rJ_{l-1}(\alpha_{ln}r)drd\varphi + \frac{i\mu\hat{k}_{nl}}{8\pi c N_{l+1}}\int_0^{r_0}\int_0^{2\pi} e^{-il\varphi}\left(\frac{\partial (P_{NL})_r}{\partial t} + i\frac{\partial}{\partial t}(P_{NL})_\varphi\right)$$
$$\times rJ_{l+1}(\alpha_{ln}r)drd\varphi\Big). \tag{9.169}$$

The final result presents a complex equation that describes implicitly the interaction between pulse polarized components through nonlinearity. The obtained final equation is a version of the results presented in papers [15–18] obtained by the projecting technique. It is dedicated to the ultra-short pulses propagation theory that was derived directly with explicit projection operators. The presented result was derived in the same unit system as the one used by Leble and Reichel in their works. Thus, it is possible to use formulas for nonlinear coefficients that were calculated in the Leble and Reichel works. Such a program is realized via the application of a symbolic program that picks up the components of the electromagnetic field (9.143), expressed in terms of the mode fields by (9.123) taken in t-domain:

$$\begin{aligned} \mathscr{A}_{ln}(z,\omega) &= \frac{i\alpha_{ln}^2}{2\hat{k}_{nl}}(\Lambda_1(z,\omega) - \Pi_1(z,\omega)), \\ \mathscr{C}_{ln}(z,\omega) &= \tfrac{1}{2}(\Lambda_1(z,\omega) + \Pi_1(z,\omega)) \\ \mathscr{K}_{ln}(z,\omega) &= \frac{i\alpha_{ln}^2}{2\hat{k}_{nl}}(\Lambda_2(z,\omega) - \Pi_2(z,\omega)) \\ \mathscr{M}_{ln}(z,\omega) &= \tfrac{1}{2}(\Lambda_2(z,\omega) + \Pi_2(z,\omega)). \end{aligned} \tag{9.170}$$

This should be plugged into expressions for the nonlinear polarization terms (9.150). The algorithm is quite transparent; such work "by hands" is also possible, but very complicated.

9.7 APPENDIX

Here, we build the projecting matrix operators for the evolution operator of the form typical for electrodynamics, see Section 9.1. During calculations, we will use temporary notations for shorthand, rewriting the result in terms of the matrices of Section 9.1.

Let the evolution operator have the structure

$$\mathscr{L} = \begin{pmatrix} 0 & L \\ M & 0 \end{pmatrix}, \tag{9.171}$$

then it is convenient to introduce two-component columns. For α it is

$$\begin{pmatrix} 1 \\ \gamma \end{pmatrix}, \tag{9.172}$$

which guarantees the standard normalization whereas the column β has components to be expressed in terms of α. The eigen problem

$$\begin{pmatrix} 0 & L \\ M & 0 \end{pmatrix} \begin{pmatrix} \alpha \\ \beta \end{pmatrix} = \lambda \begin{pmatrix} \alpha \\ \beta \end{pmatrix} \tag{9.173}$$

may be reformulated as the system

$$\begin{aligned} L\beta &= \lambda\alpha \\ M\alpha &= \lambda\beta \end{aligned} \tag{9.174}$$

or, having the direct link

$$\beta = \lambda^{-1} B\alpha, \tag{9.175}$$

we write

$$LM\alpha = \lambda^2\alpha. \tag{9.176}$$

Denoting the 2×2 matrix as $C = LM$ and its eigenvalue as $\tilde{\mu} = \lambda^2$, we write the eigen problem as

$$\begin{pmatrix} a & b \\ c & d \end{pmatrix} \begin{pmatrix} 1 \\ \gamma \end{pmatrix} = \tilde{\mu} \begin{pmatrix} 1 \\ \gamma \end{pmatrix}, \tag{9.177}$$

or

$$\begin{aligned} a + b\gamma &= \tilde{\mu}, \\ c + d\gamma &= \tilde{\mu}\gamma. \end{aligned} \tag{9.178}$$

So, we arrive at expression for the component γ and the spectral condition.

$$\begin{aligned} \gamma &= b^{-1}(\tilde{\mu} - a), \\ bc + d(\tilde{\mu} - a) &= \tilde{\mu}(\tilde{\mu} - a). \end{aligned} \tag{9.179}$$

Finally, the eigenvalues of the whole problem are evaluated as

$$\begin{aligned} \lambda_i^{\pm} &= \pm\sqrt{\tilde{\mu}_i}, \\ \tilde{\mu}_i &= \tfrac{a+d}{2} \pm \sqrt{\tfrac{(a+d)^2}{4} - \det C}. \end{aligned} \tag{9.180}$$

The eigenvectors are constructed by two-component columns. For the four-component eigenvectors $\psi_i^{\pm}$ it is

$$\psi_i^{\pm} = \begin{pmatrix} \alpha_i \\ \frac{1}{\lambda_i^{\pm}} B\alpha_i \end{pmatrix}, \tag{9.181}$$

where

$$\alpha_i = \begin{pmatrix} 1 \\ b^{-1}(\tilde{\mu}_i - a) \end{pmatrix}. \tag{9.182}$$

So, to determine all the necessary ingredients of the algorithm for the projecting operators calculation, we start from the 2×2 submatrices L, M and the matrix LM, arriving at the eigenvectors formula (9.180) construct eigenvectors by the formulas (9.181) and (9.182).

For the rectangular waveguide in the notations of the Section 9.1 we write

$$\hat{L} = \begin{pmatrix} 0 & B \\ C & 0 \end{pmatrix}. \tag{9.183}$$

The column vectors

$$\Xi_1 = \begin{pmatrix} 1 \\ \Omega_2 \end{pmatrix}, \Xi_2 = \begin{pmatrix} \Omega_3 \\ \Omega_4 \end{pmatrix}$$

The bi-quadratic equation

$$\lambda^4 + b\lambda^2 + d = 0,$$

and all necessary matrix elements are found by a symbolic computation program, for example, SWP. The results are as follows:

$$(BC)_{11} = -\frac{q\omega^2 + qrs + s\mu\nu\varepsilon}{p^2 s\nu}$$

$$(BC)_{12} = q\mu\varepsilon\frac{iq\omega^2 - ips\nu - ps\omega}{p^3 s\nu\omega}$$

$$(BC)_{21} = \frac{-ip\omega^3 + q\omega^2 + qrs - iprs\omega - is\mu\nu\varepsilon\omega^3 + iprs\nu\omega}{p^3 s\nu^2}$$

$$(BC)_{22} = -\frac{iq^2\mu\varepsilon\omega + pq\mu\varepsilon\omega^2 + ip^2 s\mu\varepsilon\omega - pqs\mu\varepsilon + p^2 qrs\nu - pqs\mu\nu\varepsilon\omega^2}{p^4 s\nu^2}$$

The eigenvalues of the matrix BC are given by the roots of quadratic equation (9.7) for λ^2, the coefficients b,d to be found by the Wiet theorem. The matrix of eigenvectors Ω are calculated by the relations

$$\Omega_2 = (\lambda^2 - (BC)_{11})/(BC)_{12}, \tag{9.184}$$

and, as

$$\Xi_2 = \lambda^{-1} C \Xi_1. \tag{9.185}$$

plugging for each vector $i = 1,2,3,4$ the corresponding eigenvalue λ_i. Evaluation the projectors it is convenient to use the direct product of columns of Ω and rows of Ω^{-1} as prescribed by the formula

$$P_{ik}^{(j)} = \Omega_{ij} \Omega_{jk}^{-1}, \quad i,j,k = 1,2,3,4. \tag{9.186}$$

REFERENCES

1. Hondros, D., and P. Debye. 1910. Elektromagnetische wellen an dielektrischen drahten. *Annals of Physics* 32: 465.
2. Leble, S.B. *Waiveguide Propagation of Nonlinear Waves in Stratified Media* (in Russian), Leningrad University Press, Berlin, 1988. Extended Ed. in Springer-Verlag, 1990.
3. Kuszner, M., and S. Leble. 2011. Directed electromagnetic pulse dynamics: projecting operators method. *Journal of the Physical Society of Japan* 80: 024002.
4. Kuszner, M., and Leble S. 2014. Ultrashort opposite directed pulses dynamics with Kerr effect and polarization account. *Journal of the Physical Society of Japan* 83: 034005.
5. Ampilogov, D., and S. Leble. 2015. General equation for directed electromagnetic pulse propagation in 1D metamaterial: projecting operators method. arXiv:1512.01682v1 [math-ph].
6. Kuszner, M., and S. Leble. 2015. Waveguide electromagnetic pulse dynamics: projecting operators method. In: Porsezian, K., and Ganapathy, R. (Eds), *Odyssey of Light in Nonlinear Optical Fibers: Theory and Applications*. CRC Press Boca Raton, FL.
7. Fock, V.A. 1978. *Fundamentals of Quantum Mechanics*. Mir Publishers, Moscow.
8. Schäfer, T., and C.E. Wayne. 2004. Propagation of ultra-short optical pulses in cubic nonlinear media *Physica D*. 196: 90–105.
9. Chung, Y., and T. Schäfer. 2007. Stabilization of ultra-short pulses in cubic nonlinear media. *Physics Letters A*: 63–69.
10. Boyd, R.W. 1992. *Nonlinear Optics*. Academic Press, Boston.
11. Pietrzyk, M., I. Kanattsikov, and U. Bandelow. 2008. On the propagation of vector ultrashort pulses. *Journal of Nonlinear Mathematical Physics* 15: 2.
12. Galitskii, V.M., and Ermachenko, V.M. 1988. *Macroscopic Electrodynamics*. Moscow High School, Moscow.
13. Kielich, S. 1977. *Nonlinear Molecular Optics*. PWN, Warsaw.
14. Carlos Montes and all. 2013. *Without Bounds: A Scientific Canvas of Nonlinearity and Complex Dynamics*. Springer, Berlin.
15. Leble, S., and B. Reichel. 2009. Coupled nonlinear schrodinger equations in optical fibers theory: from general aspects to solitonic ones. *The European Physical Journal Special Topics* 173: 5–55.

16. Leble, S.B., and B. Reichel. 2007. Mode interaction in few-mode optical fibres with Kerr effect. *Journal of Modern Optics* 55: 1–11.
17. Leble, S.B., and B. Reichel. 2008. The equations for interaction of polarization modes in optical fibres including the Kerr effect. *Journal of Modern Optics*. doi:10.1007/s10910-008-9457-5
18. Leble, S.B., and Reichel, B. 2008. On convergence and stability of a numerical scheme of coupled nonlinear Schrodinger equations. *Computers and Mathematics with Applications* 55: 745–759.
19. Kuszner, M., S. Leble, and B. Reichel. 2011. Multimode systems of nonlinear equations: derivation, integrability, and numerical solutions. *Theoretical and Mathematical Physics* 168(1): 977.

10 Waves in 3D space

10.1 INTRODUCTORY NOTE

The main obstacle in electromagnetic phenomena modeling is the multicomponent character of the basic Maxwell system. It has six fields to be found in most simple cases without moving the charges account; see the main source for this chapter [1]. Being in the last chapter, we would repeat the main algorithm of the basic problem to which this book is devoted. The solution of a problem of time or the coordinate evolution of a system with a multicomponent state includes a classification of basic states as eigenstates of the evolution operator or modes; see the introduction and, as in the preceded book of one of the authors [2]. Thinking about a mode content on basis of the following algorithm, applicable in the case of problems that may be formulated as a system of t-evolution differential equations with constant coefficients for unknown variables organized in a vector of state ψ (see e.g. [2–4]):

1. Each component of the state vector ψ is subjected to the Fourier transformation to $\vec{k}$-representation (transform)

$$\psi \to \mathscr{F}\psi$$

2. The evolution operator L then is transformed as

$$L \to \mathscr{L} = \mathscr{F}L\mathscr{F}^{-1}$$

   resulting in a matrix dependent on $\vec{k}$.
3. The eigenvectors of the matrix $\mathscr{L}(\vec{k})$ form the matrix $\Psi(\vec{k})$. The fact that the eigenproblem is a homogeneous system of equations allows us to choose one of the components as 1.
4. This matrix defines the projecting operators; see Chapter 2 [5]

$$(\tilde{P}^s)_{ij} = \Psi_{is}\Psi^{-1}_{sj}.$$

5. Its Fourier transforms to x-domain, named a dynamic projecting operators allow to split the evolution problem. The same happens with space of initial condition (and in any moment of time!) that defines evolution in each mode subspace [5].

The algorithm is transparent and effective in the case of low dimensions for a wide class of Cauchy problems at infinite space. Even in two- or three-dimensional cases it can lead to the appearance of complicated integral operators as the projectors matrix elements. A big number of a vector components also include a cumbersome technique or approximations when the eigenvalues are evaluated approximately. Special efforts are necessary if boundary conditions are taken into account as, for example, for an electromagnetic field in waveguides [2].

An alternative idea to carry through the splitting of such addresses is the direct manipulation of the evolution operator without the Fourier transformations application [6–8].

This chapter continues such studies, considering the important example of full Maxwell equations system for waves in a space without boundaries. The complete system of Maxwell equations is splitting into independent subsystems by means of the dynamic projecting technique. The previously mentioned formalism, based on Fourier transformations, is used at the initial stage, but at the next step, the operator relations are used. Finally, the technique relies upon a direct connection between the field components that determine corresponding subspaces. The links are effectively used in conditions of some symmetry, and, next some approximations by a small parameter may be developed. It is illustrated by examples of spherical symmetry and quasi-one-dimensional ones.

Section 10.2 of the current chapter contains the complete system of Maxwell equations and the matrix form of equations, which will be further considered. We also define a linear operator **L**, which contains spatial derivatives. In Section 10.3, we determine an operator's eigenvalues and eigenvectors, which, in turn, by standard formula give projecting operators for the operator **L** eigenspaces. Afterwards, we derive equations resulting from the application of projector operators in matrix form of a considered part of Maxwell equations. In Section 10.4, we use the projection technique for the Maxwell equations for the case of isotropic media: the linear dependence of electric induction on the electric field and the magnetic induction on the magnetic field. Section 10.5 contains examples of equations that have a symmetry and the solutions obtained by means of the projecting operators application built with the symmetry account.

10.2 BASIC EQUATIONS AND STARTING POINTS

Our starting point is Maxwell's equations for a medium in Lorentz-Heaviside's unit system. Its evolution part is

$$\frac{1}{c}\frac{\partial \vec{B}}{\partial t} + \vec{\nabla} \times \vec{E} = \vec{0}. \tag{10.1}$$

originated from Faraday's law, and

$$\frac{1}{c}\frac{\partial \vec{D}}{\partial t} - \vec{\nabla} \times \vec{H} = -\frac{4\pi}{c}\vec{j}_f. \tag{10.2}$$

arisen from Ampere-Maxwell ones. The stationary differential links between components are

$$\vec{\nabla} \cdot \vec{B} = 0, \tag{10.3}$$

$$\vec{\nabla} \cdot \vec{D} = 4\pi\rho_f, \tag{10.4}$$

where the inductions $\vec{B}, \vec{D}$ may be expressed in terms of polarization and magnetization as

$$\vec{B} = \vec{H} + 4\pi\vec{M},$$

$$\vec{D} = \vec{E} + 4\pi\vec{P}.$$

Now, developing the method, we include sources functions such as the charge density in (10.4) and the current density in (10.2). Let us introduce the following form of the evolution operator $\mathbf{L}$:

$$\mathbf{L} = \begin{pmatrix} 0 & 0 & 0 & 0 & -\frac{\partial}{\partial z} & \frac{\partial}{\partial y} \\ 0 & 0 & 0 & \frac{\partial}{\partial z} & 0 & -\frac{\partial}{\partial x} \\ 0 & 0 & 0 & -\frac{\partial}{\partial y} & \frac{\partial}{\partial x} & 0 \\ 0 & \frac{\partial}{\partial z} & -\frac{\partial}{\partial y} & 0 & 0 & 0 \\ -\frac{\partial}{\partial z} & 0 & \frac{\partial}{\partial x} & 0 & 0 & 0 \\ \frac{\partial}{\partial y} & -\frac{\partial}{\partial x} & 0 & 0 & 0 & 0 \end{pmatrix}. \tag{10.5}$$

The system of Eqs (10.1, 10.2) can be written in the matrix notation:

$$\frac{1}{c}\frac{\partial \phi}{\partial t} + \mathbf{L}\psi = -\frac{4\pi}{c}\vec{j}_{ex}, \tag{10.6}$$

where

$$\phi = \begin{pmatrix} B_x \\ B_y \\ B_z \\ D_x \\ D_y \\ D_z \end{pmatrix} = \begin{pmatrix} \vec{B} \\ \vec{D} \end{pmatrix}, \ \psi = \begin{pmatrix} H_x \\ H_y \\ H_z \\ E_x \\ E_y \\ E_z \end{pmatrix} = \begin{pmatrix} \vec{H} \\ \vec{E} \end{pmatrix}, \ \vec{j}_{ex} = \begin{pmatrix} 0 \\ 0 \\ 0 \\ j_{f,x} \\ j_{f,y} \\ j_{f,z} \end{pmatrix} = \begin{pmatrix} \vec{0} \\ \vec{j}_f \end{pmatrix}.$$

For given operator $\mathbf{L}$ we can determine the system of projecting operators $\mathbf{P}^i$, $i = 1, \ldots, 6$, which possesses standard properties (see Chapter 2).

10.3 DETERMINATION OF OPERATOR EIGENVALUES, EIGENVECTORS AND PROJECTING OPERATORS FOR A FULL SYSTEM OF MAXWELL'S EQUATIONS

Let us determine projection operators for **L**. For this purpose, we use the spatial direct and inverse Fourier transformations. The Fourier transform representing all fields is nothing but a superposition of planar waves:

$$f(\vec{r},t) = \int_{R^3} \widetilde{f}(\vec{k},t)\exp(-i\vec{k}\cdot\vec{r})d\vec{k},$$

$\widetilde{f}(\vec{k},t)$ denotes the Fourier-transforms of $f(\vec{r},t)$; the inverse one is calculated as $\widetilde{f}(\vec{k},t) = \frac{1}{(2\pi)^3}\int_{R^3} f(\vec{r},t)e^{i\vec{k}\cdot\vec{r}}d\vec{r}$ and $\vec{r} = (x, y, z), \vec{k} = (k_x, k_y, k_z)$.

The resulting formulas may be written via operator eigenvalues of the operator **L** taking the following form

$$\begin{aligned} \lambda_{1,2} &= 0, \\ \lambda_{3,4} &= \sqrt{\Delta}, \\ \lambda_{5,6} &= -\sqrt{\Delta}, \end{aligned} \tag{10.7}$$

where the designation $\sqrt{\Delta}$ (the square root of the Laplacian) is an integral operator, which corresponds to $-i\sqrt{k_x^2+k_y^2+k_z^2}$ in the space of the Fourier transforms. In the quasi-one-dimensional geometry, for example, when we consider the propagation of an electromagnetic wave in the form of the Gaussian beam along axis OX, components of $\vec{k}$ satisfy the condition $k_x^2 \gg k_y^2 + k_z^2$, which allows us to introduce small parameter $\varepsilon = \frac{k_y^2+k_z^2}{k_x^2} \ll 1$ as the diffraction parameter. We can interpret the operator $\sqrt{\Delta}$ in the following way:

$$\sqrt{\Delta} \approx \partial/\partial x + 0.5\varepsilon\Delta_\perp \int dx,$$

where $\Delta_\perp = \frac{\partial^2}{\partial y^2} + \frac{\partial^2}{\partial z^2}$ and $\varepsilon = \frac{k_y^2+k_z^2}{k_x^2} \ll 1$ is the diffraction parameter.

In other models, when we know that a considered function f has spherical symmetry with respect to point $r_0 = (x_0, y_0, z_0)$ and depends only on radial coordinates, $r = \sqrt{(x-x_0)^2+(y-y_0)^2+(z-z_0)^2}$. Then

$$\sqrt{\Delta}f = \frac{1}{r}\frac{\partial}{\partial r}(rf).$$

To simplify the projection operators view, let us introduce the notation:

$$\mathbf{P}_d = \frac{1}{\Delta}\begin{pmatrix} \frac{\partial^2}{\partial x^2} & \frac{\partial^2}{\partial x\partial y} & \frac{\partial^2}{\partial x\partial z} \\ \frac{\partial^2}{\partial y\partial x} & \frac{\partial^2}{\partial y^2} & \frac{\partial^2}{\partial y\partial z} \\ \frac{\partial^2}{\partial z\partial x} & \frac{\partial^2}{\partial z\partial y} & \frac{\partial^2}{\partial z^2} \end{pmatrix}, \tag{10.8}$$

$$\mathbf{P}_r = \frac{1}{\sqrt{\Delta}} \begin{pmatrix} 0 & -\frac{\partial}{\partial z} & \frac{\partial}{\partial y} \\ \frac{\partial}{\partial z} & 0 & -\frac{\partial}{\partial x} \\ -\frac{\partial}{\partial y} & \frac{\partial}{\partial x} & 0 \end{pmatrix}, \tag{10.9}$$

$$\mathbf{0} = \begin{pmatrix} 0 & 0 & 0 \\ 0 & 0 & 0 \\ 0 & 0 & 0 \end{pmatrix}. \tag{10.10}$$

Designations $\frac{1}{\sqrt{\Delta}}$ and $\frac{1}{\Delta}$ denote integral operators that correspond to $\frac{i}{\sqrt{k_x^2+k_y^2+k_z^2}}$ and $\frac{1}{k_x^2+k_y^2+k_z^2}$ in the space of the Fourier transforms. With such markings the projection operators $\mathbf{P}_1$ and $\mathbf{P}_2$ are rewritten as

$$\mathbf{P}_1 = \begin{pmatrix} \mathbf{0} & \mathbf{0} \\ \mathbf{0} & \mathbf{P}_d \end{pmatrix}, \tag{10.11}$$

$$\mathbf{P}_2 = \begin{pmatrix} \mathbf{P}_d & \mathbf{0} \\ \mathbf{0} & \mathbf{0} \end{pmatrix}. \tag{10.12}$$

Every pair of operators $\mathbf{P}_3$, $\mathbf{P}_4$ and $\mathbf{P}_5$, $\mathbf{P}_6$ generate a two-dimensional subspace. In this connection, their appearance depends on the choice of eigenvectors within their subspace, but their sum ($\mathbf{P}_3+\mathbf{P}_4$ and $\mathbf{P}_5+\mathbf{P}_6$) will always take the same form:

$$\mathbf{P}_+ = \mathbf{P}_3 + \mathbf{P}_4 = \frac{1}{2} \begin{pmatrix} -\mathbf{P}_r^2 & \mathbf{P}_r \\ -\mathbf{P}_r & -\mathbf{P}_r^2 \end{pmatrix}, \tag{10.13}$$

$$\mathbf{P}_- = \mathbf{P}_5 + \mathbf{P}_6 = \frac{1}{2} \begin{pmatrix} -\mathbf{P}_r^2 & -\mathbf{P}_r \\ \mathbf{P}_r & -\mathbf{P}_r^2 \end{pmatrix}. \tag{10.14}$$

The form of eigenvectors we get from equality:

$$\mathbf{P}_i \psi = \psi_i = \begin{pmatrix} \vec{H}_i \\ \vec{E}_i \end{pmatrix}. \tag{10.15}$$

Let us also use the notation

$$\mathbf{P}_i \phi = \phi_i = \begin{pmatrix} \vec{B}_i \\ \vec{D}_i \end{pmatrix}. \tag{10.16}$$

Indices $i = 1,2,3,4,5,6$ at individual components will mark vectors that were obtained after applying to them projection operators $\mathbf{P}_i$.

10.3.1 PROJECTION BY OPERATOR $\mathbf{P}_1$ APPLICATION

Applying the operator $\mathbf{P}_1$ to the system (10.6) and by using Eq. (10.4) after simplification, we get the equation of continuity:

$$\frac{\partial \rho_f}{\partial t} = -\vec{\nabla} \cdot \vec{J}_f. \tag{10.17}$$

The eigenvector will have the form

$$\psi_1 = \begin{pmatrix} H_{x,1} \\ H_{y,1} \\ H_{z,1} \\ E_{x,1} \\ E_{y,1} \\ E_{z,1} \end{pmatrix} = \frac{1}{\Delta} \begin{pmatrix} 0 \\ 0 \\ 0 \\ \frac{\partial}{\partial x} \\ \frac{\partial}{\partial y} \\ \frac{\partial}{\partial z} \end{pmatrix} \vec{\nabla} \cdot \vec{E}. \tag{10.18}$$

Next, for the vector ϕ_1, the following equality will be fulfilled:

$$\phi_1 = \begin{pmatrix} B_{x,1} \\ B_{y,1} \\ B_{z,1} \\ D_{x,1} \\ D_{y,1} \\ D_{z,1} \end{pmatrix} = \frac{1}{\Delta} \begin{pmatrix} 0 \\ 0 \\ 0 \\ \frac{\partial}{\partial x} \\ \frac{\partial}{\partial y} \\ \frac{\partial}{\partial z} \end{pmatrix} \vec{\nabla} \cdot \vec{D}.$$

Other properties:

$$\vec{\nabla} \cdot \vec{E}_1 = \vec{\nabla} \cdot \vec{E},$$
$$\vec{\nabla} \cdot \vec{D} = \vec{\nabla} \cdot \vec{D}_1 = \vec{\nabla} \cdot \vec{E}_1 + 4\pi \vec{\nabla} \cdot \vec{P}_1,$$
$$\vec{\nabla} \cdot \vec{J}_f = \vec{\nabla} \cdot \vec{J}_{f,1},$$
$$\vec{B}_1 = \vec{H}_1 = \vec{0},$$

arriving at conditions of the zero rotations,

$$\vec{\nabla} \times \vec{E}_1 = \vec{0},$$
$$\vec{\nabla} \times \vec{D}_1 = \vec{0},$$

having potential fields, where $\vec{D}_1 = \mathbf{P}_d \vec{D}$ and $\vec{E}_1 = \mathbf{P}_d \vec{E}$.

10.3.2 PROJECTION WITH OPERATOR $\mathbf{P}_2$

Applying operator $\mathbf{P}_2$ on system (10.6) we obtain the equality

$$\frac{\partial \vec{B}_2}{\partial t} = \vec{0}; \tag{10.19}$$

therefore, the corresponding projection of magnetic induction vector $\vec{B}_2$ does not change over time, where $\vec{B}_2 = \mathbf{P}_d\vec{B}$ and $\vec{H}_2 = \mathbf{P}_d\vec{H}$. Taking into account equation (10.3) we get identity equation.

The eigenvector ψ_2 will have form

$$\psi_2 = \begin{pmatrix} H_{x,2} \\ H_{y,2} \\ H_{z,2} \\ E_{x,2} \\ E_{y,2} \\ E_{z,2} \end{pmatrix} = \frac{1}{\Delta} \begin{pmatrix} \frac{\partial}{\partial x} \\ \frac{\partial}{\partial y} \\ \frac{\partial}{\partial z} \\ 0 \\ 0 \\ 0 \end{pmatrix} \vec{\nabla} \cdot \vec{H}.$$

And for vector ϕ_2 we will have that

$$\phi_2 = \begin{pmatrix} \vec{B}_2 \\ \vec{D}_2 \end{pmatrix} = \frac{1}{\Delta} \begin{pmatrix} \vec{\nabla} \\ \vec{0} \end{pmatrix} \vec{\nabla} \cdot \vec{B}.$$

Other properties:

$$\vec{\nabla} \cdot \vec{H}_2 = \vec{\nabla} \cdot \vec{H},$$

$$\vec{\nabla} \cdot \vec{B} = \vec{\nabla} \cdot \vec{B}_2 = \vec{\nabla} \cdot \vec{H}_2 + 4\pi \vec{\nabla} \cdot \vec{M}_2,$$

$$\vec{D}_2 = \vec{E}_2 = \vec{J}_{f,2} = \vec{0},$$

$$\vec{\nabla} \times \vec{H}_2 = \vec{0},$$

$$\vec{\nabla} \times \vec{B}_2 = \vec{0}.$$

10.3.3 RESULTS FOR OTHER PROJECTOR OPERATOR

We cannot give the form of other projector operators until we specify how eigenvectors unbutton adequate subspaces. Not knowing the form of eigenvectors, however, we can specify the general properties of the subspace generated by operators $\mathbf{P}_+$ and $\mathbf{P}_-$. We can choose such eigenvectors and adequate operators $\mathbf{P}_3$, $\mathbf{P}_4$, $\mathbf{P}_5$ and $\mathbf{P}_6$, which will generate subspaces with the same properties as operators $\mathbf{P}_+$ and $\mathbf{P}_-$.

For any choice of eigenvectors $i = 3,4,5,6$ will fulfill the following dependences:

$$\frac{\partial \vec{B}_2}{\partial t} = \vec{0}, \vec{\nabla} \times \vec{H}_i = -\lambda_i \vec{E}_i, \tag{10.20}$$

$$\frac{\partial \vec{B}_2}{\partial t} = \vec{0}, \vec{\nabla} \times \vec{E}_i = \lambda_i \vec{H}_i. \tag{10.21}$$

From these equalities, it follows that

$$\Delta \vec{H}_i = -\vec{\nabla} \times (\vec{\nabla} \times \vec{H}_i),$$

$$\Delta \vec{E}_i = -\vec{\nabla} \times (\vec{\nabla} \times \vec{E}_i);$$

therefore,

$$\vec{\nabla}(\vec{\nabla} \cdot \vec{H}_i) = \vec{0} \text{ we can choose such } \vec{H}_i \text{ for which } \vec{\nabla} \cdot \vec{H}_i = 0$$

and

$$\vec{\nabla}(\vec{\nabla} \cdot \vec{E}_i) = \vec{0} \text{ we can choose such } \vec{E}_i \text{ for which } \vec{\nabla} \cdot \vec{E}_i = 0.$$

In addition we list:

$$\vec{\nabla} \cdot \vec{B}_+ = \vec{\nabla} \cdot (\vec{B}_3 + \vec{B}_4) = 0,$$
$$\vec{\nabla} \cdot \vec{B}_- = \vec{\nabla} \cdot (\vec{B}_5 + \vec{B}_6) = 0,$$
$$\vec{\nabla} \cdot \vec{H}_+ = \vec{\nabla} \cdot (\vec{H}_3 + \vec{H}_4) = 0,$$
$$\vec{\nabla} \cdot \vec{H}_- = \vec{\nabla} \cdot (\vec{H}_5 + \vec{H}_6) = 0,$$
$$\vec{\nabla} \cdot \vec{M}_+ = \vec{\nabla} \cdot (\vec{M}_3 + \vec{M}_4) = 0,$$
$$\vec{\nabla} \cdot \vec{M}_- = \vec{\nabla} \cdot (\vec{M}_5 + \vec{M}_6) = 0,$$

$$\vec{\nabla} \cdot \vec{D}_+ = \vec{\nabla} \cdot (\vec{D}_3 + \vec{D}_4) = 0,$$
$$\vec{\nabla} \cdot \vec{D}_- = \vec{\nabla} \cdot (\vec{D}_5 + \vec{D}_6) = 0,$$
$$\vec{\nabla} \cdot \vec{E}_+ = \vec{\nabla} \cdot (\vec{E}_3 + \vec{E}_4) = 0,$$
$$\vec{\nabla} \cdot \vec{E}_- = \vec{\nabla} \cdot (\vec{E}_5 + \vec{E}_6) = 0,$$
$$\vec{\nabla} \cdot \vec{P}_+ = \vec{\nabla} \cdot (\vec{P}_3 + \vec{P}_4) = 0,$$
$$\vec{\nabla} \cdot \vec{P}_- = \vec{\nabla} \cdot (\vec{P}_5 + \vec{P}_6) = 0.$$

In this book, we don't pay much attention to inhomogeneous (with sources) systems of equations. In the following, however, we do some steps in this interesting direction.

$$\vec{\nabla} \cdot \vec{j}_{f,+} = \vec{\nabla} \cdot (\vec{j}_{f,3} + \vec{j}_{f,4}) = 0,$$
$$\vec{\nabla} \cdot \vec{j}_{f,-} = \vec{\nabla} \cdot (\vec{j}_{f,5} + \vec{j}_{f,6}) = 0.$$

Let's introduce the notation

$$\lambda_+ = \lambda_3 = \lambda_4,$$
$$\lambda_- = \lambda_5 = \lambda_6,$$
$$\psi_+ = \psi_3 + \psi_4,$$
$$\psi_- = \psi_5 + \psi_6,$$
$$\phi_+ = \phi_3 + \phi_4,$$
$$\phi_- = \phi_5 + \phi_6.$$

Then, applying operator $\mathbf{P}_+$ on the system (10.6), we obtain the equations with sources account:

$$\frac{\partial \phi_+}{\partial t} + c\lambda_+\psi_+ = -4\pi\mathbf{P}_+\left(\frac{\partial}{\partial t}\begin{pmatrix}\vec{M}\\ \vec{P}\end{pmatrix} + \vec{j}_{ex}\right) = -4\pi\left(\frac{\partial}{\partial t}\begin{pmatrix}\vec{M}_+\\ \vec{P}_+\end{pmatrix} + \mathbf{P}_+\vec{j}_{ex}\right), \tag{10.22}$$

$$\mathbf{P}_+\vec{j}_{ex} = \frac{1}{2}\begin{pmatrix}\mathbf{P}_r\\ -\mathbf{P}_r^2\end{pmatrix}\vec{j}_f. \tag{10.23}$$

Similarly, acting by the operator $\mathbf{P}_-$ on the system (10.6), we obtain

$$\frac{\partial \phi_-}{\partial t} + c\lambda_-\psi_- = -4\pi\mathbf{P}_-\left(\frac{\partial}{\partial t}\begin{pmatrix}\vec{M}\\ \vec{P}\end{pmatrix} + \vec{j}_{ex}\right) = -4\pi\left(\frac{\partial}{\partial t}\begin{pmatrix}\vec{M}_-\\ \vec{P}_-\end{pmatrix} + \mathbf{P}_-\vec{j}_{ex}\right), \tag{10.24}$$

$$\mathbf{P}_-\vec{j}_{ex} = \frac{1}{2}\begin{pmatrix}-\mathbf{P}_r\\ -\mathbf{P}_r^2\end{pmatrix}\vec{j}_f. \tag{10.25}$$

In the vacuum, the preceding equation will take the form

$$\frac{\partial \psi_+}{\partial t} + c\lambda_+\psi_+ = \begin{pmatrix}\vec{0}\\ \vec{0}\end{pmatrix}, \tag{10.26}$$

$$\frac{\partial \psi_-}{\partial t} + c\lambda_-\psi_- = \begin{pmatrix}\vec{0}\\ \vec{0}\end{pmatrix}. \tag{10.27}$$

10.4 THE CASE OF THE LINEAR DEPENDENCE OF ELECTROMAGNETIC INDUCTION ON THE ELECTRIC FIELD AND THE MAGNETIC INDUCTION ON THE MAGNETIC FIELD

Consider the case where the relation between $\vec{B}$ and $\vec{H}$ and $\vec{D}$ and $\vec{E}$ are linear:

$$\vec{B} = \mu\vec{H}, \tag{10.28}$$

$$\vec{D} = \varepsilon\vec{E} \tag{10.29}$$

then, we can write the right side of equations (10.1), (10.2) using the vector

$$\tilde{\psi} = \begin{pmatrix} B_x \\ B_y \\ B_z \\ E_x \\ E_y \\ E_z \end{pmatrix}, \tag{10.30}$$

For operator $\tilde{\mathbf{L}}$ we arrive at

$$\tilde{\mathbf{L}} = \begin{pmatrix} 0 & 0 & 0 & 0 & -\frac{\partial}{\partial z} & \frac{\partial}{\partial y} \\ 0 & 0 & 0 & \frac{\partial}{\partial z} & 0 & -\frac{\partial}{\partial x} \\ 0 & 0 & 0 & -\frac{\partial}{\partial y} & \frac{\partial}{\partial x} & 0 \\ 0 & \frac{1}{\varepsilon\mu}\frac{\partial}{\partial z} & -\frac{1}{\varepsilon\mu}\frac{\partial}{\partial y} & 0 & 0 & 0 \\ -\frac{1}{\varepsilon\mu}\frac{\partial}{\partial z} & 0 & \frac{1}{\varepsilon\mu}\frac{\partial}{\partial x} & 0 & 0 & 0 \\ \frac{1}{\varepsilon\mu}\frac{\partial}{\partial y} & -\frac{1}{\varepsilon\mu}\frac{\partial}{\partial x} & 0 & 0 & 0 & 0 \end{pmatrix}. \tag{10.31}$$

The system of equations (10.1), (10.2) can be rewritten in matrix form:

$$\frac{1}{c}\frac{\partial \tilde{\psi}}{\partial t} + \tilde{\mathbf{L}}\tilde{\psi} = -\frac{4\pi}{c\varepsilon}\vec{j}_{ex}. \tag{10.32}$$

10.4.1 PROJECTION OPERATORS

In the considered case the operator eigenvalues of operator $\tilde{\mathbf{L}}$ take the form

$$\tilde{\lambda}_{1,2} = 0,$$

$$\tilde{\lambda}_{3,4} = \frac{1}{\sqrt{\varepsilon\mu}}\sqrt{\Delta}, \tag{10.33}$$

$$\tilde{\lambda}_{5,6} = -\frac{1}{\sqrt{\varepsilon\mu}}\sqrt{\Delta}.$$

With such notations, projector operators take the form

$$\tilde{\mathbf{P}}_1 = \mathbf{P}_1 = \begin{pmatrix} \mathbf{0} & \mathbf{0} \\ \mathbf{0} & \mathbf{P}_d \end{pmatrix}, \tag{10.34}$$

$$\tilde{\mathbf{P}}_2 = \mathbf{P}_2 = \begin{pmatrix} \mathbf{P}_d & \mathbf{0} \\ \mathbf{0} & \mathbf{0} \end{pmatrix}, \tag{10.35}$$

$$\tilde{\mathbf{P}}_+ = \tilde{\mathbf{P}}_3 + \tilde{\mathbf{P}}_4 = \frac{1}{2}\begin{pmatrix} -\mathbf{P}_r^2 & \sqrt{\varepsilon\mu}\mathbf{P}_r \\ -\frac{1}{\sqrt{\varepsilon\mu}}\mathbf{P}_r & -\mathbf{P}_r^2 \end{pmatrix}, \tag{10.36}$$

$$\tilde{\mathbf{P}}_- = \tilde{\mathbf{P}}_5 + \tilde{\mathbf{P}}_6 = \frac{1}{2}\begin{pmatrix} -\mathbf{P}_r^2 & -\sqrt{\varepsilon\mu}\mathbf{P}_r \\ \frac{1}{\sqrt{\varepsilon\mu}}\mathbf{P}_r & -\mathbf{P}_r^2 \end{pmatrix}. \tag{10.37}$$

All the results obtained for the operators $\tilde{\mathbf{P}}_1$ and $\tilde{\mathbf{P}}_2$ will be the same as those previously obtained for $\mathbf{P}_1$ and $\mathbf{P}_2$.
For elements of eigenvectors generated by $\tilde{\mathbf{P}}_i$ from vector $\tilde{\psi}$, where $i = 3,4,5,6$, the following relationships are fulfilled:

$$\vec{\nabla} \times \vec{\tilde{E}}_i = \tilde{\lambda}_i \vec{\tilde{B}}_i, \tag{10.38}$$

$$4\vec{\nabla} \times \vec{\tilde{B}}_i = -\varepsilon\mu\tilde{\lambda}_i \vec{\tilde{E}}_i.$$

From these equalities it follows that

$$\Delta\vec{\tilde{B}}_i = -\vec{\nabla} \times (\vec{\nabla} \times \vec{\tilde{B}}_i),$$

$$\Delta\vec{\tilde{E}}_i = -\vec{\nabla} \times (\vec{\nabla} \times \vec{\tilde{E}}_i),$$

And for operators $\tilde{\mathbf{P}}_+$ and $\tilde{\mathbf{P}}_-$ we obtain the following equations:

$$\frac{\partial \tilde{\psi}_+}{\partial t} + \frac{c}{\sqrt{\varepsilon\mu}}\lambda_+ \tilde{\psi}_+ = -2\pi \begin{pmatrix} \sqrt{\varepsilon\mu}\mathbf{P}_r \\ -\mathbf{P}_r^2 \end{pmatrix} \frac{\vec{J}_f}{\varepsilon}, \tag{10.39}$$

$$\frac{\partial \tilde{\psi}_-}{\partial t} + \frac{c}{\sqrt{\varepsilon\mu}}\lambda_- \tilde{\psi}_- = 2\pi \begin{pmatrix} \sqrt{\varepsilon\mu}\mathbf{P}_r \\ \mathbf{P}_r^2 \end{pmatrix} \frac{\vec{J}_f}{\varepsilon}, \tag{10.40}$$

where $\tilde{\psi}_+ = \tilde{\mathbf{P}}_+ \tilde{\psi}$ and $\tilde{\psi}_- = \tilde{\mathbf{P}}_- \tilde{\psi}$. Besides, for the subspace generated by $\tilde{\mathbf{P}}_+$ and $\tilde{\mathbf{P}}_-$ following equalities will be satisfied:

$$\vec{\nabla} \cdot \vec{\tilde{B}}_+ = \vec{\nabla} \cdot (\vec{\tilde{B}}_3 + \vec{\tilde{B}}_4) = 0,$$

$$\vec{\nabla} \cdot \vec{\tilde{E}}_+ = \vec{\nabla} \cdot (\vec{\tilde{E}}_3 + \vec{\tilde{E}}_4) = 0,$$

$$\vec{\nabla} \cdot \vec{\tilde{B}}_- = \vec{\nabla} \cdot (\vec{\tilde{B}}_5 + \vec{\tilde{B}}_6) = 0,$$

$$\vec{\nabla} \cdot \vec{\tilde{E}}_- = \vec{\nabla} \cdot (\vec{\tilde{E}}_5 + \vec{\tilde{E}}_6) = 0.$$

10.5 EXAMPLES WITH A SYMMETRY ACCOUNT

As in the first example, let's consider equations in a region with no currents ($\vec{j}_g = \vec{0}$). For this case, from equation (10.17), we can see that charges for the first mode are invariant with respect to time.

10.5.1 SPHERICAL GEOMETRY

Now let us consider a problem in which we have no sources of electromagnetic field: no currents, no charges, and the electric fields changes support the spherical symmetry; then, equations (10.26) and (10.27) take the form

$$\frac{\partial E_+}{\partial t} + c\sqrt{\Delta}E_+ = 0, \tag{10.41}$$

$$\frac{\partial E_-}{\partial t} - c\sqrt{\Delta}E_- = 0. \tag{10.42}$$

The integral operator for such a problem simplifies to the differental operator form $\sqrt{\Delta}f = \frac{1}{r}\frac{\partial}{\partial r}(rf)$, where $r = \sqrt{(x-x_0)^2+(y-y_0)^2+(z-z_0)^2}$ and $r_0 = (x_0, y_0, z_0)$ is the point, where the source of electromagmetic field is located. An analytical solution of these equations by the chatacteristics method is built via

$$E_+(r,t) = \frac{F(t-r/c)}{r},$$

$$E_-(r,t) = \frac{F(t+r/c)}{r}.$$

10.5.2 QUASI-ONE-DIMENSIONAL GEOMETRY

Spherical geometry cannot be used when we consider the propagation of X-rays. In such a case, we need to consider the propagation of a Gaussian beam along a specified direction (for examplle, axis OX) and during propagation, the beam will slowly expand in a direction perpendicular to the direction of propagation of X-rays. To describe such a problem, we need to usc the approximate form of integral operator:

$$\sqrt{\Delta} \approx \partial/\partial x + 0.5\Delta_\perp \int dx$$

and then find solutions of equations:

$$\frac{\partial E_+}{\partial t} + c\frac{\partial E_+}{\partial x} + 0.5c\Delta_\perp \int E_+ dx = 0, \tag{10.43}$$

$$\frac{\partial E_-}{\partial t} - c\frac{\partial E_-}{\partial x} - 0.5c\Delta_\perp \int E_- dx = 0. \tag{10.44}$$

Let's consider Eq. (10.43). Differentiating it by x, we can rearrange it to the form

$$\frac{\partial^2 E_+}{\partial t \partial x} + c\frac{\partial^2 E_+}{\partial x^2} + 0.5c\Delta_\perp E_+ = 0. \tag{10.45}$$

Using anzatz

$$E_+(\vec{r},t) = A(\vec{r},t)\exp\left(i\left(k_0 x - \omega_0 t\right)\right), \qquad k_0 = \frac{\omega_0}{c}, \tag{10.46}$$

we get an equation for function $A(\vec{r},t)$ that varies slowly with all argument variables $\vec{r}$ and t. Leaving only expressions that have the greatest impact on changes of function $A(\vec{r},t)$ and assuming that function $A(\vec{r},t)$ does not change in time leads us to equation

$$\frac{\partial A}{\partial x} = \frac{ic}{2\omega_0}\left(\frac{\partial^2}{\partial y^2} + \frac{\partial^2}{\partial z^2}\right)A. \tag{10.47}$$

This equation has important applications in optic science, where it provides solutions that describe a propagation of electromagnetic waves in the form of either paraboloidal waves or Gaussian beams. Solutions for this equation can be obtained with help of the so-called Kirchhoff propagator when the boundary condition $A(0,y,z)$ is known

$$A(x,y,z) = \iint_S A(0,\xi,\eta)\, G(x,y,z,\zeta,\eta)\, d\xi\, d\eta, \tag{10.48}$$

and where $G(x,y,z,\zeta,\eta)$ is a Kirchhoff propagator, which has the form

$$G(x,y,z,\zeta,\eta) = -\frac{i}{2\pi\sigma^2}\cdot\exp\left(i\cdot\frac{(y-\zeta)^2+(z-\eta)^2}{2\sigma^2}\right), \qquad \sigma^2 = \frac{xc}{\omega_0}. \tag{10.49}$$

For the boundary condition $A(0,y,z) = \exp\left(-\alpha(y^2+z^2)\right)$ the solution obtained via Eq. (10.48) is presented here:

$$A(x,y,z) = \frac{1}{1+i\cdot 2\alpha\sigma^2}\exp\left(-\alpha\cdot\frac{y^2+z^2}{1+i\cdot 2\alpha\sigma^2}\right).$$

A theory and examples of numerical solution of Eq. (10.47) with other boundary conditions can be found in publications [9,10].

10.6 CONCLUDING REMARKS

A development of the theory are first of all in more advanced material relations compared to (10.28, 10.29) of this chapter, in the spirit of Chapters 8 and 9. An obvious option is in a different geometry that changes the action of the operators used in the calculations of Sections 10.3 and 10.5.

REFERENCES

1. Wojda, P., and S. Leble. 2017. General dynamic projecting of Maxwell equations. *TASK Quarterly* 21(2): 73–84.
2. Leble, S.B. 1988. *Waveguide Propagation of Nonlinear Waves in Stratified Media (in Russian).* Leningrad University Press, Leningrad. Extended Ed. in Springer-Verlag 1990.
3. Perelomova, A. 1993. Construction of directed disturbances in one-dimensional isothermal atmosphere model. *Izvestiya, Atmospheric and Oceanic Physics* 29(1): 4750.
4. Kuszner, M., and S. Leble. 2011. Directed electromagnetic pulse dynamics: Projecting operators method. *Journal of the Physical Society of Japan* 80: 024002.
5. Leble, S. 2016. General remarks on dynamic projectors method. *TASK Quarterly* 20(2): 113–130.
6. Perelomova, A., and S. Leble. 2002. ollmien-Shlichting and sound waves, interaction: Nonlinear resonances. In: Rudenko, O.V. (Ed), *Nonlinear Acoustics at the Beginning of the 21st Century*, vol. 1. MSU, Moscow, p. 203.
7. Perelomova, A., and S. Leble. 2005. Vortical and acoustic waves interaction. From general equations to integrable cases. *Theoretical and Mathematical Physics* 144: 10301039.
8. Belov, V.V., S.Y. Dobrohotov, and T.Y. Tudorovskiy. 2006. Operator separation of variables for adiabatic problems in quantum and wave mechanics. *Journal of Engineering Mathematics* 55(14): 183237.
9. Kshevetskii, S., P. Wojda, and V. Maximov. 2016. A high-accuracy complex-phase method of simulating X-ray propagation through a multi-lens system. *Journal of Synchrotron Radiation* 23: 1305–1314.
10. Kshevetskii, S., and P. Wojda. 2015. Efficient quadrature for the fast oscillating integral of paraxial optics. *Mathematica Applicanda* 43(2): 253–267.

Index

D

E

F

G

H

I

K

L

M

N

O

P

U

V

W

X

Z

PGMO 04/05/2018